The Engineering Approach to Winter Sports

Francesco Braghin • Federico Cheli
Stefano Maldifassi • Stefano Melzi
Edoardo Sabbioni
Editors

The Engineering Approach to Winter Sports

Springer

Editors
Francesco Braghin
Department of Mechanical Engineering
Politecnico di Milano
Milano, Italy

Federico Cheli
Department of Mechanical Engineering
Politecnico di Milano
Milano, Italy

Stefano Maldifassi
AC Milan SpA
Milano, Italy

Stefano Melzi
Department of Mechanical Engineering
Politecnico di Milano
Milano, Italy

Edoardo Sabbioni
Department of Mechanical Engineering
Politecnico di Milano
Milano, Italy

ISBN 978-1-4939-3019-7 ISBN 978-1-4939-3020-3 (eBook)
DOI 10.1007/978-1-4939-3020-3

Library of Congress Control Number: 2015950192

Springer New York Heidelberg Dordrecht London

Printed on acid-free paper

Springer Science+Business Media LLC New York is part of Springer Science+Business Media (www.springer.com)

Preface

Why did we decide to write this book? In fact, we had never thought of writing a book on winter sports since our competences were focused just on bobsleigh, skeleton and luge (except Stefano Maldifassi who, being responsible for the research of the Italian Winter Sports Federation, had carried out research in all disciplines but had no time for writing a book). Too less material for writing a book.

One day, however, we took part to a congress with a special session on sports equipment, and there was a great interest in the sensors we had developed and the test campaign we had carried out to characterize the dynamics of the sled coupled with the actions by the athletes. Our surprise was to see that what we had done within our discipline was very similar to what colleagues had done in different disciplines, e.g. in skating. In fact, all winter disciplines have one thing in common: the sliding on a slippery surface (either ice or snow). Thus, to maximize performances, one has to minimize (or control, such as in curling) the friction at equipment—ice/snow interface.

The following question arose: Why not putting the experiences gained in all the various disciplines together to cross-fertilize and provide new ideas? Looking around, however, there was no common archive of the engineering studies done on winter sport disciplines. You had to go around the internet and the various public or private databases trying to collect all the literature. Even more problematic, several colleagues, especially in the past, had published their researches at conferences and the corresponding proceedings were almost lost. Not to talk about brilliant researches done within Ph.D. thesis that had not been published elsewhere.

Thus, we decided to collect all the relevant literature on the engineering studies done in the various disciplines and to write a book summing them up. However, a mere review would have been not that interesting. We therefore tried to highlight the research paths, to critically analyse the obtained results, to show the technological trends, to draw the attention to the open questions, and, even more important, to stress the similarities and the differences between the different disciplines. We decided to split the book into two main sections: the first focusing on the common denominator of all winter sports, i.e. ice and snow as well as their interaction with skis and runners, respectively, and the second focusing on the various disciplines.

We decided to limit our analysis to the most popular Olympic disciplines but interesting hints for disciplines not considered in the book can easily be obtained. At the end, we added a concluding chapter that bridges engineering with the other most relevant science in sports: medicine. In fact, you may have the best equipment but you will win or lose according to your training and your psychology. We hope you will enjoy the reading of this book.

Milano, Italy

Francesco Braghin
Federico Cheli
Stefano Maldifassi
Stefano Melzi
Edoardo Sabbioni

Contents

Chapter 1
Ice and Snow for Winter Sports

Norikazu Maeno

In the past winter sports were played only in cold areas during winter, but the development of techniques to artificially produce snow and ice has changed the situation. Now winter sports are enjoyed even in summer and almost all over the world regardless of season and area. Ice and snow are indispensable to winter sports because they cover a variety of ground surfaces, mountains, forests, lakes and so on, and provide us surfaces on which we can walk and play. Moreover ice and snow prepare slippery surfaces necessary for various kinds of winter sports such as ski, skate, sledge, curling, etc. The slipperiness is the most important property of ice and snow for winter sports. It may be stupid to make a question why ice and snow are so slippery, but this inquiry gives an important key to understand the essential property of ice and snow. In physical sense slipperiness is not an intrinsic property of ice and snow. The reason is as follows. Figure 1.1 shows the homologous temperatures of three familiar materials. The homologous temperature is defined as T/T_m where T is the temperature and T_m is the melting point expressed in the absolute temperature (K). The homologous temperature 100 % means the highest temperature any solid material can attain because it melts away at T_m; it is 1809 K (1536 °C) for iron and 273 K (0 °C) for ice.

For instance, the homologous temperature 80 % is −55 °C for ice and 1174 °C for iron. −55 °C is quite a cold temperature for human beings, but it is extremely high temperature for ice and corresponds to iron heated to 1174 °C. We should remember that ice and snow we see are just like iron heated to red-hot and white-hot above 1000 °C. Another example in Fig. 1.1; imagine an iron pot and knife on a table at room temperature 25 °C. The room temperature corresponds to the homologous

N. Maeno (✉)
Professor Emeritus Hokkaido University, Hanakawa Minami 7-2-133, Ishikari, Hokkaido 061-3207, Japan
e-mail: maenony@ybb.ne.jp

F. Braghin et al. (eds.), *The Engineering Approach to Winter Sports*,
DOI 10.1007/978-1-4939-3020-3_1

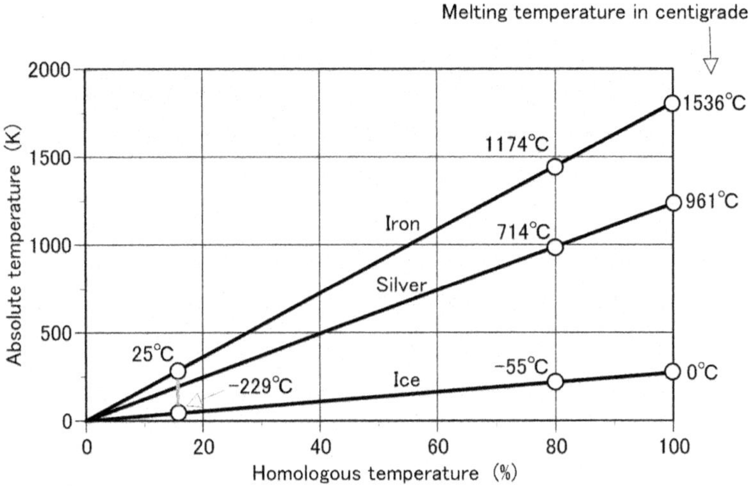

Fig. 1.1 Homologous temperature. Homologous temperature is the fraction of melting temperature of a material, that is, T/T_m where T and T_m represent, respectively, the temperature and the melting temperature of the material in Kelvin

temperature of iron 16 %, which is −229 °C for ice, that is, the iron pot and knife at the room temperature are equivalent to ice cooled to −229 °C.

In winter sports we should carefully understand that we are treating ice and snow at such high homologous temperatures. Slipperiness of ice and snow may be caused by pre-melting near the melting point, and many other processes may be attributed to high homologous temperatures. An example is shown in three sequential pictures in Fig. 1.2, which gives small ice particles in contact at −8 °C. As seen in the pictures bonds between the particles grow with time. The solid-state process below the melting point is known as "sintering," which has been studied extensively in metallurgy and ceramics. As we see in the following discussion, structures and various physical properties of ice and snow vary dramatically with time depending on temperature, humidity, and other factors. In winter sports we should recognize that ice and snow are always changing with time. This is because the situation is just like the iron pot and knife or the ceramic pottery works kept in a firing kiln at homologous temperatures above 80 %. In this sense we may call ice and snow as *pottery works in a kiln* [1]. In the following we briefly review the structure of ice and snow, and then their friction properties.

1.1 Structure and Friction Properties of Ice

1.1.1 Structure of Ice

Hexagonal ice is the crystal form of all natural snow and ice on Earth, on which we play winter sports. It is a stable phase of ice at atmospheric pressures and called

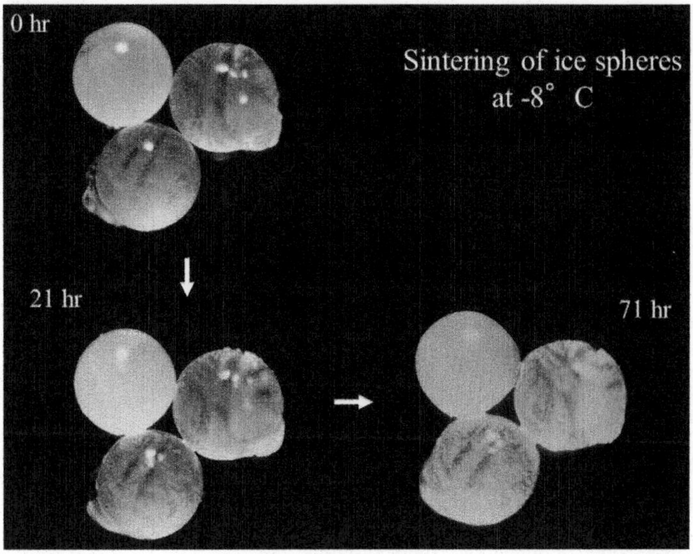

Fig. 1.2 Sintering of ice particles. The diameter of ice particle at the *top left* is 0.9 mm and temperature is −8 °C. The bond growth process below the melting point is called sintering

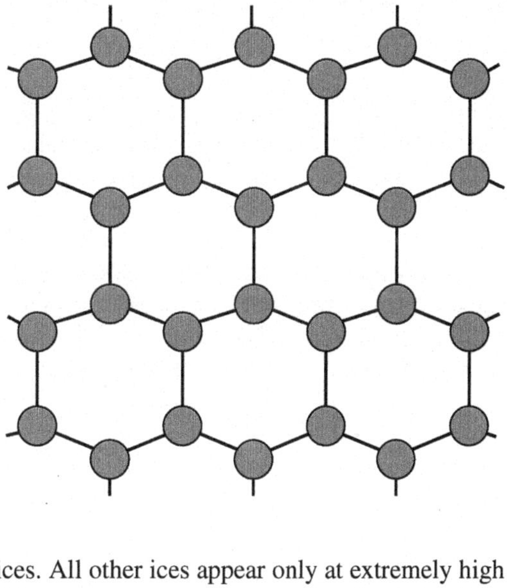

Fig. 1.3 Arrangement of water molecules in a basal plane of ice Ih. *Dark balls* are water molecules and *solid lines* are hydrogen bonds. The layer of six-fold symmetric rings is the "basal plane" and the normal direction to the plane is the *c*-axis

"ice Ih" to distinguish from other ices. All other ices appear only at extremely high pressures or low temperatures [2]. The arrangement of water molecules in Ice Ih is schematically illustrated in Fig. 1.3.

Water molecules, shown by dark balls, are arranged in a layer of hexagonal or six-fold symmetric rings. Lines connecting water molecules are hydrogen bonds. The crystallographic layer is called "basal plane," and the normal to the basal plane is the *c*-axis or the optical axis of the crystal. A single crystal or monocrystalline

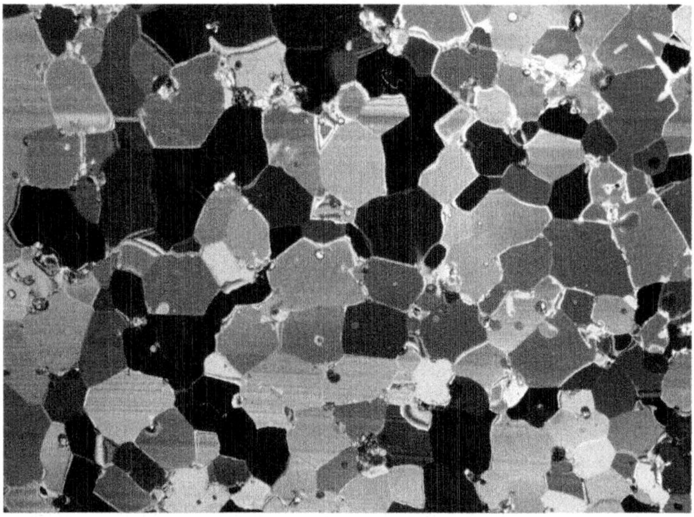

Fig. 1.4 Cross section of ordinary ice (polycrystalline ice)

ice can be thought of as consisting of these basal planes. Normally single crystal deforms by gliding on its basal planes just as a pack of cards is deformed by sliding cards each other. This means that the mechanical property of a single crystal of ice is different in different directions, or anisotropic, and the anisotropy can be described by the orientation of the *c*-axis. Anisotropy is also found in other physical properties such as optical refractive index, thermal conductivity, and so on.

Actual ice in our daily life, however, is usually polycrystalline, that is, a block of ice is composed of several single crystals. Figure 1.4 shows an example, a thin section of an ice cube made in a home-refrigerator. We see that the block is composed of small single crystals with different sizes, shapes, and *c*-axis orientations. The difference of orientations gives the different colors to individual crystals; the picture was taken by placing the thin section between two crossed polarization filters. Strictly speaking, the physical properties of the ice cube are dependent on sizes, shapes, and orientations of composing crystals, but we consider that usual polycrystalline ice we treat is composed of many single crystals with *c*-axes in different orientations so that its physical properties are isotropic in average.

1.1.2 Compressive Strength of Ice

Mechanical strength of ice can be tested by applying compressive, tensile or shear force on an ice specimen. We consider here only the compressive strength. The compressive strength of polycrystalline ice has been measured by many

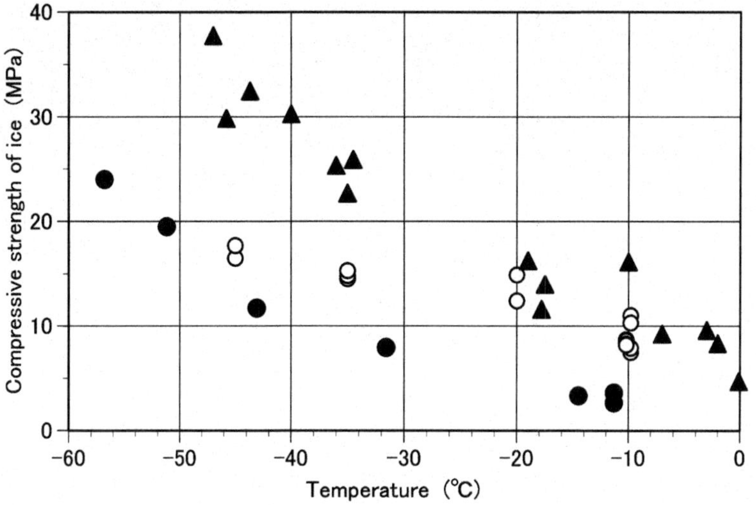

Fig. 1.5 Compressive strength of ice versus temperature. Data obtained by three different research groups are shown. *Filled triangle*: [3], strain rate $10^{-2}\,s^{-1}$; *open circle*: [10], $10^{-3}\,s^{-1}$; *filled circle*: [11], $10^{-5}\,s^{-1}$

investigators but reported data show wide range of scatter mainly because the strength depends on many variables such as temperature, strain rate, ice grain size, and tested volume (see [3–12]).

Figure 1.5 shows the temperature dependence of compressive strength measured at three strain rates from 10^{-5} to $10^{-2}\,s^{-1}$. We can see that the compressive strength is larger at larger strain rates and increases with decreasing temperature from about 5 MPa at 0 °C to 20–40 MPa at −50 °C.

Compressive stress–strain curves show ductile behavior at lower strain rates than about $10^{-4}\,s^{-1}$, and brittle behavior at higher strain rates at −10 °C [13]. The relation between the strength and grain size has not been studied in detail, but according to the measurement by Currier and Schulson [5] at $10^{-6}\,s^{-1}$ and −10 °C the tensile strength of ice decreases with increasing diameter. We may assume that the compressive strength is similarly smaller for ice composed of larger grains.

1.1.3 Friction Coefficient of Ice

Slipperiness is of the most important characteristics of ice for winter sports. The degree of slipperiness is expressed by the friction coefficient, μ, which is defined as the ratio of the force of friction and the force pressing the two surfaces. Compared to other materials μ of ice is much smaller, as small as 0.01, and a number of friction measurements of ice have been carried out as found in references of [14–16]. However, most of them were made using sliding surfaces of ice and

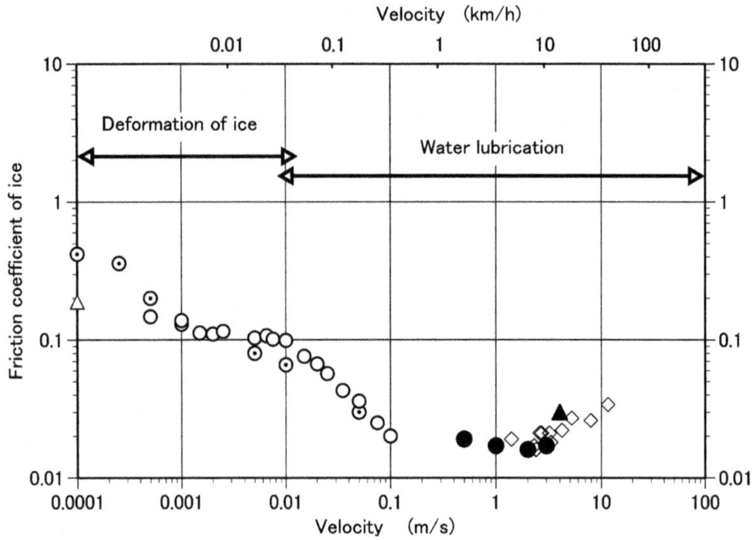

Fig. 1.6 Friction coefficient of ice versus sliding velocity. Data of ice–ice friction coefficient are shown. *Filled triangle*, [17]; *filled circle*, [18]; *dotted circle*, [19]; *open diamond*, [20]; *open circle*, [21]; *open triangle*, [22]. From [23] and [24] with changes

different materials mainly because of practical application needs to skates, skis, and other various structural interactions in ice environments. But ice surfaces show different friction behaviors to different materials, e.g. wood, polymer, metal, etc., and at different sliding velocities. Accordingly each result gave different explanations and could not lead to the systematic understanding of the intrinsic friction property of ice.

Figure 1.6 shows the friction coefficient of ice measured in a wide range of sliding velocity, from 0.0001 to 100 m/s. In the figure only the data obtained by sliding an ice block on ice at about −10 °C are selected to look into the intrinsic friction property of ice. It is clear that the velocity dependence of μ of ice is not simple but quite complicated, but we can explain the behavior by two physical mechanisms working in two different regions of sliding velocity, as indicated in the figure. One is the water lubrication mechanism above roughly 0.01 m/s, and the other is the plastic deformation of ice at the friction interface, which works below roughly 0.01 m/s. Here we discuss briefly the implication of the ice friction behavior to winter sports before we make more detailed explanation of the two mechanisms in the next section.

Ice friction in the velocity region above 0.01 m/s is more important for winter sports. In the region as the sliding velocity increases from 0.01 m/s the magnitude of μ decreases due to the increase of production of frictional heat to form water film as lubricant, and at higher velocities it increases again with increasing velocity. It is noted that there is a somewhat constant value region in between. The V-shaped relation between μ and velocity (U) can be expressed with the following theoretical equations discussed in the next section:

$$\mu = \frac{A}{\sqrt{U}} \tag{1.1}$$

at lower velocities and

$$\mu = B\sqrt{U} \tag{1.2}$$

at higher velocities where A and B are numerical constants.

Winter sports are played at various sliding velocities, so that depending on the kinds of winter sports μ at different regions become important. In curling, for instance, Eq. (1.1) may be useful because stones are normally delivered at velocities less than a few m/s, but for speed skates Eq. (1.2) may be more important. It should be noted that Fig. 1.6 shows the overall characteristics of the intrinsic friction of ice. For specific materials sliding on ice the absolute magnitudes of μ are different depending on their chemical and physical properties, but the general behaviors are similar [14–16].

The response of μ on temperature is important. According to the various measurements [14, 25, 26] it decreases linearly with temperature rise and expressed as follows:

$$\mu = C - D(T_m - T) \tag{1.3}$$

where T is the temperature, T_m is the melting temperature of ice (273 K), and C and D are numerical constants.

1.1.4 Friction Mechanisms of Ice

Slipperiness of ice, or low value of friction coefficient of ice, has long been explained by water lubrication. It is quite convincing because many things became slippery by water film as evidenced by wet tiles and pavements. Furthermore we know that water appears easily on the ice surface because we live at extremely high homologous temperatures of ice (Fig. 1.1). Any blocks of ice in front of us are just below the melting point and are threatened to melt away, just as a ceramic plate kept in a firing kiln at high temperatures as 1000 °C or higher.

Figure 1.7 illustrates schematically the friction of a material sliding on ice. The slipperiness of ice is realized by the appearance of some lubricant between the two surfaces. The lubricant is a thin water film at sliding velocities above 0.01 m/s, or plastically deformed ice layer below 0.01 m/s. The thickness of the film or layer has never been directly measured but it may be of order of 100 nm and is almost impossible to recognize with naked eyes. What is the physical mechanism to produce water on sliding ice surface? The first proposal was "pressure

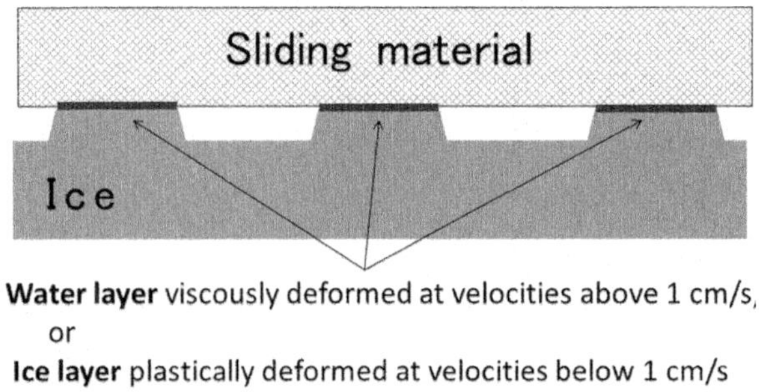

Fig. 1.7 Schematic description of ice friction. Slipperiness of ice can be explained by the existence of a thin layer which deforms with small shear force and behaves as lubricant. At sliding velocities above 0.01 m/s the layer is thin water film produced by frictional heating, which flows viscously. At velocities below 0.01 m/s the layer is a thin top surface of ice, which deforms plastically

melting" originated in the nineteenth century [27, 28]. It was a reasonable concept in those days when the thermodynamics just established showed that the melting point of ice (T_m) decreases with increasing pressure (p). However, the explanation is qualitatively correct but quantitatively incorrect because the effect of pressure melting is too small to explain the slipperiness of ice; the decrease of T_m with p is $dT_m/dp = -0.074$ K/MPa.

In the twentieth century the water lubrication mechanism was proposed by Bowden and Hughes [17]. The physical processes involved were extensively studied by Evans et al. [25] and then a reasonable quantitative expression for the friction coefficient was formulated by Oksanen and Keinonen [18] as a function of sliding velocity, temperature, and normal stress. The model of Oksanen and Keinonen [18] explains fairly well the measured V-shaped relation of ice friction coefficients at sliding velocities above 0.01 m/s (Fig. 1.6) that is, the decrease ($U^{-1/2}$) and increase ($U^{1/2}$) with the increasing sliding velocity, Eqs. (1.1) and (1.2).

The water lubrication model cannot explain the small value of μ at low sliding velocities below 0.01 m/s and at lower temperatures because heat produced by friction is not enough to warm and melt ice. At these low velocities and temperatures the mechanism of plastic deformation of ice is predominant. As discussed earlier in this chapter ice is easily deformed by sliding of basal planes. The shear deformation takes place only in a thin layer of the ice surface and adhesion and sintering between the sliding material and ice are also significant due to relatively long time of contact [24].

1.2 Structure and Friction Properties of Snow

1.2.1 Structure of Snow

Snow crystals form and grow by condensation of water vapor in clouds. Their outer forms are essentially hexagonal prisms reflecting the crystal symmetry of ice Ih. But actually snow crystals take a variety of forms, hexagonal plates, columns, needles, or dendrites, depending on the temperature, degree of water vapor saturation, and other conditions in clouds. Furthermore they often collide with each other to form snowflakes in the falling process.

Figure 1.8a shows an example of dendritic snow crystals a few hours after snowfall. All side branches have disappeared and the main branches are deformed and decaying horribly. We cannot see any delicate beautiful surface features characteristic of snow crystals, frequently introduced in many literatures. However it should be noted that these are actual snow crystals and snowflakes more commonly and frequently observed in wintertime. Once snow crystals and snowflakes reach the ground they immediately begin changing the forms. As the convex points tend to evaporate they lose their sharp corners and edges, but at the same time the concave parts grow by the condensation of water vapor so that with the passage of time they become round grains in average. An example is shown in Fig. 1.8b, which is a picture of a thin section of settled snow taken under a crossed polarized light. We see that complex shapes and features with high surface energy disappeared and most snow grains are of simpler shape. By estimating the mean masses of snow crystals and grains we find that a grain corresponds to roughly ten snow crystals, that is, ten snow crystals evaporated to grow a grain.

We should notice one more important fact in Fig. 1.8b that individual grains are connected to each other, that is, ice bonds grow between grains in contact. Needless to say, the physical mechanism of the bond growth is sintering mentioned in the first section of this chapter. Due to high homologous temperatures sintering is always progressing in snow. This is one of the most important characteristics in discussing snow for winter sports because bonds between grains alter the physical properties such as mechanical, thermal, and other properties.

The change of snow structure and properties may include other physical processes; if there exists a temperature gradient in snow the diffusional transport of water vapor produces faceted or hoar grains, and if the snow is wet the changing process will undergo faster due to liquid water. Anyhow snow continues to change its structure and properties forever if it does not disappear. The transformation process is called variously, e.g., metamorphism, densification, firnification, and so on.

The transformation of structure of snow is summarized as follows. Snow is essentially a mixture of ice grains and air. New snow is fluffy and full of air; its density is normally about 100 kg/m^3 or less. This means that 89 % of its volume is air. As the structure of snow changes with time, grains become larger, bonds connecting them become thicker and their packing becomes more compact, so that

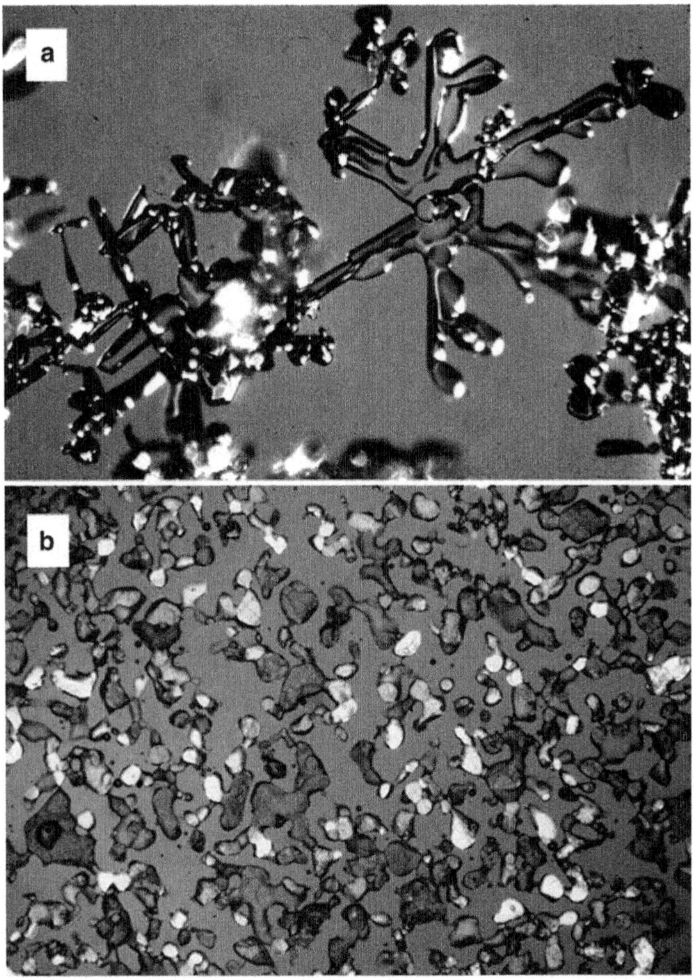

Fig. 1.8 New snow (**a**) and settled snow (**b**). (**a**) Snow crystals a few hours after snowfall ($-10\,^\circ$C). Horizontal width of picture 3 mm. (**b**) Cross section of settled snow of density 330 kg/m^3 under the crossed polarized light. Horizontal width 20 mm

the density increases and most of various physical properties also increase. In the case of seasonal snow, however, it melts away in spring when its density becomes about 300 kg/m^3, but in cold regions having glaciers the transformation continues long time and snow changes finally to ice. Generally the critical density at which snow transforms to ice is 830 kg/m^3, at which all air passages are sealed off so that air only exits as isolated bubbles in ice [29]. The fraction of air is 10 %. If we compare the two pictures in Figs. 1.4 and 1.8a, we can describe the transformation of snow to ice as follows: in new snow ice grains as "solute" are dispersed in air as 'solvent', but in ice air bubbles as 'solute' are dispersed in ice as "solvent."

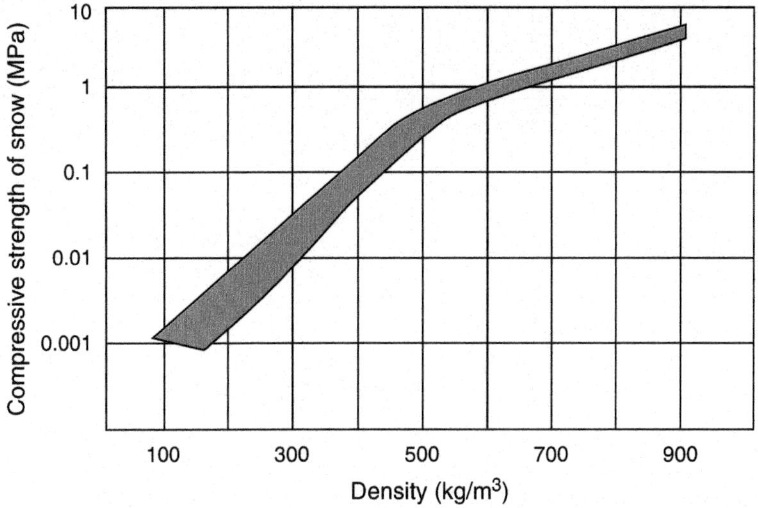

Fig. 1.9 Compressive strength of snow versus density. Range of data measured by 11 independent research groups. Strain rates are between 10^{-4} and $10^{-2}\ s^{-1}$ and temperature is between −7 and −25 °C. From [30] with changes

1.2.2 Compressive Strength of Snow

The knowledge of mechanical strength of snow is essential for winter sports. However, it involves many difficult problems as expected from the characteristic structure of snow. First of all snow is deformed by a minimum mechanical force, that is, snow creeps or shows viscoelastic behavior. The irreversible volumetric change makes measurements and interpretations quite difficult [30]. Here we consider the compressive strength of snow at relatively high strain rates, which may be more useful for winter sports.

Figure 1.9 is the measured compressive strength of snow plotted against the density. Hasty conclusion may be misleading since elastic and viscous properties of snow are very sensitive to density change. The data in Fig. 1.9 were obtained at high strain rates so that creep strain and density change are relatively small, i.e. the strength is regarded as characterizing brittle fracture. The increase of compressive strength with the increasing density is reasonable if we remember the densification process accompanied by bond growth due to sintering.

1.2.3 Friction Coefficient of Snow

Numerous measurements have been carried out and reported on the friction coefficients of snow using a variety of materials of practical importance as reviewed by Colbeck [31]. We will discuss only the results of ski sliding on snow. Figure 1.10

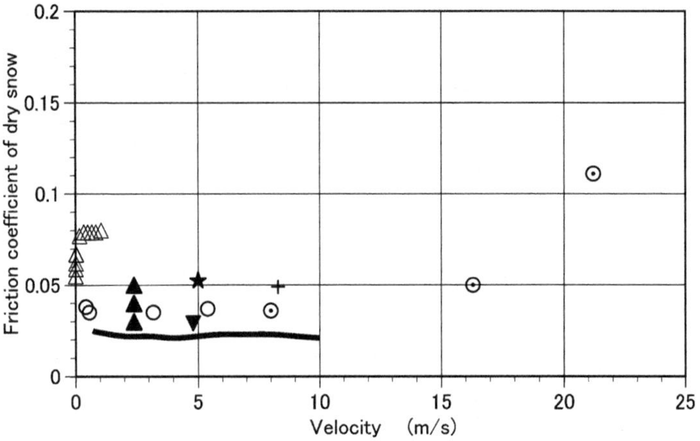

Fig. 1.10 Friction coefficient of dry snow versus sliding velocity. Data obtained by friction measurements of polyethylene ski on dry snow. *Filled triangle*: [32] at −10.5 and −13 °C; *filled inverted triangle*: [33] at −10 °C; *dotted circle*: [34] at −2.5 and −1.6 °C; *open circle*: [35] at −8 °C; *plus symbol*: [36] at −2.7 and −4.2 °C; *filled star*: [37] at −10 °C; *open triangle*: [38] at −10 °C; *solid line*: [39] at −7.5 °C

shows friction coefficient of snow obtained by eight research groups using a popular polyethylene ski without waxing. Though the density of snow used in each measurement is different and the temperature ranges from −1.6 to −13 °C, we can recognize clearly a general tendency. At the sliding velocities around 2–8 m/s the friction coefficient is almost constant, about 0.03, but it is 0.05 or even larger at velocities below 1 m/s, and it increases rapidly with velocity at velocities above 10 m/s. This general form looks very much like that of ice (Fig. 1.6), and we can regard this as a modified V-shaped relation of the friction behavior.

It is obvious from the discussion in Sect. 1.1.4 that the main friction mechanism of ski sliding at these velocities is the water lubrication, which is the viscous flow of thin water film produced between the ski base and snow grains. The magnitude of friction coefficient and the range of velocity variation may be different depending on several factors such as snow density, grain sizes, temperature, etc., and lead to a modified shape of V-shaped dependence as compared to that of the intrinsic ice friction (Fig. 1.6).

Two processes may be responsible for the modification; one is the viscoelastic deformation of snow and the other is compaction and plowing resistance of snow. As discussed earlier the compressive strength of snow is very sensitive to strain rates or loading time. If a ski slides at low velocities its bottom compresses the snow surface at small strain rates and the composing snow grains are deformed viscoelasitically and add a drag force, leading to larger friction coefficients below 1 m/s (Fig. 1.10). The effect at the low sliding velocities will be more significant for snow of low density or high temperature. The other process is the compaction and plowing resistance of snow, which becomes significant at higher velocities and its

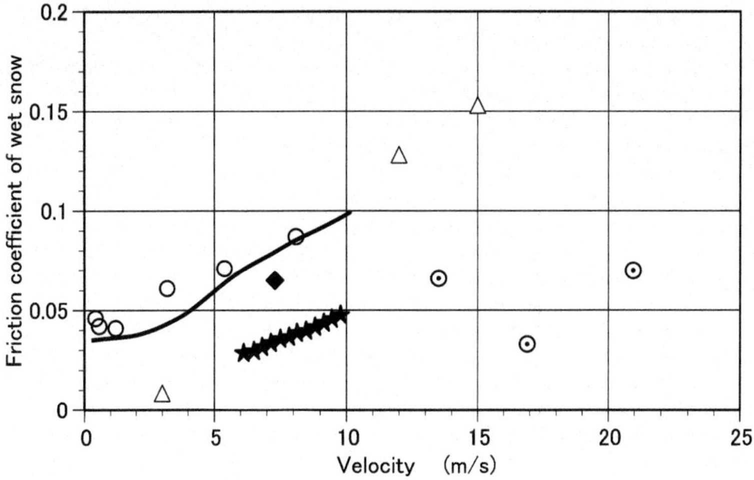

Fig. 1.11 Friction coefficient of wet snow versus sliding velocity. Data obtained by friction measurements of polyethylene ski on wet snow at 0 °C. *Dotted circle*: [34]; *filled star*: [43]; *open circle*: [35]; *open triangle*: [44]; *filled diamond*: [36]; *solid line*: [39]

velocity dependence may be expressed as U^2 just as air drag [40, 41]. The resistance is also larger for lower density snow and at higher temperatures. It is reasonable to assume that the effects caused by the two characteristic processes of snow are added to the friction property of intrinsic ice and lead to the modified V-shaped dependence of the friction coefficient of snow.

If snow is wet, composing snow grains are covered with thin water films, and if snow is very wet fairy amount of water is held between grains by capillary force. It is expected that such water in snow changes the mechanical properties and produces a drag effect to a ski through capillary attraction [31, 42]. Figure 1.11 shows the friction coefficient of wet snow. Since the water content and density of snow are different in each measurement the obtained data points spread in a wide range, but it is certain that the magnitudes are generally larger than those of dry snow.

References

1. N. Maeno, *Ice Science* (Hokkaido University Press, Sapporo, 2004)
2. V.F. Petrenko, R.W. Whitworth, *Physics of Ice* (Oxford University Press, Oxford, 1999)
3. F.D. Haynes, Effect of temperature on the strength of snow-ice, Cold Regions Research and Engineering Laboratory. CRREL Report 78-27 (1978)
4. R.L. Hooke et al., Mechanical properties of polycrystalline ice: an assessment of current knowledge and priorities for research. Cold Reg. Sci. Technol. **3**, 263–275 (1980)
5. J.H. Currier, E.M. Schulson, The tensile strength of ice as a function of grain size. Acta Metall. **30**, 1511–1514 (1982)

6. R.W. Lee, E.M. Schulson, The strength and ductility of ice under tension. J. Offshore Mech. Arct. Eng. **110**, 187–191 (1988)
7. E.M. Schulson, S.G. Hoxie, W.A. Nixon, The tensile strength of cracked ice. Philos. Mag. **A59**, 303–311 (1989)
8. J. Druez, P. Mccmber, Transactions of the Canadian Society for Mechanical Engineering. Tech. rep. **13**, 59 (1989)
9. X. Xian et al., An experimental evaluation of the tensile strength of impact ice. J. Mater. Sci. Lett. **8**, 1205–1208 (1989)
10. E.M. Schulson, The brittle compressive fracture of ice. Acta Metall. Mater. **30**, 1963–1976 (1990)
11. M. Arakawa, N. Maeno, Mechanical strength of polycrystalline ice under uniaxial compression. Cold Reg. Sci. Technol. **26**, 215–229 (1997)
12. J.J. Petrovich, Mechanical properties of ice and snow. J. Mater. Sci. **38**, 1–6 (2003)
13. L.J. Gold, Engineering properties of fresh water ice. J. Glaciol. **19**(81), 197–223 (1977)
14. M. Akkok, C.M. McC. Ettles, S.J. Calabrese, Parameters affecting the kinetic friction of ice. J. Tribol. Trans. ASME **109**, 552–561 (1987)
15. A.M. Kietzig, S.G. Hatzikiriakos, P. Englezos, Physics of ice friction. J. Appl. Phys. **107**, 08110 (2010)
16. C. Stamboulides, P. Englezos, S.G. Hatzikiriakos, The ice friction of polymeric substrates. Tribol. Int. **55**, 59–67 (2012)
17. F.P. Bowden, T.P. Hughes, The mechanism of sliding on ice and snow. Proc. R. Soc. Lond. **A172**, 280–298 (1939)
18. P. Oksanen, J. Keinonen, The mechanism of friction of ice. Wear **78**, 315–324 (1982)
19. D.E. Jones, An experimental investigation of low-speed ice-ice friction. MA thesis. ME thesis, Thayer School of Engineering, Dartmouth College, Hanover, 1989
20. G. Casassa, H. Narita, N. Maeno, Shear cell experiment of snow and ice friction. J. Appl. Phys. **69**, 3745–3756 (1991)
21. A. Yasutome, M. Arakawa, N. Maeno, Measurements of ice-ice friction coefficients. Seppyo **61**(6), 443–447 (1999)
22. S. Kanazawa, M. Arakara, N. Maeno, Measurements of ice-ice friction coefficients at low sliding velocities. Seppyo **65**(4), 389–397 (2003)
23. N. Maeno et al., Ice-ice friction measurements, and water lubrication and adhesion-shear mechanisms. Can. J. Phys. **81**, 241–249 (2003)
24. N. Maeno, M. Arakawa, Adhesion shear theory of ice friction at low sliding velocities, combined with ice sintering. J. Appl. Phys. **95**(1), 134–139 (2004)
25. D.C.B. Evans, J.F. Nye, K.J. Cheeseman, The kinetic friction of ice. Proc. R. Soc. Lond. **A347**, 493–512 (1976)
26. D. Slotfeldt-Ellingsen, L. Torgersen, Water on ice; influence on friction. J. Phys. D **16**, 1715–1720 (1983)
27. J. Joly, The phenomena of skating and Professor J. Thomson's thermodynamic relation. Sci. Proc. R. Dublin Soc. **5**, 453 (1887)
28. O. Reynolds, On the slipperiness of ice. Mem. Proc. Manch. Lit. Philos. Soc. **43**(5), 2 (1899)
29. W.S.B. Patterson, *The Physics of Glaciers*, 3rd edn. (Elsevier, New York, 1994)
30. M. Mellor, A review of basic snow mechanics. IAHS-AISH Pub. **114**, 251–291 (1975)
31. S.C. Colbeck, A review of the processes that control snow friction. CRREL Monograph 92–2, 1–40 (Cold Regions Research and Engineering Laboratory, Hanover, 1993)
32. M. Shimbo, Sliding on snow surface and friction. Seppyo **22**(5), 147–156 (1960)
33. F.P. Bowden, D. Tabor, *The Friction and Lubrication of Solids* (Oxford University Press, Oxford, 1964)
34. D. Kuroiwa et al., The coefficient of sliding friction between ski and chemically treated or mechanically compressed snow surface. Low Temp. Sci. **A27**, 229–245 (1969)
35. R. Tanahashi, Mechanism of ski-sliding: effects of materials and vibration on friction. J. Jpn. Soc. Mech. Eng. **95**(888), 1001–1004 (1992)

36. D.A. Moldestad, Some aspects of ski base sliding friction and ski base structure, Engineering thesis, Department of Structural Engineering, Norwegian University of Science and Technology, Trondheim, 1999
37. D. Buhl, M. Fauve, H. Rhyner, The kinetic friction of polyethylen on snow: the influence of the snow temperature and the load. Cold Reg. Sci. Technol. **33**, 133–140 (2001)
38. M. Takeda et al., Friction of the short model ski at low velocity. J. Phys. Conf. Ser. **258**, 1–9 (2010)
39. M.A.H. Leino, E. Spring, Determination of the Coefficient of Kinetic Friction Between Ski and Snow from Gliding Velocity of a Skier. Rep. Ser. Geophys. **19**, 1–11 (University of Helsinki, Helsinki, 1984)
40. R.I. Perla, Avalanche release, motion, and impact, in *Dynamics of Snow and Ice Masses*, ed. by S.C. Colbeck (Academic, New York, 1980), pp. 397–462
41. B. Glenne, Sliding friction and boundary lubrication of snow. J. Tribol. **109**(4), 614–617 (1987)
42. S.C. Colbeck, Kinetic friction of snow. J. Glaciol. **34**(116), 78–86 (1988)
43. J. Erkkila et al., A cinematographic method for determination of the kinetic friction of skis on snow. Rep. Ser. Geophys. Univ. Hels. **21**, 1–15 (1985)
44. P. Kapps, W. Nachbauer, M. Mossner, Determination of kinetic friction and drag area in alpine skiing. Ski Trauma Skiing Saf. **10**, 1–17 (1996)

Chapter 2
Friction Between Ski and Snow

Werner Nachbauer, Peter Kaps, Michael Hasler, and Martin Mössner

2.1 Introduction

The first skis were developed to improve locomotion across the natural, wind packed snow surface in the European northern countries. The skis were made of flat planks with shovels at the tips. Under load, the tips and ends of the skis bended up causing resistance against forward movement. An improvement of the gliding of skis was the invention of the bow-shaped cambered ski, arched up towards his center. Under load, the ski lies flat on the snow surface and the load is more evenly distributed along the ski. With the appearance of downhill skiing, the turning properties of skis became more important. In 1928, Lettner (AT) invented steel edges to give the skis more grip. During the first half of the twentieth century, the technique was developed to produce laminated skis composed of a wooden core with different bottom and upper layers. In 1955, Kofler (AT) introduced the first ski with a polyethylene base, which remarkably improved the gliding properties. In addition, the repair of minor scratches was easily possible. In the recent past, the gliding properties of skis were further developed by special grinding techniques for the ski base and by the development of special waxes.

W. Nachbauer (✉)
Department of Sport Science, University of Innsbruck, Innsbruck, Austria

Centre of Technology of Ski and Alpine Sports, University of Innsbruck, Innsbruck, Austria
e-mail: werner.nachbauer@uibk.ac.at

P. Kaps
Department of Engineering Mathematics, University of Innsbruck, Innsbruck, Austria

M. Hasler • M. Mössner
Centre of Technology of Ski and Alpine Sports, University of Innsbruck, Innsbruck, Austria

F. Braghin et al. (eds.), *The Engineering Approach to Winter Sports*,
DOI 10.1007/978-1-4939-3020-3_2

Besides practical ski development, a lot of knowledge about the friction processes was gained. The initial considerations of friction on snow regarded dry friction. Soon the reason for the low friction on snow and ice was presumed to lie in a thin water film. This water film was first supposed to be generated by pressure melting. Ice melts at lower temperature when additional pressure is applied [e.g., 1]. Energy considerations led to another theory. At snow temperatures relevant for skiing, the heat generated by friction is sufficient to melt snow [2]. This process causes lubricated friction, which is now considered as the dominant friction regime of skis on snow. In the last decade, an additional theory came up. Quasi-liquid films were observed due to premelting even below the melting point of crystals [3] which supposedly have an effect on snow friction.

2.2 Theory of Snow Friction

2.2.1 *Dry Friction*

For friction between a solid and its supporting surface the friction force, F_F is given by $F_F = \mu F_N$. F_N is the normal load and μ is the coefficient of friction. One has to distinguish between static and kinetic friction, depending on whether the object is resting or moving. Static coefficients usually are larger than kinetic ones.

Leonardo da Vinci (1452–1519) already stated the two basic laws of friction: (1) the area of contact has no effect on friction and (2) if the load of an object is doubled, its friction will be doubled. Two centuries later Amonton (1663–1705) investigated friction. According to Amonton's first law the friction force is proportional to the applied load, and due to Amonton's second law the constant of proportionality is independent of the contact area. Coulomb (1736–1806) continued Amonton's studies and found that the coefficient of kinetic friction is independent of speed. After another century, [4, 5] stated that the friction between solid bodies arises from the contact of the asperities on the surface of a solid body. As contact pressure increases, more asperities come into contact, and so frictional resistance increases.

In skiing dry friction is usually modeled as solid-to-solid interaction. The frictional resistance originates from shearing off the tips of the snow asperities and/or wax or running base asperities. The coefficient of friction is given according to [6] as

$$\mu = \frac{\tau}{\sigma} \tag{2.1}$$

(which is equivalent to $F_F = \mu F_N$). τ represents the shear stress and σ the normal stress acting between ski base and snow. For the shear stress, τ one has to use the shear strength of the snow asperities or the ski base asperities including wax. The smaller value is the appropriate one. At a temperature near 0 °C, the snow asperities are softer than the ski base but harder as most ski waxes [7, Fig. 10]. For snow/ice temperatures from 0 down to $-32\,^{\circ}$C the shear strength of the ice asperities increases about 15 % and the compression strength roughly about 400 %. Consequently, the coefficient of dry friction drops for decreasing temperatures [8, Table 1].

2.2.2 *Wet Friction*

A lubricating water film between ski and snow causes reduced friction compared to dry friction. Bowden and Hughes [2] related the formation of the water film to frictional heating. Friction between ski and snow generates heat, which dissipates into snow and ski. If the melting temperature of snow is reached, water acts as lubricant between ski and snow. When a complete water film exists between snow asperities and ski base, the friction force for a laminar flow is given by

$$F_{F,\text{wet}} = \frac{\eta_W A_C v}{h}. \tag{2.2}$$

A_C denotes the real contact area between ski and snow asperities, h the thickness of the water film, and v the speed of the ski. The dynamic viscosity of water is given by $\eta_W = 1.79 - 0.054\vartheta_S$ mPa s with ϑ_S the snow temperature in °C.

Depending on the properties of the snow and the texture of the ski base, the real contact area between snow and ski may be quite small. The snow is composed of ice crystals with various shapes and sizes, which are packed and sintered together. The structure of the ski's running surface is given by its texture. The whole load applied by the ski is supported by a small fraction of the snow surface. For small contact pressures, the ski will run on some spikes. For larger contact pressures, these spikes collapse and so the real contact area gets larger. Let A be the area of the ski base and $A_C = aA$ be the real contact area. According to [8], the fraction of the real contact area, a ranges between 0.001 and 0.015 and to [9] between 0.01 and 0.1.

The critical Reynolds number gives the transition from laminar to turbulent flow. For a rectangular cross section the flow remains laminar as long as

$$R = \frac{\rho_W h v}{\eta_W} < R_{\text{crit}} = 1500 \quad \text{or if} \quad hv < 2 \cdot 10^{-3}\,\text{m}^2\,\text{s}^{-1}. \tag{2.3}$$

In skiing speeds are below 40 m s^{-1} and the water film thickness is clearly below 50 μm [10]. Therefore, the flow in the water film is laminar.

2.2.3 *Mixed Friction*

Dry friction occurs when meltwater lubrication is absent and wet friction when the ski-snow contacts are completely covered by a layer of water. In many cases, snow asperities and ski base may not be fully separated by a water film and so solid-to-solid contacts as well as lubricated contacts occur, i.e. mixed lubrication conditions prevail. Due to frictional heating the water film thickens and the number of lubricated contacts increases. Squeezing the water away from the snow asperities reduces the water film thickness and increases the number of solid-to-solid contacts.

Along the ski there may exist solid-to-solid contacts at the front of the ski and towards the tail of the ski, due to frictional heating, mixed conditions, and finally solely wet friction. Therefore, the friction coefficient varies along the ski too. To calculate the overall coefficient of friction for a ski, one may partition the ski into several segments. Then the coefficient of friction for the whole ski is given by

$$\mu = \frac{1}{F_N} \sum_{i=1}^{n} F_{F,i}. \tag{2.4}$$

The friction force ($F_{F,i}$, $i = 1, \ldots, n$) has to be calculated for every segment separately by using the appropriate friction model (dry, mixed, or wet friction) with the appropriate parameters (water film thickness, normal load, contact area, etc.).

In an advanced approach, one has additionally to consider contributions due to ploughing and compression [8] specifically in the front region of a ski, and for turns, skidding and carving processes.

2.2.4 Quasi-Liquid Layer

At snow temperatures below 0 °C, no melt water is produced. Yet, there is a boundary layer, called quasi-liquid layer, with the thickness of some few water molecules [3, 11]. It is in an intermediate state between solid and liquid. Above −80 °C the crystalline structure of the ice begins to alter and for temperatures between −30 and 0 °C the ice surface is covered with a 1–10 nm thick quasi-liquid layer of quasi-liquid water [3, Fig. 9]. This tiny layer rounds the edges of the ice asperities and reduces the solid-to-solid interactions between the ice crystals. It is presumed that this makes the ice asperities slippery and contributes to the small coefficients of friction at moderate snow temperatures below 0 °C.

2.3 Heat Considerations

2.3.1 Basic Considerations

The solid state of water exists in nature as snow and ice. The amount of heat per unit mass necessary to raise the temperature of a certain material by 1 K is called specific heat capacity c_p. Table values for ice, snow, and water are 2.110, 2.009, and 4.186 $\mathrm{kJ\,kg^{-1}\,K^{-1}}$, respectively. The heat to raise the temperature of the snow mass m_S

with volume V and density ρ_S by ΔT degrees of Kelvin is given by $Q = c_{p,S} m_S \Delta T$. When the snow temperature reaches the melting point of snow any further supplied heat is used to melt snow. When the whole snow is molten, the water temperature rises according to the specific heat capacity of water. If water is cooled, the water temperature drops until 0 °C and freezing starts. The water-snow mixture keeps the temperature of 0 °C until all water is frozen. Once the water is fully frozen, the temperature of the snow continues to fall. The specific melting heat or enthalpy of fusion is the heat per unit mass added during the phase transition from snow to water and vice versa. Its value is $H_{\mathrm{fus},W} = 335\,\mathrm{kJ\,kg^{-1}}$. The heat required to melt snow is given by $Q = H_{\mathrm{fus},W} m_S$. This amount of energy is needed to release the water molecules from the lattice structure of the snow crystals. The same amount of energy is released when the water freezes again.

Because of frictional heating, the temperature of the tips of the snow asperities below the ski is increased. When the temperature reaches 0 °C, snow asperities melt causing a layer of water. We consider a particular location on the snow surface. The passage of a 1.75 m long and 0.06 wide ski loaded with 400 N produces an amount of 35 J of heat ($Q = \mu F_N L$) when presuming a friction coefficient of 0.05. The heat flows partly into snow and ski and when the snow temperature is high enough, snow melts. The energy needed to heat snow is much lower than the energy needed to melt snow. In the following two extreme cases are examined. Firstly, it is assumed that all 35 J of heat is used to heat snow. The snow mass of 1.74 g can be heated from −10 to 0 °C ($Q = c_{p,S} m_S \Delta T$). This mass of snow has a volume of 3480 mm^3 if one assumes a snow density of 500 kg m^{-3}. This corresponds to a 0.033 mm thick layer of snow between ski and snow ($A = 1.75 \times 0.06 = 0.105\,\mathrm{m^2}$). In reality, the snow is not heated at once and one has to solve the heat equation to study the propagation of heat. Nevertheless, one can conclude that during the passage of a ski the heat transfer is restricted to a quite thin layer of snow, whose thickness is approximately the magnitude of the snow asperities. Secondly, it is assumed that all 35 J of heat is used to melt snow. The snow mass of 0.11 g is converted to water ($Q = H_{\mathrm{fus},W} m_W$). If the amount of melt water is uniformly distributed along the complete running surface of the ski ($A = 0.105\,\mathrm{m^2}$), the thickness of the water film is 1.0 µm. The real contact area is not known, since the melt water is concentrated on the snow asperities with contact to the ski. Published results differ by two orders of magnitude (see Sect. 2.2). If only a tenth of the running surface of the ski is covered by water, the water film thickness is 10.0 µm. Experimentally, the water film thickness was measured between 5 and 20 µm using a dielectric device [10]. Because their measurement device had a hydrophilic coating, the measured values are likely too large. Strausky et al. [12] used fluorescence spectroscopy to detect the water film. The lower limit of their measurement range was 0.05 µm. They did not detect any water in their measurement range. This may be reasonable for the measured speed below 0.1 m s^{-1}. Bäurle et al. [9] presented simulations of the heat flow with water film thicknesses considerably lower than 1.0 µm. Their results suggest that a large part of the produced heat flows into ski and snow and only a small fraction is used to melt snow.

2.3.2 Modeling of Heat Flow Between Ski and Snow

In this section, the location along the ski is determined, where the transition from dry to mixed/wet friction occurs. For this, it is not necessary to know the exact amount of the produced melt water. For simplicity, we focus on the heat flow into the snow and neglect the flow into the ski. Thus, the melting temperature at the snow surface is reached faster, since the complete produced heat is transmitted into the snow.

At a particular location on the snow surface a Cartesian coordinate system is introduced with the x-coordinate along the ski's longitudinal direction, the y-coordinate transverse to the ski, and the z-coordinate normal to the snow surface. The snow surface is given by the plane $z = 0$ and the positive z-coordinate points towards the snow. The heat flow in z-direction is computed by solving the one-dimensional heat equation. During the passage of the ski, heat is produced with the rate $\frac{\Delta Q}{\Delta t} = F_F v$. Here F_F is the friction force and v the speed of the ski. This causes a constant heat flux

$$J = \frac{1}{A}\frac{\Delta Q}{\Delta t} = \frac{F_F v}{A} \tag{2.5}$$

into the snow. A denotes the contact area of the whole ski with the snow. The friction force along the ski is assumed to be constant. $T(z,t)$ is the snow temperature at the time t and at the penetration depth z. The start time ($t = 0$) is given with the first contact and the end time ($t = L/v$) with the last contact of the ski. A complete set of equations to calculate the snow temperature, T is given by the heat equation for the snow, the initial temperature of the snow, T_0, and the heat flux into the snow surface, J:

$$\frac{\partial T}{\partial t} = \alpha\frac{\partial^2 T}{\partial z^2}, \qquad T(z,0) = T_0, \qquad -\lambda\frac{\partial T}{\partial z}(0,t) = J. \tag{2.6}$$

Because the ski is only supported by snow asperities, for the thermal diffusivity, α and the thermal conductivity, λ table values for ice are used: $\alpha_I = 0.93\,\text{mm}^2\,\text{s}^{-1}$ and $\lambda_I = 1.8\,\text{W}\,\text{m}^{-1}\,\text{K}^{-1}$. The solution for the heat equation is [13]:

$$T(z,t) = T_0 + \frac{2J}{\lambda}\sqrt{\frac{\alpha t}{\pi}}f\left(\frac{z}{\sqrt{4\alpha t}}\right) \qquad \text{with} \qquad f(q) = e^{-q^2} - \sqrt{\pi}q\,\text{erfc}(q). \tag{2.7}$$

Here, erfc denotes the complementary error function. From this equation the temperature rise on the snow surface ($z = 0$) can be computed. We get

$$\Delta T = \frac{2J}{\lambda}\sqrt{\frac{\alpha t}{\pi}}. \tag{2.8}$$

The time with dry friction is defined by $t_{\text{dry}} = L_{\text{dry}}/v$ with L_{dry} the length of the section in the frontal part of the ski with dry friction. In this time, the temperature of the tips of the snow asperities raises from T_S to 0 °C. We set $\Delta T = \vartheta_I$, with ϑ_I the temperature of the snow asperities in degree Celsius. Putting all together, one gets

$$L_{\text{dry}} = \frac{\pi}{\alpha_I v}\left(\frac{\lambda_I A_C \vartheta_I}{2\mu_{\text{dry}} F_N}\right)^2. \tag{2.9}$$

As numerical example we consider a cross-country ski of the size $A = 1.75 \times 0.06 = 0.105\,\text{m}^2$, with a real contact area of a tenth ($a = 0.1$), loaded with $F_N = 400\,\text{N}$, and moving at a speed of $v = 7\,\text{m}\,\text{s}^{-1}$. For dry friction $\mu = 0.25$ and 0.22 are used for snow temperatures of $\vartheta_I = -5$ and -10 °C [8, Table 2]. In this case, the sections with dry friction have a length $L_{\text{dry}} = 0.11$ and 0.55 m, respectively. If the real contact area is only a twentieth ($a = 0.05$), or if the speed is $v = 28\,\text{m}\,\text{s}^{-1}$ (e.g. downhill skiing), the corresponding numbers are $L_{\text{dry}} = 0.03$ and 0.14 m. The calculated L_{dry} is a lower bound for the real length with dry friction, since some heat also flows into the ski.

Note, the sections with dry friction are longer for lower temperatures due to the lower temperature of the snow as well as due to the smaller coefficient of dry friction. The smaller coefficient of friction causes lower frictional heating. This contributes to the well-known fact that the total friction for the whole ski increases with decreasing temperature.

The presented modeling approach has two major limitations. The heat production rate along the ski is not constant and heat is not only transmitted into the snow but also into the ski. To obtain a more realistic approximation of the amount of produced melt water, the heat equation for the heat flow into snow and ski has to be solved. The major difficulties are the rate of heat production for the different kinds of friction (dry, mixed, wet), the heat consumption for melting of snow, and the thinning of the water film by squeezing. Such models were presented for ice [14, 15] and for snow [9]. All these approaches needed the numeric solution of a complicated set of equations.

2.4 Experiments

Friction between skies and snow is affected by numerous factors, e.g. speed, load, snow properties (temperature, density, liquid-water content, hardness, texture, etc.), and ski properties (stiffness, thermal conductivity, base material, base roughness, etc.). Most of the knowledge about ski-snow friction is from laboratory measurements. The results of the laboratory studies are important for explaining the basic processes of snow friction; however, the results are restricted to the conditions in which the experiments were conducted. Load, speed, slider characteristics, and snow properties used in the laboratory do usually not correspond to ski gliding, where long, highly loaded skies move at high speed under a diversity of man-made and natural snow conditions. Furthermore, friction varies along a sliding ski, which cannot be mirrored with short probes.

Tests under ski-specific conditions have to be carried out in the field, e.g. on long gliding tracks with skiers. The results of these field tests are regrettably impaired by changing environmental conditions and variation of the skier's motion. To determine snow friction, air and gliding resistance have to be separated which constitutes a major problem, especially when measuring at high speeds where drag is overwhelming. In addition, the conduction of outdoor tests is time-consuming and cumbersome. Hence, only few measurements have been published. They are presented in the next chapter.

To investigate the factors influencing snow friction one by one in a controlled environment, various experimental devices were developed. Among these, pin-on-disk tribometers are the most common types. In these rotational tribometers, a stamp carrying the probe is pressed against a snow-filled rotating disk and the frictional force acting on the probe is measured. Tribometers with rotational disks have serious drawbacks. One is the repeated contact of the probe with the snow or ice on each rotation. Temperature and water film can subsequently build up and the snow surface is polished with each passage of the probe. A second drawback is the centrifugal force at high speeds. Melt water might be moved radially outwards altering the measured friction force. Consequently, at the Centre of Technology of Ski and Alpine Sports at the University of Innsbruck a linear tribometer was developed on which snow friction can be measured for a whole ski at speeds up to $30\,\mathrm{m\,s^{-1}}$. A description of this tribometer follows in a separate chapter below.

2.4.1 Field Tests

In competitive alpine and cross-country skiing, every year numerous gliding tests are conducted by ski manufacturers and national ski federations to select the fastest skies for their athletes. Results of these tests are usually not published. For alpine skiing, experienced ski testers glide straight down the test courses in the tucked downhill position. The test courses are about 350 m long; it is intended to reach maximal speeds above $30\,\mathrm{m\,s^{-1}}$. Besides split and end times using photo cells some environmental conditions like air and snow temperature, air humidity, and snow moisture are taken. Running times and corresponding speeds are used to assess the sliding friction of different base material, base texture, and waxes.

On a permanent 342 m long test course located in Seefeld (AT) an experienced ski tester performed 175 test runs with the same equipment during a winter season. Air and snow conditions were measured for each run [16]. As an example running time vs air temperature for snow with and without liquid water content is presented in Fig. 2.1. At about 6 °C liquid water was partly measured in the snow. One can see different tendencies: Decreasing friction with increasing air temperature on dry snow and increasing friction with increasing temperature on wet snow. It may be speculated that the decreasing friction on dry snow is due to increased lubrication by increased water film thickness and the increasing friction on wet snow is caused by an increase of capillary attachments between ski and snow.

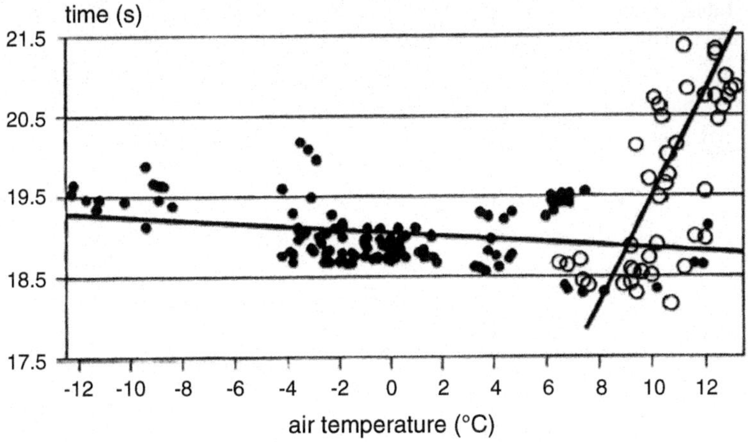

Fig. 2.1 Run time versus air temperature in a straight-run gliding test. The *filled circles* correspond to snow without and the *open circles* to snow with liquid water content [16]. Reprinted, with permission, from ASTM STP1266 Skiing Trauma and Safety: Tenth Volume, copyright ASTM International, 100 Barr Harbor Drive, West Conshohocken, PA 19428

In Table 2.1 the published field friction measurements for straight gliding, traversing, and turning are summarized. In the following the field experiments conducted by our group are briefly outlined. On the permanent test course in Seefeld (AT) the coefficient of friction and the drag area of skiers during straight gliding were determined [17]. Nine photocells to measure split times were installed. By geodetic surveying slope inclination and the locations of the photocells were determined. The motion of the skier was formulated as differential-algebraic equation and an optimization method was used to separate drag and friction force. The results revealed a drag area of $0.22\,\mathrm{m}^2$ and a coefficient of friction of 0.0085 (Table 2.1). The value for the drag area compares to wind tunnel experiments, whereas the value for the coefficient of friction is lower than the value found in other studies. A reason might be that the coefficient of friction may depend on a term proportional to v^2, too, but such a term cannot be separated from drag.

Coefficients of friction and drag areas for traversing and turning are published by Kaps and Nachbauer [18] and Kaps et al. [17, 19]. Position-time of the center of mass of the skiers was determined by two- or three-dimensional video analysis. The skier's equation of motion was formulated as differential-algebraic and by dynamic optimization friction coefficients and drag area were determined. In these cases, snow friction is significantly affected by ploughing, compression, and skidding and carving actions of the ski. As one can see from the coefficients of friction (Table 2.1), the frictional interaction at the ski-snow interface plays a minor role especially for turning.

Table 2.1 Overview of friction measurements during straight gliding, traversing, and turning

μ	Speed (m s^{-1})	Method	Source
Straight running			
0.03–0.04	2.2	Towing method	[20]
0.1	8–27	Run-out, light cells	[20]
0.02–0.1	8–22	Light cells	[21]
0.02 (dry snow)	0.5–10	Run-out, light cells	[22]
0.03–0.1 (wet snow)	0.5–10	Run-out, light cells	[22]
0.02–0.05	5.5–10.5	Run-out, light cells	[23]
0.0085	3–29	Light cells	[17]
0.041	0–4	Towing method	[24]
Traversing			
0.064	1–11	Video	[17]
0.128	11–13	Video	[17]
0.153	15–17	Video	[17]
Turning			
0.2–0.3	Carving	Video	[18]
0.45	Downhill racing turn	Video	[19]
< 0.3	Stem and parallel turn, Wedeln, and snow plow	Serial images	[25]

2.4.2 *Laboratory Tests*

Bowden and Hughes [2] first investigated in the laboratory the frictional resistance when gliding on snow or ice. They developed a friction apparatus to analyze friction on snow for different materials, loads, and speeds. Based on their observations, they presumed the generation of a water film by frictional heating between slider and snow. At present, this is a generally accepted theory. In the following, many different laboratory tribometers were developed. Figure 2.2 gives an overview of the publications in the field. Studies on snow were carried out to a maximum speed of 5 m s^{-1} and a sample size of 0.1 m. Even on ice, a much easier manageable surface than snow, probe size never was in regions of interest for snow sports.

The pin-on-disk tribometer is the most common device type used in the research on friction on snow and ice. It consists of a rotating disk covered with ice or snow and a probe pushed with given normal force to the surface. The force against the rotating direction parallel to the disk is measured. It has been built in a wide range of sizes from a diameter of several mm [26] to 1.8 m [27]. Evans et al. [28] chose a design with a pendulum hanging around a revolving drum of ice to imitate the movement of an ice skating blade on the ice. Jones et al. [29] determined the friction coefficient using a conveyor belt and measuring the friction force on a fixed ice sample. Several studies have been carried out using ring-on-disk or disk-on-disk tribometers, where instead of a pin a ring or a disk is pressed against the snow or

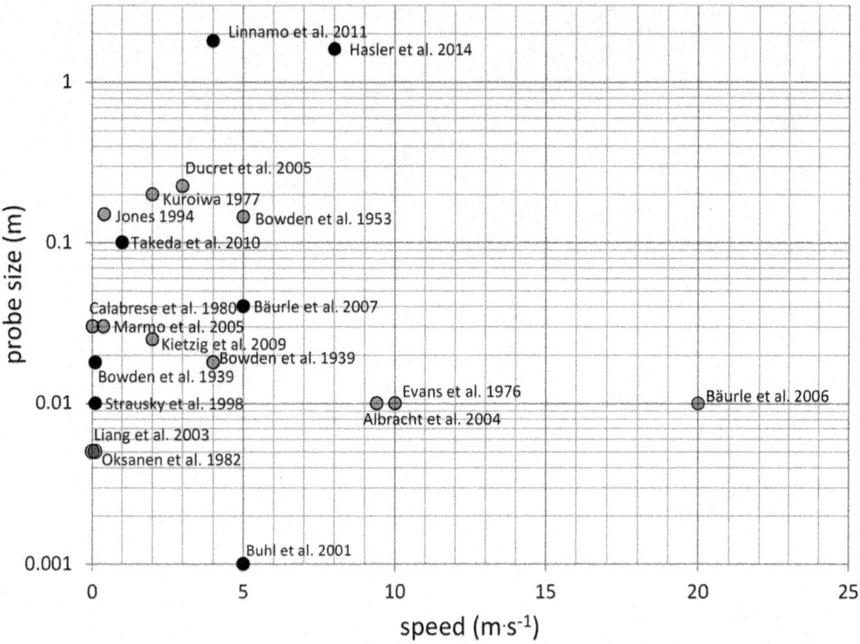

Fig. 2.2 Overview of laboratory experiments on snow (*black circles*) and ice (*grey circles*) [34–41]

ice surface [e.g., 21, 30, 31]. Temperatures ranged from −150 °C [26] to the melting point of 0 °C.

The first laboratory measurements were carried out on ice and on snow [2]. Later, mainly ice friction measurements were done because of the easier setup of the testing surface and the better reproducibility of the experiments. Furthermore, on ice mechanisms subject to friction are easier to model since the contact area is flat and hence well defined. Especially with rotating measurement devices snow is difficult to handle due to its graininess and the change of the contact area by the abrasion that occurs with each rotation.

2.4.3 Friction Measurements with the Linear Tribometer

At the Centre of Technology of Ski and Alpine Sports of the University of Innsbruck a linear tribometer (Fig. 2.3) was developed for investigating friction between snow and skies. The device consists of a linear beam, a trough for the snow, and carriage for the probe or ski. The trough can be moved laterally in order to have up to ten fresh tracks of the same snow for the measurements available. The carriage is pulled by an 80 kW high torque electric-motor. At the end of a run, the carriage is decelerated by the electric motor and an eddy current brake. Each run consists

Fig. 2.3 Linear tribometer

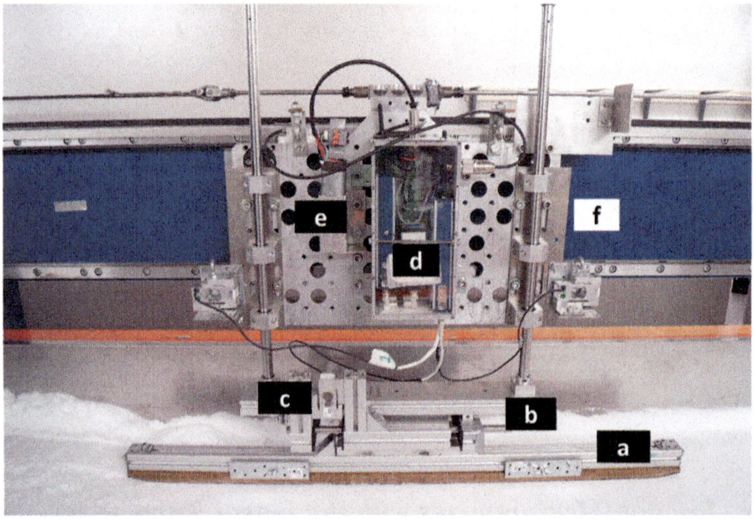

Fig. 2.4 Carriage of the linear tribometer. (*a*) Ski base probe, (*b*) lower carriage, (*c*) load cell horizontal force, (*d*) data acquisition and transmission system, (*e*) battery, (*f*) linear beam

of three phases: acceleration, measurement, and deceleration. In the measurement section, the carriage runs with almost constant speed. Available speeds range between 0.1 and 30 m s^{-1} resulting in a measurement length of 20–5 m. The skies or probes are attached to the carriage (Fig. 2.4). With vertical springs, normal loads are applied to the ski between 10 N and 700 N. Normal and horizontal forces acting on the ski are measured with load cells. The signals of the load cells are transmitted wireless to a computer. The speed of the carriage is determined by an inductive length measurement system. The whole tribometer is located in a cooling chamber in which air can be cooled down to approximately −20 °C.

Snow is produced in a separate cooling chamber using a snow lance. The snow is placed in the through, compacted and leveled using a stainless steel blade. Cooling coils positioned on the bottom of the trough are used to control snow temperature. Snow properties (temperature, density, hardness, water content, grain form, and grain size), air temperature, and air humidity are determined before each test.

In the following, the effects of snow temperature, speed, and normal load on snow friction are presented. The probes with a size of 1.03×0.05 m were mounted below a stiff aluminum profile to get a uniform load distribution. Snow temperature was set between −11 and −3 °C, speed to 5, 10, 15, and 20 m s^{-1}, and normal load between 10 and 600 N. Each measurement consisted of ten consecutive runs on the same track. For runs four to ten the means of the measured horizontal and vertical force were calculated and from these the mean coefficient of friction was determined. The horizontal force was adjusted to account for air drag. The measurements are highly reliable with an accuracy, which decreases with increasing speed to approximately 1.3 % at 10 m s^{-1}. The mean coefficients of friction with the 95 % confidence intervals are presented in Figs. 2.5 and 2.6.

2.4.3.1 Effect of Snow Temperature

At the low speed of 5 m s^{-1} the friction coefficient decreased with increasing snow temperature (Fig. 2.5). This behavior is in accordance with experimental results of Buhl et al. [32] on snow and with a manifold of studies on ice [e.g., 27, 31]. Buhl and colleagues found the minimal friction between −5 and 0 °C. It is assumed that the decrease of the friction coefficient with increasing temperature is due to enhanced lubrication at higher snow temperatures.

In contrast to the measurements at low speed, there was no clear temperature effect observed on the coefficient of friction at a high speed of 20 m s^{-1} (Fig. 2.5).

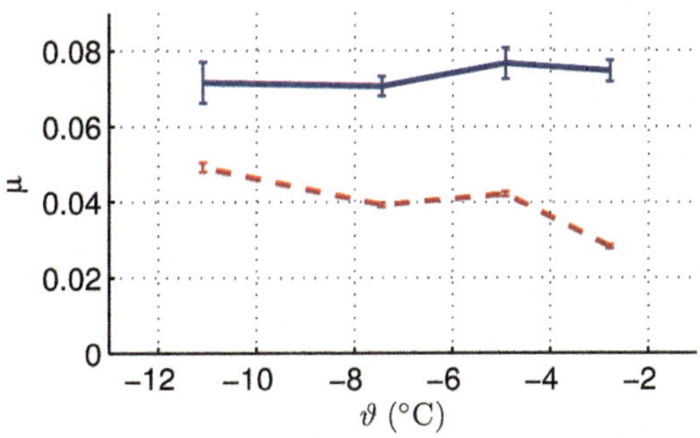

Fig. 2.5 Coefficient of friction, μ with 95 % confidence intervals versus snow temperature, ϑ at a speed of 5 m s^{-1} (*dashed line*) and 20 m s^{-1} (*solid line*). Normal load $F_N = 146$ N, snow density $\rho = 400$ kg m^{-3}

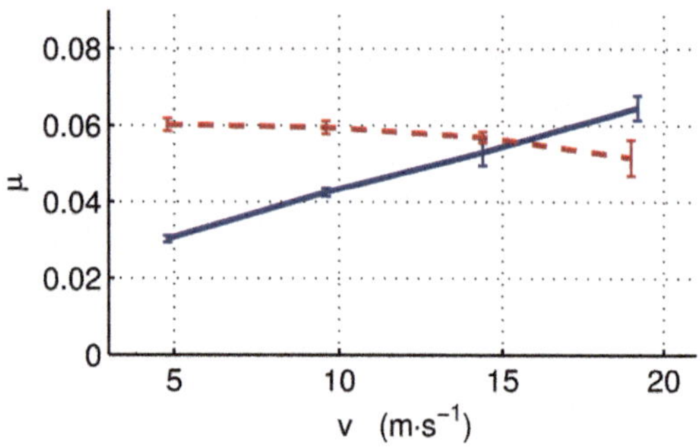

Fig. 2.6 Coefficient of friction, μ with 95 % confidence intervals versus speed v at snow temperature of $-3\,^\circ$C (*solid line*) and $-11\,^\circ$C (*dashed line*). Normal load F_N =146 N, snow density $\rho = 400\,\mathrm{kg\,m^{-3}}$

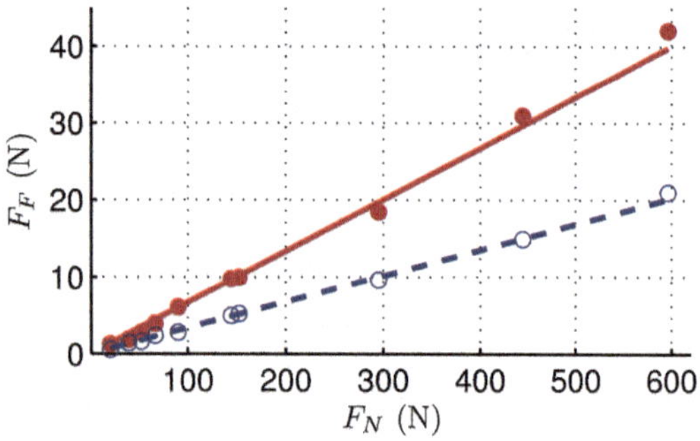

Fig. 2.7 Friction force, F_F versus normal load, F_N at speed of $5\,\mathrm{m\,s^{-1}}$ (*open circles*) and $15\,\mathrm{m\,s^{-1}}$ (*filled circles*). Snow temperature ϑ = $-4\,^\circ$C, snow density $\rho = 390\,\mathrm{kg\,m^{-3}}$

2.4.3.2 Effect of Speed

Contrary to the measurement at the snow temperature of $-11\,^\circ$C, at $-3\,^\circ$C the friction coefficient increased from 0.03 to 0.07 with increasing speed (Fig. 2.6). This increase at high temperatures was also found by Kuroiwa [21]. The increase of friction for the higher snow temperature may be explained by the increase of speed assuming wet friction $F_{F,\mathrm{wet}} = \frac{\eta_W A_C v}{h}$.

2.4.3.3 Effect of Normal Load

Friction force was directly proportional to the normal force for the measured speeds of 5 and $15\,\mathrm{m\,s^{-1}}$ (Fig. 2.7). Thus, the friction coefficient was independent of the normal force as it is stated for Coulomb type friction. For snow, load-independent static [33] and dynamic [32] friction coefficients were measured, which is in agreement with our findings. For ice, it was repeatedly documented that the friction coefficient decreases at a certain load [e.g., 27, 28]. This is explained by the higher friction force with increasing loads, which causes more meltwater due to higher frictional heating and a reduction of the coefficient of friction by improved lubrication. The friction coefficient on snow might not be affected by the load because additional meltwater after a certain level is squeezed out or absorbed by the snow.

References

1. O. Reynolds, *Papers on Mechanical and Physical Subjects*, vol. 2 (Cambridge University Press, Cambridge, 1901)
2. F.P. Bowden, T.P. Hughes, The mechanism of sliding on ice and snow. Proc. R. Soc. Lond. **A172**, 280–298 (1939). doi:10.1098/rspa.1939.0104
3. Y. Li, G.A. Somorjai, Surface premelting of ice. J. Phys. Chem. **111**(27), 9631–9637 (2007)
4. F.P. Bowden, D. Tabor, The friction and lubrication of solids. Technical Report, Oxford University Press, Oxford, 1950
5. F.P. Bowden, D. Tabor, Friction, lubrication, and wear: a survey of work during the last decade. Br. J. Appl. Phys. **17**, 1521–1544 (1966)
6. B. Glenne, Sliding friction and boundary lubrication of snow. J. Tribol. **109**(4), 614–617 (1987)
7. F.P. Bowden, Friction on snow and ice and the development of some fast-running skis. Nature **176**, 946–947 (1955)
8. S.C. Colbeck, A review of the processes that control snow friction. Technical Report 92-2 Cold Regions Research and Engineering Laboratory (CRREL), Hanover (1992). http://www.ntis.gov/index.html
9. L. Bäurle et al., Sliding friction of polyethylene on snow and ice: contact area and modeling. Cold Reg. Sci. Technol. **47**, 276–289 (2007). doi:10.1016/j.coldregions.2006.10.005
10. W. Ambach, B. Mayr, Ski gliding and water film. Cold Reg. Sci. Technol. **5**, 59–65 (1981)
11. G. Sazaki et al., Quasi-liquid layers on ice crystal surfaces are made up of two different phases. Proc. Natl. Acad. Sci. USA **109**(4), 1052–1055 (2012)
12. H. Strausky et al., Sliding plastics on ice: fluorescence spectroscopic studies on interfacial water layers in the μm thickness regime. Appl. Phys. B **66**(1), 599–602 (1998)
13. B.N.J. Persson, *Sliding Friction. Physical Principles and Applications* (Springer, Berlin, 2000). ISBN: 978-3-642-08652-6. doi:10.1007/978-3-662-04283-0
14. E.P. Lozowski, K. Szilder, S. Maw, A model of ice friction for a speed skate blade. Sports Eng. **16**, 239–253 (2013). doi:10.1007/s12283-013-0141-z
15. E.P. Lozowski, K. Szilder, FAST 2.0 derivation and new analysis of a hydrodynamic model of speed skate ice friction. J. Biomech. Eng. **134**, 8 p (2012)
16. W. Nachbauer, P. Schröcksnadel, B. Lackinger, Effects of snow and air conditions on ski friction, in *Skiing Trauma and Safety*, ASTM STP 1266, vol. 10, ed. by C.D. Mote et al. (American Society for Testing and Materials, Philadelphia, 1996), pp. 178–185
17. P. Kaps, W. Nachbauer, M. Mössner, Determination of kinetic friction and drag area in alpine skiing, in *Skiing Trauma and Safety*, ASTM STP 1266, vol. 10, ed. by C.D. Mote et al. (American Society for Testing and Materials, Philadelphia, 1996), pp. 165–177. doi:10.1520/STP37926S
18. P. Kaps, W. Nachbauer, Snow friction during skiing turns, in *Abstracts Volume II, Second World Congress of Biomechanics*, Amsterdam, NL, ed. by L. Blankevoort, J.G.M. Kooloos (1994), p. 142
19. P. Kaps, W. Nachbauer, M. Mössner, Snow friction and drag in alpine downhill racing, in *Abstract Book, 4th World Congress of Biomechanics*, Calgary, 2002
20. B. Habel, Über die Bestimmung von Luftwiderstand und Gleitreibung beim Skilauf. Europa-Sport **20**, 950–955 (1968)
21. D. Kuroiwa, The kinetic friction on snow and ice. J. Glaciol. **19**, 141–152 (1977)
22. M.A.H. Leino, E. Spring, Determination of the coefficient of kinetic friction between ski and snow from gliding velocity of a skier. Report Series in Geophysics, vol. 19. University of Helsinki, FI, 1984
23. J. Erkkilä et al., A cinematographic method for determination of the kinetic friction of skis on snow. Report Series in Geophysics, vol. 21. University of Helsinki, FI, 1985

24. P. Miller et al., Development of a prototype that measures the coefficient of friction between skis and snow, in *The Engineering of Sport 6*, vol. 1, ed. by E.F. Moritz, S.J. Haake (International Sports Engineering Association, Munich, 2006), pp. 305–310. doi:10.1007/978-0-387-46050-5_24
25. T. Sahashi, S. Ichino, Coefficient of kinetic friction of snow skis during turning descents. Jpn. J. Appl. Phys. **37**, 720–727 (1998)
26. H. Liang, J.M. Martin, T.L. Mogne, Experimental investigation of friction on low-temperature ice. Acta Mater. **51**(9), 2639–2646 (2003)
27. L. Bäurle et al., Sliding friction of polyethylene on ice: tribometer measurements. Tribol. Lett. **24**(1), 77–84 (2006). doi:10.1007/11249-006-9147-z
28. D.C.B. Evans, J.F. Nye, K.J. Cheeseman, The kinetic friction of ice. Proc. R. Soc. Lond. **A347**, 493–512 (1976)
29. S.J. Jones et al., Friction of melting ice. Ann. Glaciol. **19**, 7–12 (1994)
30. S.J. Calabrese, R. Buxton, G. Marsh, Frictional characteristics of materials sliding against ice. Lubr. Eng. **36**(5), 283–289 (1980)
31. A.M. Kietzig, S.G. Hatzikiriakos, P. Englezos, Ice friction: the effects of surface roughness, structure and hydrophobicity. J. Appl. Phys. **106**(2), 024303 (2009). doi:10.1063/1.3173346
32. D. Buhl, M. Fauve, H. Rhyner, The kinetic friction of polyethylene on snow: the influence of the snow temperature and the load. Cold Reg. Sci. Technol. **33**, 133–140 (2001)
33. F.P. Bowden, Friction on snow and ice. Proc. R. Soc. Lond. A **217**, 462–478 (1953). doi:10.1098/rspa.1953.0074
34. F. Albracht et al., On the influences of friction on ice. Mater. Werkst. **35**, 620–625 (2004). doi:10.1002/mawe.200400822
35. V. Linnamo et al., Sports technology, science and coaching, in *Proceedings of the 2nd International Congress on Science and Nordic Skiing*, 2011, pp. 28–31
36. M. Hasler et al., Kinetic friction of boardercross-snowboards. Procedia Eng. **72**, 310–314 (2014)
37. S. Ducret et al., Friction and abrasive wear of UHWMPE sliding on ice. Wear **258**, 26–31 (2005)
38. M. Takeda et al., Friction of the short model ski at low velocity. J. Phys. Conf. Ser. **258**, 012007 (2010)
39. L. Bäurle, Sliding friction of polyethylene on snow and ice, Ph.D. thesis, Swiss Federal Institute of Technology Zurich, 2006
40. B.A. Marmo, J.R. Blackford, C.E. Jeffree, Ice friction, wear features and their dependence on sliding velocity and temperature. J. Glaciol. **51**(174), 391–398 (2005)
41. P. Oksanen, J. Keinonen, The mechanism of friction of ice. Wear **78**, 315–324 (1982)

Chapter 3
Friction Between Runner and Ice

Francesco Braghin, Edoardo Belloni, Stefano Melzi, Edoardo Sabbioni, and Federico Cheli

The scientific literature on ice friction has a long history, going back to [39] and beyond. It involves many areas of science and engineering, including physics, chemistry, tribology, fluid dynamics, thermodynamics. In this chapter we will focus on only those research that were focused on the interaction of skate and runners with ice. The low ice friction coefficient in competitive sliding sports is the result of a liquid water lubricating layer separating the blade and ice. Two different mechanisms contribute to this liquid layer: frictional heating and pressure melting. The thickness of the meltwater layer varies with ice temperature and sliding speed and is of the order of 1 μm. Conduction of heat into the ice and slider and lateral squeeze flow of the lubricating liquid from under the slider tend to reduce the thickness of the lubricating layer and hence increase the friction coefficient. Friction is due to the energy dissipated at the contact, i.e. to the ploughing of ice (ice permanent deformation, crushing and extrusion), and to the shear stress in the liquid layer. To be able to predict the sled dynamics as well as to optimize its layout in order to maximize performances, both aerodynamic and ice-runner contact forces [1–3] have to be accurately predicted. While assessing aerodynamic forces is reasonably well defined (e.g. through wind tunnel tests), ice-runner friction forces are much more difficult to predict and measure since they depend on various parameters, such as ice surface temperature, pressure, humidity, runner material, knife design, etc. The challenge for all sled and runner designers is to understand the relationships between runner and ice, and runner and sled. There are two possibilities of assessing

F. Braghin (✉) • S. Melzi • E. Sabbioni • F. Cheli
Department of Mechanical Engineering, Politecnico di Milano, Milano, Italy
e-mail: francesco.braghin@polimi.it; stefano.melzi@polimi.it; edoardo.sabbioni@polimi.it; federico.cheli@polimi.it

E. Belloni
Dipartimento di Meccanica, Politecnico di Milano, Via Giuseppe La Masa, 1, Milano, Italy
e-mail: edoardo.belloni@polimi.it

F. Braghin et al. (eds.), *The Engineering Approach to Winter Sports*,
DOI 10.1007/978-1-4939-3020-3_3

the kinetic friction coefficient between runner and ice: either tribometer tests with simplified specimen or indoor/outdoor tests with real runners. The first approach is very helpful [4] for understanding the physics behind the low value of the measured friction coefficient due to the possibility of precisely controlling the different variables involved (air and ice temperature, relative humidity, water quality, normal pressure, sliding orientation with respect to crystallographic plane, etc.) but the real working conditions of runners (sliding speed, normal pressure) as well as the ice surface conditions (specimen sliding on virgin ice) cannot be easily reached [5]. In this paragraph we will therefore focus on indoor/outdoor tests carried out with real runners and thus affected by unpredictable variability ("measurement noise").

3.1 Measurement of Runner–Ice Interaction

The first attempts to measure the friction between ice and sled runners date back to 1987–1989 [6–8][1] and deduce friction indirectly: by measuring the velocity of a free-sliding sled (Fig. 3.1), the influence of runner material, construction and sharpening of the runners, and surface conditions on the kinetic friction coefficient was assessed. A similar approach had already been proposed in skating by Furushima [9] and in sledding by Ericksson [10] and Ulman and Cross [11]. However, by making reference to single velocity measurement placed in multiple locations on a track, the measured kinetic friction coefficient values were affected by great dispersion due to the presence of several different events between two measuring stations.

By the measurements of initial V_i and final V_f, velocity between the optical gates separated by distance L, an average coefficient of friction μ was calculated as

$$\mu = (V_i^2 - V_f^2)/2gL \tag{3.1}$$

g being the acceleration of gravity. Unfortunately, only a limited range of velocities could be achieved through the developed experimental setup and only straight track running was taken into account. As expected, the coefficient of friction was higher at the lower temperatures than at higher ones and smooth runners showed lower friction at around −1 °C than around −10 °C. The friction of rough runners showed little temperature dependence. Moreover, a minimum of the kinetic friction coefficient, around 0.006, was observed at −3 °C for highly polished runners and materials having lower thermal conductivity showed the least resistance, as predicted by the frictional melting theory.

Also in [12] an empirical method for assessing the frictional performances of sleds is presented. However, as for the previously presented measurement campaigns, these analyses only assessed a limited range of runner working conditions and were affected by many uncertainties, not last the aerodynamic effects on the

[1]Ice friction measurements with real skate blades mounted to a frame are reported in a previous work by Kuroiwa dating back in 1977.

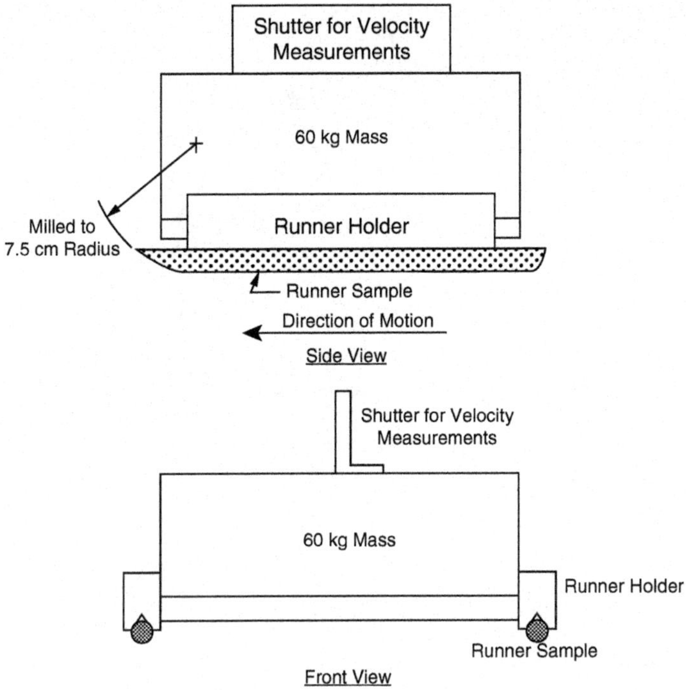

Fig. 3.1 Scheme of the model sled used in [6, 8] to measure the kinetic friction coefficient of runners on ice

sled. The first attempt to directly measure the kinetic friction coefficient between ice and runners/skates is reported in [13]: special skates for speed skating were instrumented through strain gauges and calibrated to determine both vertical and longitudinal skate-ice contact forces. Experiments were made on different indoor and outdoor ice skating rinks with an experienced ice skater. It was observed that the coefficient of friction has a minimum (equal to approx. 0.0042) at around −8 °C which is lower than the temperature observed by Itagaki et al. [6]. There are many differences between the two tests, but the main one is that in [13] the higher sliding speed is higher than in [6] (8 m/s vs. less than 2 m/s). Moreover, it was found that the coefficient of friction increases with increasing speed. This was explained by De Koning et al. [13] by making reference to the liquid layer that is generated at skate–ice interface and that, at increasing speed, hasn't enough time to grow up till the "optimal" thickness. To validate this assertion and to verify which is the dominant contribution for the liquid layer at runner–ice interface, Colbeck et al. [14] carried out a similar experiment inserting a thermocouple in the skates (Fig. 3.2). In fact, in [15] the presence of the liquid layer is explained by making reference to pressure melting and frictional heating. Pressure melting (alone) cannot be responsible for the low friction of ice since the pressure needed to reach the melting temperature is above the compressive failure stress of ice and, if it did occur, high squeeze losses would result in very thin films.

Fig. 3.2 Instrumented skate for direct temperature measurement [14]

The tests were performed both indoors under controlled conditions and outdoors. The direct (thermocouples in the skates) and indirect (ice surface temperature behind the skate) temperature measurements carried out in [14] showed that frictional heating is the dominant mechanism because the temperature of the skates increases with speed and with thermal insulation. According to [16–18], instead, contact pressure is considered to be the most influencing factor of the kinetic friction coefficient between steel and ice, not only because of the pressure melting phenomenon but especially due to the ploughing of the ice track: it is experimentally shown that the friction coefficient reaches a minimum value for a given pressure value (optimum pressure value p_{opt}, Fig. 3.3). Thus, runners should be designed so to achieve this optimum pressure value in any working condition.

The first attempt to scientifically design optimum runners is proposed in [19–21]: low friction runners were designed and manufactured based on elastic-plastic contact analysis, so to obtain the minimum contact area between the runner and ice within the shape that prevents ploughing. The developed bobsleigh runner based on the analysis was named "Nagano Special" and was able to reduce the total run time by 0.42 s to 1.23 s with respect to standard Japanese runners.

In [16] the design of the optimum runner is carried out based on the optimum normal pressure since the other parameters that influence the kinetic friction coefficient cannot be easily modified or their influence is monotonic (and thus the optimum value is already known; think of the runner surface roughness). For example, both temperature and sliding speed have a high impact on friction but the ways to influence T and v in bobsleigh are limited. In fact, since 2006 the international body for bobsleigh and skeleton, the Fédération International de Bobsleigh et de Tobogganing (FIBT), has required that runners be fabricated

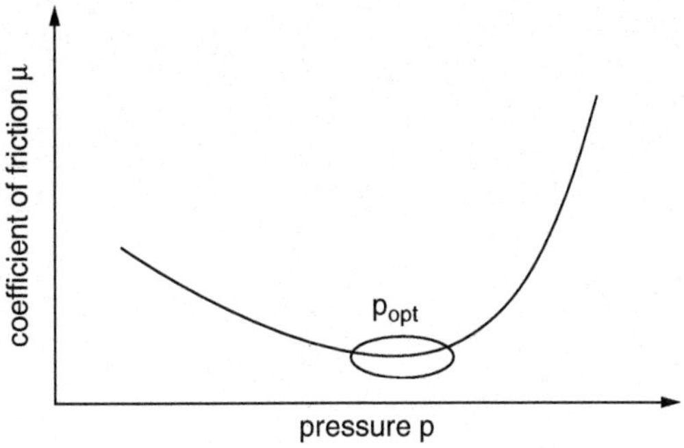

Fig. 3.3 Pressure vs. kinetic friction coefficient [16]

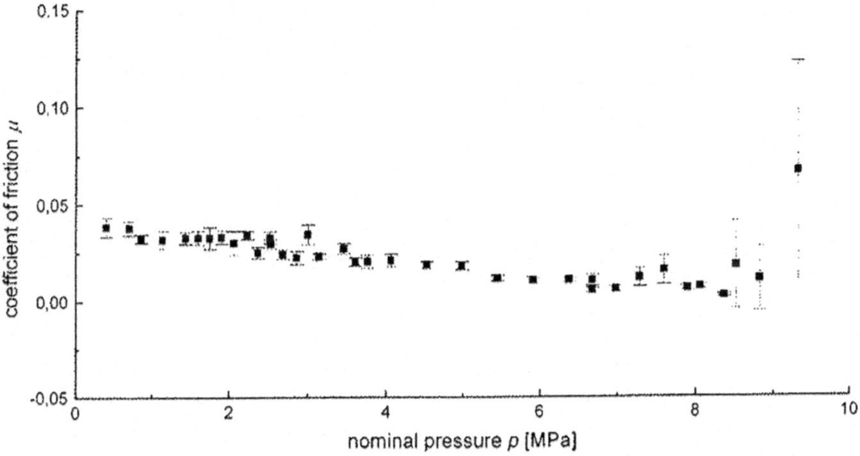

Fig. 3.4 Pressure vs. kinetic friction coefficient for a sliding speed of 1 m/s and a testing temperature of −2 °C [16]

from a standardized steel, any heating of the runners is forbidden, the effect of thermal conductivity depends on the weather situation and sliding speed can only be influenced by the initial push-off phase that is in any case optimized. Any change in the runner's geometry directly affects the pressure distribution between runner and ice and, according to the rules, there are no restrictions on the curvature of the runner and only upper or lower limits for many of the other design parameters, e.g. the radius of the running surface, the thickness and the height of the runner. Thus, a wide range of pressure distributions can be achieved. By looking at the traces of runners on the track, one can conclude that in straight parts the actual pressure p is lower than p_{opt} (slight traces) while in curved parts the actual pressure p is higher than p_{opt} (deep traces). Consequently, the friction coefficient can be decreased either by increasing the normal pressure in the straight parts of the track or by decreasing the normal pressure in the curves. It is assumed that most potential time saving will be realized by improving the runner behavior in curves. To determine p_{opt} both tribometer and full scale tests using instrumented runners were carried out. Figures 3.4 and 3.5 show the kinetic friction coefficient measured on a test

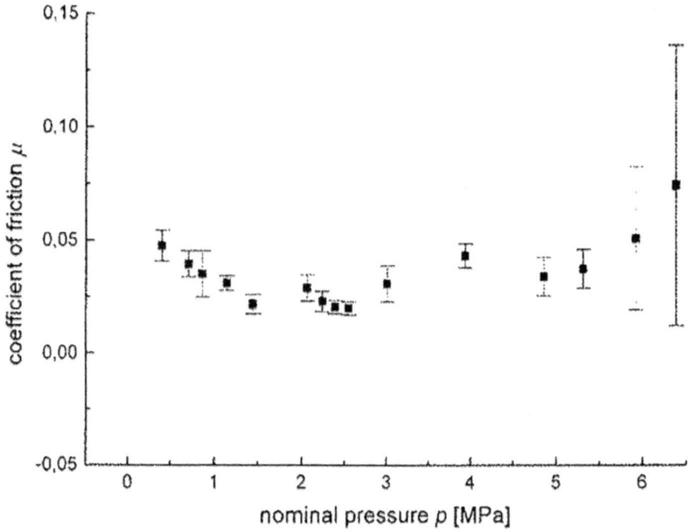

Fig. 3.5 Pressure vs. kinetic friction coefficient for a sliding speed of 1 m/s and a testing temperature of −8 °C [16]

bench at 1 m/s and at two different temperatures. It can be seen that p_{opt} shifts towards lower pressures at decreasing temperature (from approx. 7 MPa to approx. 2.5 MPa). The trend towards lower values of μ for higher pressures is stopped by deep ploughing resulting in trace depths of $d \gg 0.1$ mm. In fact, ploughing was observed to occur for lower pressures at decreasing temperature due to the increase of pre-cracks generated during the preparation of the ice surface. It should be noted that the optimum pressure p_{opt} was measured at sliding speeds considerably lower than those experienced by a bobsleigh runner. However, for the velocities measured, no change in p_{opt} was observed. This finding was explained making reference to the finding that the strength of ice is not affected by the strain rate above $10^{-2} 1/s$ for fine grained ice but may be related just to a small dependence of p_{opt} on the sliding speed at small sliding speeds. In any case, in [22] and [16] p_{opt} for high sliding speeds is assumed to be equal to p_{opt} for low sliding speeds.

For being able to measure the kinetic friction coefficient in real working conditions, runners and axles were instrumented using strain gauges (Fig. 3.6). The aim of instrumented runners was to measure the strains resulting from impacts and to determine the specific section of the running surface where the impact occurred while the aim of instrumented axles was to determine the total force applied to the runner during a run through the bending of each end of axle. According to the signal of axle bending, high vibrations of the runners were observed during the runs. These vibrations lead to impacts and thus ploughing and ice destruction that dissipate a great amount of energy resulting in higher friction. Therefore, [16] recommends to optimize the damping of the bobsleigh frame and of the suspension system. Strains measured in the runners, instead, showed that impacts were localized at

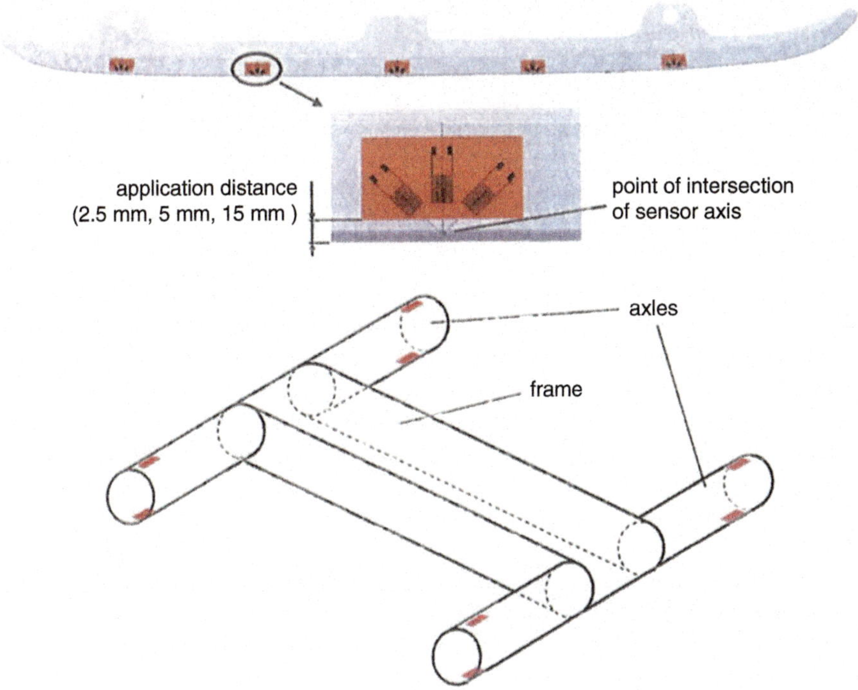

Fig. 3.6 Front runner and axles of a 2-man bobsleigh equipped with strain gauges [16, 22]

given positions on the runners and that ice temperature has an effect on this section: if temperature increases, ice becomes softer and the runner penetrates deeper into the ice; consequently, the section of the runner that contacts ice irregularities moves towards the front of the runner. For a constant impact force, the pressure p decreases with an increasing contact area and contributes to decreased ploughing during impacts. Following these considerations and measurements, based on finite element analysis an optimized runner design, called Delta 1 (Fig. 3.7), was developed. In straights, the runner slides on two bumps with a radius of 1050 mm and a height of 0.4 mm thus decreasing the contact surface and increasing the pressure. In curves, bumps penetrate into the ice surface by elastic deformation mainly and the rest of the running surface gets in contact with ice. The curvature of these parts is designed to be equal to the real curve radius of 85 m. Hence, the contacting surface in curves was increased and the pressure is reduced.

Field tests showed that Delta 1 runner was considerably slower than the reference runner mainly due to the failure of ice induced by the bumps and not correctly considered in the finite element. However, [16] concludes, the basis are set for developing an optimum runner that has both an optimized curvature and an optimized running surface radius. As rules allow only two sets of runners per bobsleigh team and per season, it is suggested to design one runner for cold temperatures and

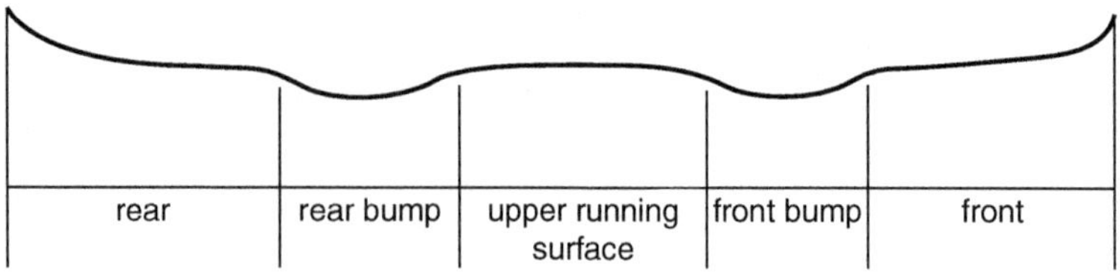

Fig. 3.7 Delta 1 optimized runner [16]

one runner for warm temperatures. What should also be investigated are the surface layers (such as ceramics and diamond-like carbon) that could be very promising. In [23] the optimization of the runner (in fact skates) cross section that minimizes friction was further evolved through specific laboratory tests. Four different steel blade cross sections were investigated concluding that friction and penetration are highly correlated and mainly influenced by the outside angle of the blade. Similar studies were carried out in [24], although on hockey skates rather than on bobsleigh runners. Compared with standard blades, the friction coefficient was lower by about 13 %, 21 %, and 22 % for flared skate blades with a blade angle of 48, 68, and 88, respectively. The reason for this was attributed to a better contact pressure distribution thus contradicting the paradigm that "thinner blades cause less friction." A direct measure of the kinetic friction coefficient was also attempted in [25, 26]: in order to measure ice-skate contact forces in real working conditions, front and rear axles of a bobsleigh have been instrumented with strain gauge bridges. The adopted strain gauge configuration (Fig. 3.8) implies:

- two half bridges to measure the moment around x (evaluation of the vertical forces),
- one full bridge to measure the axial force (evaluation of lateral forces) and
- two half bridges to measure the moment around z (evaluation of longitudinal forces).

The strain gauges are connected so to compensate for thermal effects, unwanted bending torques and axial force (for the four half bridges).

The instrumented axle was calibrated indoor by applying different combinations of external forces F_x, F_y, and F_z and measuring the corresponding strains ϵ_1, ϵ_2,…, ϵ_n. Through a least-squares approach the calibration matrix $[C]$ is thus obtained:

$$\begin{Bmatrix} F_x \\ F_y \\ F_z \end{Bmatrix} = [C] \begin{Bmatrix} \epsilon_1 \\ \epsilon_2 \\ \cdots \\ \epsilon_n \end{Bmatrix} \tag{3.2}$$

Once the calibration matrix is known, strains are measured during field runs and the corresponding contact forces are reconstructed. Besides the instrumented axles, the bobsleigh adopted for the tests on the Cesana Pariol Olympic track was equipped with:

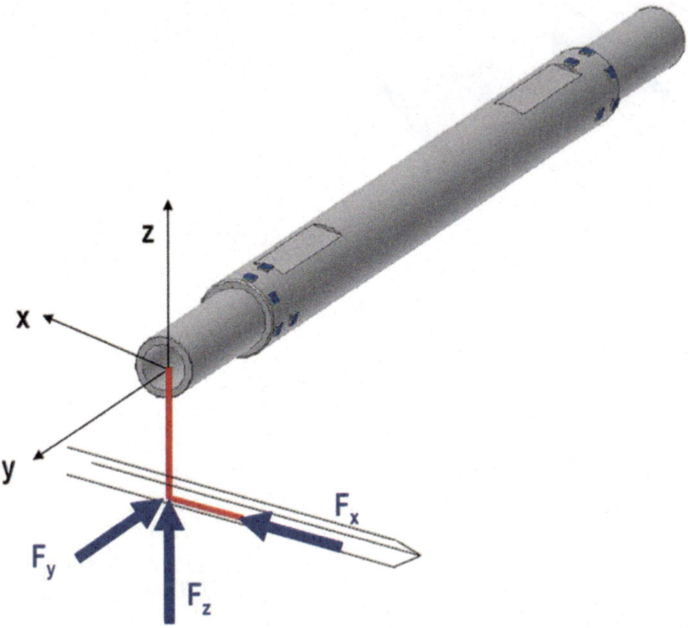

Fig. 3.8 Scheme of the instrumented axle for runner–ice contact force measurement; *blue squares* represent the strain gauges [25]

- a 9 dof inertial measurement unit;
- an optical sensor for the continuous measurement of the absolute sled speed and sideslip angle;
- four linear potentiometers for measuring the leaf spring—runner relative distance (one for each runner);
- one linear potentiometers for measuring the steer angle.

The measured friction coefficient decreases, as expected, from 0.08 to 0.025 as speed increases. If compared to the literature, the measured values are slightly higher. From the above reported literature it is clear that indoor laboratory tests with small specimen are good for assessing the friction phenomenon but do not reach the working conditions experienced by real runners during real field tests. However, field tests are affected by many unknowns and variability and thus the possibility of identifying trends is limited. Two attempts for reducing the dispersion of results and obtaining more repeatable results are available in the literature: one [27] based on the use of a special sled dragged by a car and the other [28] based on the use of a huge radial tribometer that allows to reach high sliding speeds. In [27] an approach borrowed from the one typically used for assessing steady-state tyre behavior is proposed: a special sled, that allows to change the angle of attack of the runners through a steering mechanism, is dragged by a car (Fig. 3.9) is proposed. The aim is to measure the friction forces of real runners in steady-state conditions at varying

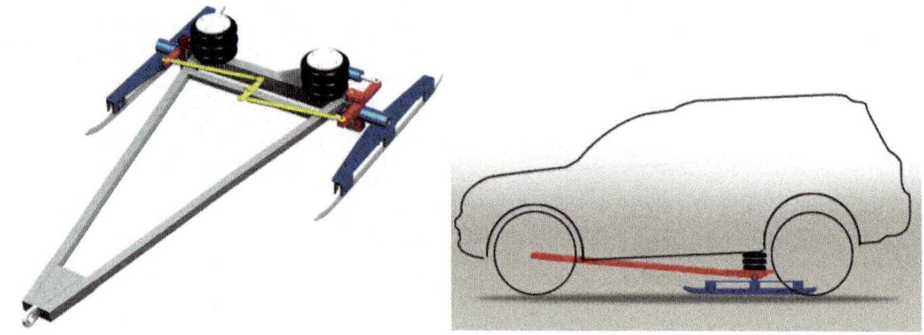

Fig. 3.9 Special sled dragged by a car for the direct measurement of steady-state friction forces at varying angle of attack, normal load, and sliding speed [27]

angle of attack (slip angle), applied normal load and imposed sliding speed. Tests can easily be repeated thus allowing to assess also the influence of ice temperature, geometry of the runners and their camber angle.

In [29, 30] a real bobsleigh equipment is used in a more controlled environment: the acceleration and deceleration of a bobsleigh training sled are measured through a radar gun located behind the sled sliding along a flat and inclined ice surfaces. From the measured accelerations/decelerations, knowing the drag force, the coefficient of kinetic friction between the runners and ice can easily be determined. Values of the coefficient of friction equal to 0.0042 ± 0.0009 were measured in a speed range of 1–10 m/s. Note that these values are considerably lower than previously published values. No dependence of μ over the sliding speed was resolvable in this speed range. Although tests should have been carried out in more controlled conditions, a large dispersion in the obtained data was obtained. A large radius drum having the inside surface covered with ice (Fig. 3.10) was instead used in [28] to carry out friction measurements under laboratory conditions with model runners. The advantage of this setup is that high sliding speeds can easily be achieved. However, the performed tests differ from natural bobsleigh conditions in several ways:

- runners pass over the same ice (same groove) with each revolution of the drum thus significantly reducing the ploughing force;
- runner rocker radius (large radius) and the radius of the ice drum are the same thus producing a flat contact length of approximately 10 cm, much larger than that obtained under standard working conditions;
- the imposed contact pressure is almost an order of magnitude less than that of real runners during real working conditions.

Measured kinetic friction coefficient values are in the order of 0.008, slightly decreasing with both sliding speed and normal load (at −5 °C).

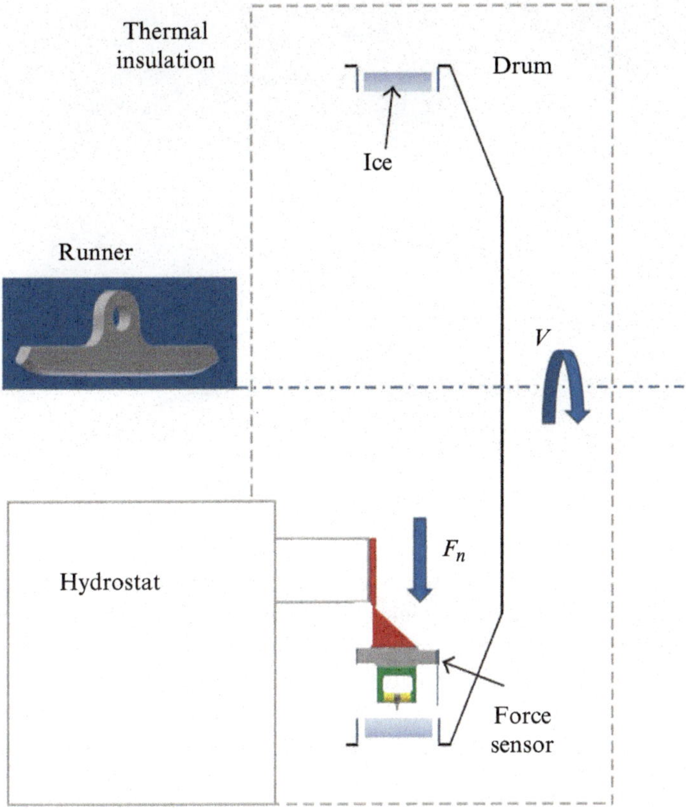

Fig. 3.10 Test cell with sensor setup and model runner used in [28] for assessing the kinetic friction coefficient

3.2 Modeling of Runner–Ice Friction

Only recently have numerical models been formulated to predict ice friction in sliding sports. The first formulas date back to [7]. However, the presented approach is purely phenomenological and does not provide any insight into the mechanism of runner–ice friction. A more theoretical approach is provided in [31]: a three-dimensional finite element static model of an existing world cup bobsled runner on an ice track is presented (Fig. 3.11). The runner is modelled using linear and quadratic tetrahedron elements while ice as a strain-softening material at −5 °C. A comparison of the runner model with bending tests confirmed accuracy to within 3 %. Resulting trace depths of the contact model of runner and ice were in good agreement with measurements taken from field-experiments. Additionally, the stress and strain state in ice was analyzed. The model was refined in [16]. It was found that bobsleigh runners must be modelled with quadratic elements rather than with linear ones due to the high runner bending values and a very refined mesh at runner–ice

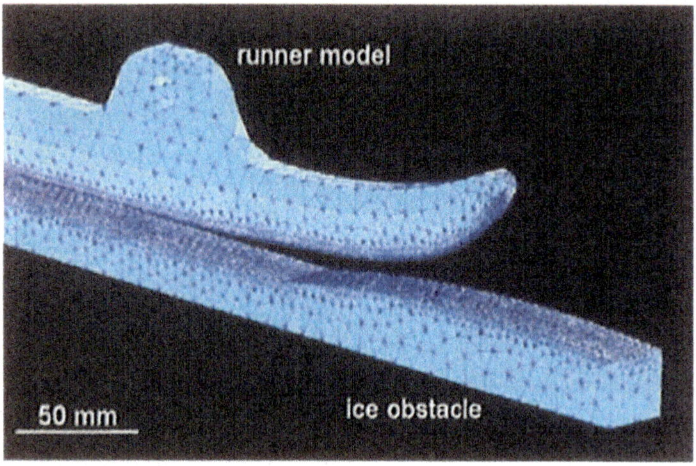

Fig. 3.11 Finite element model of a runner on an ice track [16]

interface is necessary. This however leads to considerably increased computing times due to a high number of connected nodes. Hence, dynamic models of ice friction still remain a matter of research for the future. Moreover, it is recommended to further improve the model especially for what concerns the material model of ice, to include thermic processes and fluid dynamics of the water film between runner and ice. In [25] a similar approach to that proposed in [16] is adopted. The main features of the developed model are:

- not only the runner but also the leaf spring are considered in the finite element model; their behavior is assumed to be purely elastic;
- ice is modelled as a strain-softening material;
- temperature increase of the runner is neglected;
- a mesh seed of 5 mm is used throughout the bodies except for the contacting surfaces where a mesh seed of 0.1 mm is used;
- dynamic analysis of the runner sliding on ice is carried out.

Steady state finite element simulations at varying sliding speed, yaw angle and vertical load have been carried out. A ramp of sliding speed followed by a ramp of vertical load is applied to the leaf spring and runner and the simulation is continued till steady state conditions are reached. Probably due to numerical problems, in several runs no steady state conditions were reached. However, values of the kinetic friction coefficient below 0.01 were calculated. It should be noted that each simulation requires approximately 1 week of CPU time. It is concluded that "improvements to the numerical model are required" due to the many simplifying assumptions (no temperature rise, non-phase transition of ice, no ploughing) done. The computational cost and the difficulty in determining all the parameters necessary for carrying out a realistic simulation of runner–ice interaction using finite elements has led to the development of simpler models that neglect the deformability of the runner but are able to account for physical phenomena that would otherwise be almost impossible to account for. These models were developed from 2006 on [32–34] and are named with the acronym *FAST* that stands for Friction Algorithm using Skate Thermohydrodynamics. The latest evolution is FAST 2.0i and describes ice friction in the fully lubricated, hydrodynamic regime, where a layer of meltwater completely separates the ice and slider surfaces. The effect of any contact between asperities on both surfaces is neglected. Friction results from a ploughing force, arising from (permanent) ice deformation, crushing and extrusion, and the shear stress in the lubricating Couette flow. The model takes into account frictional melting, heat conduction into the ice, and the lateral squeeze flow of the lubricating liquid. The effect of pressure on the melting temperature is also accounted for. It is assumed that the runner glides on pristine, smooth and flat ice. A bobsleigh runner has a sliding surface that can be described by two radii of curvature that are approximately constant: the longitudinal radius of curvature, parallel to the direction of motion (standard value 34 m), and the transverse radius of curvature, perpendicular to the direction of motion (standard value 4.75 mm). The ice hardness is a fundamental parameter for determining the contact area between the blade and the ice surface and thus the ploughing force and consequently the runner–ice friction coefficient. In fact, the apparent contact area is determined by the ratio of the force applied by the blade onto the ice and the hardness of the ice surface. There are several ways of assessing the dynamic ice hardness. In [29, 35] steel balls ranging from 8 to 540 g are dropped onto several different ice surfaces from heights between 0.3 and 1.2 m and the ice hardness is determined by measuring the diameter of the indentation craters. It is found that the dynamic ice hardness P, measured in MPa, varies with the temperature of the ice surface T, measured in °C, according to (Fig. 3.12):

$$P(T) = (-0.6 \pm 0.4)T + 14.7 \pm 2.1 \tag{3.3}$$

The analysis of the effect of air humidity on ice hardness was inconclusive. It should be noted, as in [36], that the ice hardness for the ploughing force could be different from the one obtained from impact tests in normal direction. In fact, since the ploughing force is largely determined by the ice hardness near the front of the contact zone, it should be higher since the ice there is largely intact. But experimental tests are more difficult to be carried out. The liquid layer is generated by the frictional melting and the pressure melting. Frictional melting is greatly influenced by the heat conduction into the runner and the ice. Moreover, layer thickness is reduced by the pressure exerted by the runner that produces a squeeze flow. It is not trivial to calculate the squeeze flow exactly. In [34] the following simplifying assumptions are made:

- the contact between runner and ice is fully lubricated (i.e. the lubrication layer thickness is more than about 0.25 μm),
- the lubricating layer thickness is a function of the longitudinal position along the runner only,

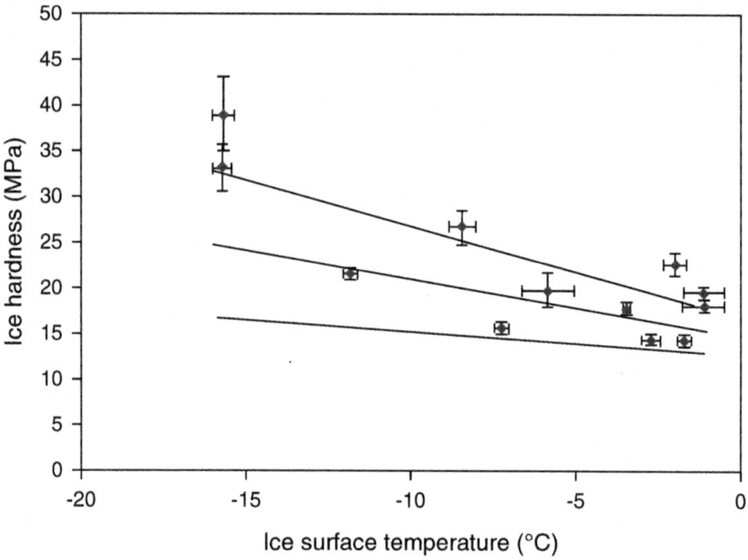

Fig. 3.12 Ice hardness of various ice surfaces at Canada Olympic Park and the Calgary Olympic Oval. The linear fit and the confidence limits are the *central* and *two outer lines*, respectively [35]

- the mean pressure in the squeeze flow, averaged over the width of the contact, is equal to the ice hardness,
- the pressure drops to zero at the lateral edges of the contact,
- the heat conduction into the runner is neglected and
- the lubricating layer, melted by the frictional energy generated by the shear stress in the lubricating layer itself, consists of bulk liquid water in a sheared, laminar Couette flow, with velocity zero at the ice/liquid interface and velocity equal to the runner velocity at the runner/liquid interface (no slip boundary condition).

The calculated contributions to the overall coefficient of friction are the following:

- the ploughing force accounts for 10–33 % of, depending on the normal load;
- the melting term contributes from about 15 % (very cold ice) to about 25 % (ice near the melting point);
- heat conduction into the ice contributes between 33 % (maximum normal load) and 40 % (static normal load) when the ice surface temperature is very cold (−40 °C) and its contribution drops to zero when the ice is at the melting point;
- the squeeze flow contribution is relatively small at −40 °C, varying from 25 % (maximum normal load) to 30 % (static normal load) and ranges from 40 % (maximum normal load) to almost 60 % (static normal load) near the melting point.

Of course, the most important output of the model is the predicted kinetic friction coefficient as a function of the ice surface temperature and the sled sliding speed

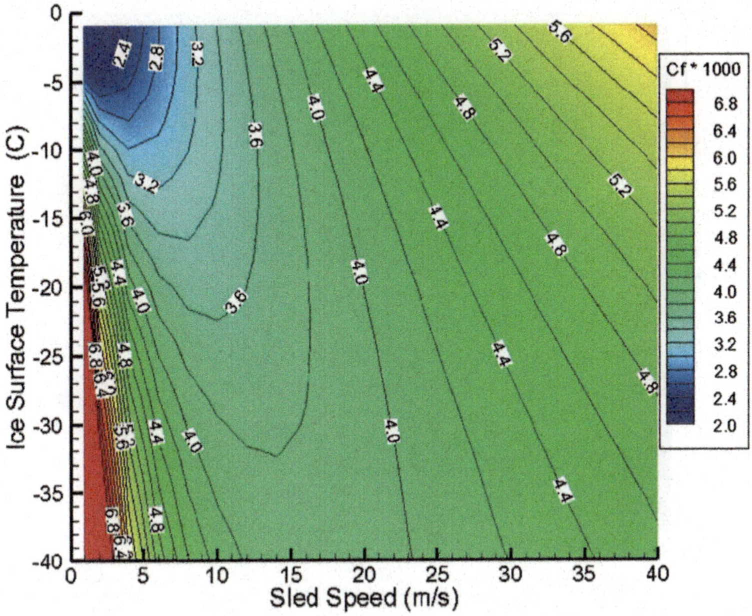

Fig. 3.13 Coefficient of ice friction as a function of ice surface temperature and sled speed [34]

(Fig. 3.13). At low speeds, not much frictional energy is generated, and cold ice conducts much of it away, giving rise to a thin film and a high shear stress force. The friction coefficient is minimized at high ice temperatures, where heat loss into the ice is also reduced. At higher speeds, ample frictional energy is generated and the friction coefficient reaches a minimum. At still higher speeds, the ploughing contribution starts to increase the friction coefficient. However, because cold ice is harder, the ploughing force and hence the total friction coefficient are reduced at low temperatures (reduced penetration depth). Unfortunately, there is little opportunity to control ice surface temperature in the sport of bobsleigh (typically it is close to ambient). In any case, according to the model, if the ice surface temperature varies with the air temperature, then runs made in the colder part of the day should have an advantage. Analyzing the influence of runner dimensions, as in [16], in [34] it is concluded that the small runner radius (transverse radius) should be maximized, especially for higher load, in order to achieve minimum ice friction, although the impact is not large. Varying the large runner radius has an even smaller effect on ice friction. The increase in friction coefficient with decreasing small runner radius arises chiefly from an increase in the ploughing force, because of greater ice penetration, coupled with an increase in the squeeze flow, because the flow path is narrower. This is partly offset by a small decrease in the frictional melting contribution with decreasing small radius. Note that, increasing the runner radii also the sled control is reduced. Thus, a reasonable trade-off should be aimed at. The developed model provides, among others, the maximum lubrication layer thickness

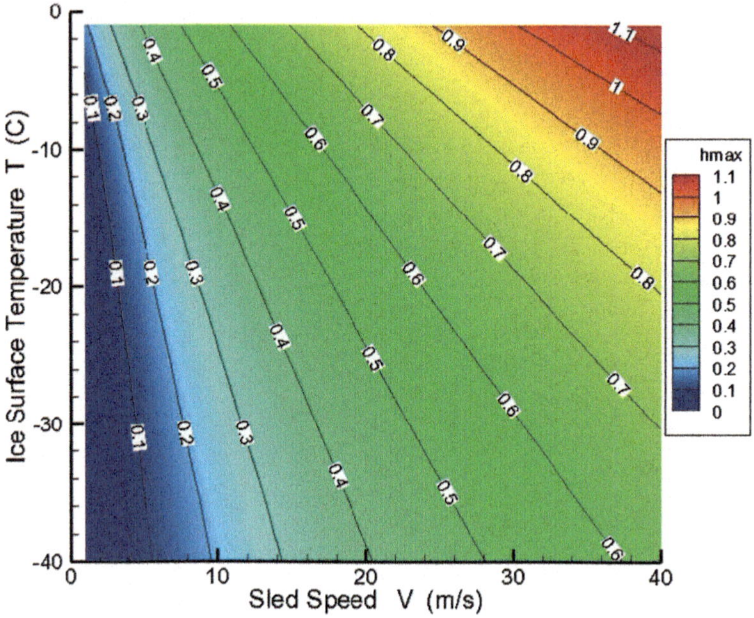

Fig. 3.14 Maximum lubrication layer thickness (μm) as a function of ice surface temperature and sled speed [34]

as a function of the ice surface temperature and the sled speed (Fig. 3.14). Note that the friction model loses its validity when *hmax* falls below the peak-to-valley height of the roughness elements, i.e. 0.25 μm. The model has been validated by comparison with measured data coming from [13] (Fig. 3.15) and [29] (Fig. 3.16). For carrying out the comparison with data provided in [13], an extension of the model is required to account for the inclination of the blades. This extension, called FAST 2.0i, is described in [37, 38]. Because it is valid in the limit as the inclination angle approaches zero, it also applies to a vertical blade. The following additional assumptions are made:

- the skater exerts zero thrust;
- the force exerted on the contacts between the side of the blade and the ice is zero;
- the thickness of the lubricating layer has no transverse variation;
- the heat conduction between the blade and the ice has been neglected.

It is shown that, for inclined blades, the ice friction coefficient could be higher. As shown in Fig. 3.15, the model and the experimental values of the kinetic friction coefficient are remarkably similar when averaged over all ice surface temperatures. However, the systematic behavior seen in the experiments, with a minimum around −7 °C, is not evident in the model at this skating speed. The model does exhibit a similar ice friction coefficient minimum, over the same temperature range, but at a slower skating speed. For what concerns the dependence on sliding speed, the

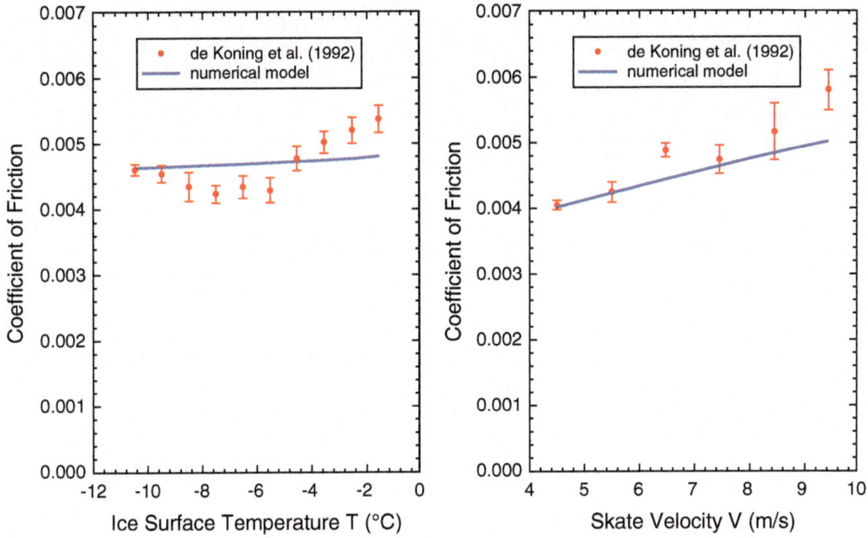

Fig. 3.15 Ice friction coefficient during a skating stroke as a function of ice surface temperature (*left*, sliding speed equal to 8 m/s) and sliding speed (*right*, ice surface temperature equal to −4.6 °C); validation of the model in [38] with experimental data in [13]

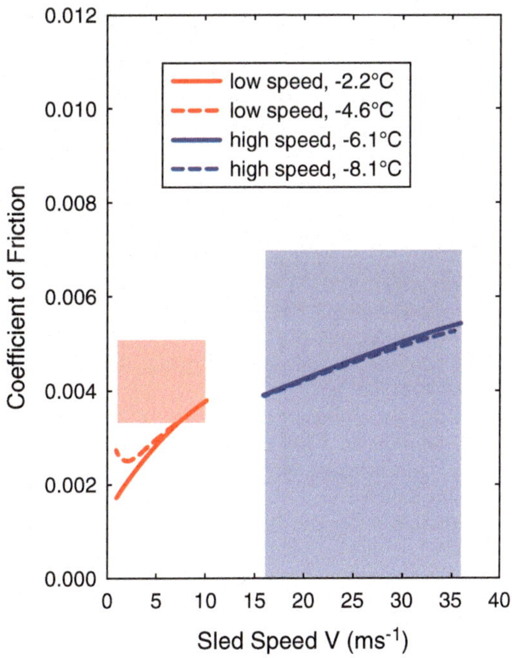

Fig. 3.16 Validation of the model presented in [34] (*continuous lines*) through comparison with experimental data provided in [29] (*coloured rectangles*)

model and the experimental values show a similar qualitative behavior, namely an increase of ice friction coefficient with skating speed. However, the rate of increase in the experiments exceeds that in the model. In Fig. 3.16, instead, experimental data for bobsleigh runners provided by Poirier [29] are affected by significant dispersion especially at higher sliding speeds. Thus, only a qualitative validation can be carried out.

References

1. K. Bromley, Factors affecting performance of skeleton bobsled. Ph.D. thesis, Mechanical Engineering, University of Nottingham, 1999
2. F. Braghin et al., Bobsleigh driver model, in *Proceedings of the Mini Conference on Vehicle System Dynamics, Identification and Anomalies* (2008), pp. 673–679
3. C. Sawade et al., Assessment of an empirical bob-skeleton steering model. Procedia Eng. **72**, 447–452 (2014)
4. K. Tusima, Tribology in skating, skiing, and curling. Toraibarojisuto/J. Jpn. Soc. Tribol. **54**(7), 470–475 (2009)
5. T. Kobayashi, Studies of the properties of ice in speed-skating rinks. ASHRAE J. **15**(1), 51–56 (1973)
6. K. Itagaki, G.E. Lemieux, N.P. Huber, Preliminary study of friction between ice and sled runners. J. Phys. (Paris) Colloque **48**, 297–301 (1987) [Chap. 3]
7. V.A. Balakin, V.N. Smirnov, O.V. Pereverzeva, Sliding of sled runner over ice. Sov. J. Frict. Wear (English translation of Trenie i Iznos) **9**(2), 57–62 (1988)
8. K. Itagaki, N.P. Huber, G.E. Lemieux, Dynamic friction of a metal runner on ice. Part 1: model sled test. Technical report, US Army Corps of Engineers, Cold Regions Research and Engineering Laboratory (1989)
9. T. Furushima, Study on the frictional resistance of skates. Seppyo **34**, 9–14 (1972)
10. R. Ericksson, Friction of runners on snow and ice. SIPRE Report TL 44, US Army Cold Regions Research and Engineering Laboratory (1955)
11. D.G. Ulman, C. Cross, Engineering a new-generation bobsled. American Society of Mechanical Engineers 79-DE-E-5 (1979)
12. V.A. Balakin, Friction performance of sports luges and bobsleds sliding along ice track. J. Frict. Wear **17**(1), 29–36 (1996)
13. J.J. De Koning, G. De Groot, G.J. van Ingen Schenau, Ice friction during speed skating. J. Biomech. **25**, 565–571 (1992)
14. S.C. Colbeck, L. Najarian, H.B. Smith, Sliding temperatures of ice skates. Am. J. Phys. **65**(6), 488–492 (1997)
15. S.C. Colbeck, Pressure melting and ice skating. Am. J. Phys. **63**, 888–890 (1995)
16. C. Hainzlmaier, A new tribologically optimized bobsleigh runner. Ph.D. thesis, Zentraliastitut fiir Medizinteclmik Teehnisehe Universitat Miinchen, 2005
17. S. Boerboom et al., Effect of runner material on ice friction in bobsleigh, in *Proceedings of the 10th Annual Congress of the European College of Sport Science*, ECSS 2005, Belgrade, July 2005
18. M. Dumm et al., The effect of pressure on friction of steel and ice and implementation to bobsleigh runners, in *The Engineering of Sport 6*, ed. by E. Moritz , S. Haake (Springer, New York, 2006), pp. 103–106
19. K. Hokkirigawa, Tribology of the bobsled: development of bobsled-runner for Nagano olympics. Jpn. J. Tribol. **43**(1), 31–38 (1998)
20. K. Hokkirigawa, Tribology in bobsleigh and skeleton-toward salt lake from Nagano. Jpn. J. Tribol. **47**(2), 69–74 (2002)

21. K. Hokkirigawa , T. Yamaguchi, Development of a low friction runner and analysis of the new start technique for bobsleigh. Hyomen Kagaku **26**(12), 762–765 (2005)
22. C. Hainzlmaier et al., Measurement of the contact force between runner and ice for bobsleigh, in *Proceedings of the 9th Annual Congress of the European College of Sport Science, ECSS 2004*, Clermont-Ferrand, March 2004
23. M. Fauve, H. Rhyner, Analysis and optimization of the sliding properties of luge steel blades on ice, in *The Engineering of Sport 7* , vol. 1 (Springer, Paris, 2011), pp. 579–586
24. P. Federolf et al., Impact of skier actions on the gliding times in alpine skiing. Scand. J. Med. Sci. Sports **18**, 790–797 (2008)
25. F. Braghin et al., Bobsleigh performance optimization through a multi-body model, in *Proceedings of Multibody Dynamics 2009 ECCOMAS Thematic Conference,* Warsav, 29 June-2 July 2009 (2009), pp. 1–12
26. F. Braghin et al., Experimental assessment of bobsleigh dynamics and ice-skate contact forces, in *Topics in Modal Analysis II, Volume 6*. Conference Proceedings of the Society for Experimental Mechanics Series (Springer, New York, 2012), pp. 487–498
27. M.R.I. Nieuwenhuijze, A set-up to investigate the steady state behavior of the iron-ice contact of bobsleigh runner. Traineeship report, Eindhoven University of Technology, 2009
28. M. Scherge et al., High-speed ice friction experiments under lab conditions: on the influence of speed and normal force. ISRN Tribol. **2013** (2013). Article ID 703202
29. L. Poirier, Ice friction in the sport of bobsleigh. Ph.D. thesis, University of Calgary, 2011
30. L. Poirier et al., Experimental analysis of ice friction in the sport of bobsleigh. Sports Eng. **14**(2–4), 67–72 (2011)
31. C. Hainzlmaier et al., Computational mechanics in bobsleigh: finite element model of runner and ice. Eng. Sport 5 **1**, 256 (2004)
32. A. Penny et al., Speedskate ice friction: review and numerical model - Fast 1.0, in *Proceedings of the 11th International Conference on the Physics and Chemistry of Ice*. Bremerhaven, 23–28 July 2006 (The Royal Society of Chemistry, Cambridge, 2007), pp. 495–504
33. E.P. Lozowski, K. Szilder, Derivation and new analysis of a hydrodynamic model of speed skate ice friction. Int. J. Offshore Polar Eng. **23**(2), 104–11 (2013)
34. E. Lozowski, K. Szilder, L. Poirier, A bobsleigh ice friction model. Int. J. Offshore Polar Eng. **24**(1), 52–60 (2014)
35. L. Poirier, E.P. Lozowski, R.I. Thompson, Ice hardness in winter sports. Cold Reg. Sci. Technol. **67**(3), 129–134 (2011)
36. E.P. Lozowski, K. Szilder, S. Maw, A model of ice friction for a speed skate blade. Sports Eng. **16**, 239–253 (2013). doi:10.1007/s12283-013-0141-z
37. E.P. Lozowski, K. Szilder, S. Maw, A model of ice friction for an inclined incising slider, in *Proceedings of the 22nd International Offshore and Polar Engineering Conference*, ISOPE 2012, Paper 2012-TPC-1018, Rhodes, June 2012, pp. 1243–1251
38. E.P. Lozowski, K. Szilder, Derivation and new analysis of a hydrodynamic model of speed skate ice friction. Int. J. Offshore Polar Eng. 23, 104–111 (2013)
39. J. Joly, The phenomena of skating, Sci. Proc. R. Soc. Dublin, New Series. **5**, 453 (1886)

Chapter 4
Alpine Ski

Stefano Melzi, Edoardo Belloni, and Edoardo Sabbioni

4.1 Introduction

Figure 4.1 shows that already in the prehistoric age (wall paintings and some antique finds strengthen this hypothesis) the use of skies existed, mainly to facilitate the mobility during hunt. During the centuries the skies were used for military and transportations in snow-covered areas, with some examples of reports about long distance travels across Europe in 1517 AD [1]. A description of ski techniques was published in 1765 AD in military manuals, while the first competitions were held in Norway in the mid-years of 1800, spreading in few years in the whole Scandinavian peninsula up to the Alps. The original disciplines ("Nordic") were only the jumping and the cross-country ones, for this reason in a first time Alps were not considered suitable for skiing. Thus, skies were initially used in the Alps only to hike up the mountains and the descent was performed without skies. As time passed, a new downhill skiing technique, the "Alpine" one, was developed by skiers, ending up in the important teaching manual "Lilienfelder Skilauftechnik" by Mathias Zdarsky in 1896 AD (for further information about him, a really peculiar and capital personality in ski history, one can refer to [2]). Along the twentieth century, the ski experienced a lot of transformations and development: in 1922, in Mürren, the first alpine ski competition took place, then in 1924 the International Ski Federation (FIS) was established, and the ski was finally included among the Olympic disciplines in 1936. Important technological and coaching improvements took place along the whole century, and probably one of the most relevant was the introduction of carving skies

S. Melzi (✉) • E. Sabbioni
Department of Mechanical Engineering, Politecnico di Milano, Milano, Italy
e-mail: stefano.melzi@polimi.it; edoardo.sabbioni@polimi.it

E. Belloni
Dipartimento di Meccanica, Politecnico di Milano, Via Giuseppe La Masa, 1, Milano, Italy
e-mail: edoardo.belloni@polimi.it

F. Braghin et al. (eds.), *The Engineering Approach to Winter Sports*,
DOI 10.1007/978-1-4939-3020-3_4

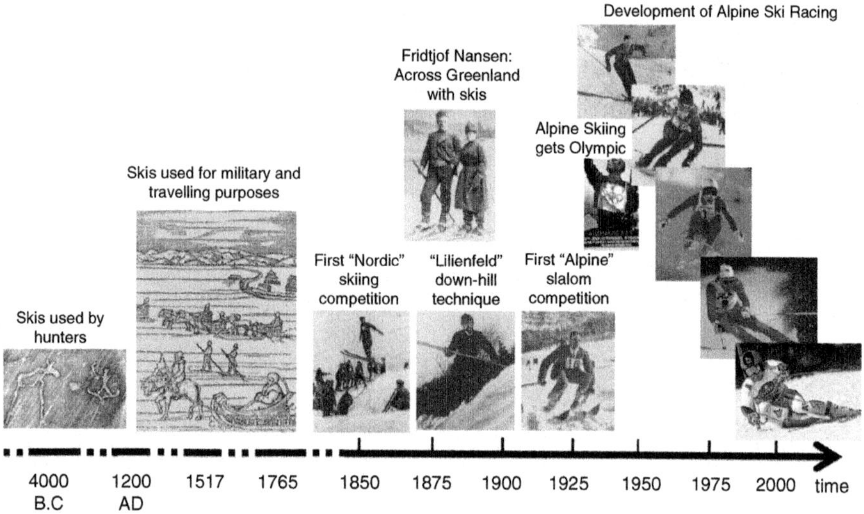

Fig. 4.1 The history of Alpine Ski by its fundamental steps (see [3])

and binding plates in 1990s: higher inclination angles and tighter curves became possible, modifying significantly the appearance of the sport [3].

The International Ski Federation website has a page entirely dedicated to Alpine Ski history [4], and many books were published regarding this topic: for example, one can refer to the works of Polednik [5] and Suttner [6].

Nowadays, the ski has a worldwide diffusion: a lot of people are involved in this sport, both doing it in first person or following the competitions as a fan. As one can argue, also the economic attractiveness for sponsors and correlated activities has experienced a remarkable boost. Following strict regulations [7], the Alpine Ski competitions have been split into four disciplines: slalom, giant slalom, super-G and downhill, with in addition the combined and team disciplines.

Our purpose in this chapter is to provide an overview of the state of the art about the scientific approach to Alpine Ski. It is common to find manuals with rules of thumbs for the skiers suggested by coaches with respect to a more technical and scientific-oriented publications. Technological and safety aspects are presented in the equipment section: focusing on the helmet, the suit, the boots and the skies, that is the most relevant part of the outfit, technological innovations from the point of view of safety, performances and ergonomics are tackled. A section is then dedicated to the experiments conducted both in open air and in wind tunnel. Different systems will be presented to obtain the desired quantities: the (3D) kinetics of a skier, the ground forces, the aerodynamic contribution, and a comparative analysis between skiing techniques and their performances. Finally, there will be presented (multibody) models for the motion of the skiers, and considerations about trajectory, style, velocity will be made. The task is not so straightforward because of the complexity of the motion and the high number of parameters involved in.

Further, a sort of "local" optimization (referred to a specific section of the track) is not sufficient, but the study must be performed on the whole track length to get meaningful results.

4.2 Equipment and Materials

As the most part of sports, also Alpine Ski uninterruptedly seeks the technology developments of the equipment. New technologies and materials allow to reach unforeseen results from the point of view of performances, guaranteeing (and sometime enhancing) at the same time safety and protections for the skiers. In this section, the most important parts of skier's equipment are described, with a particular focus on their impact on results and safety.

4.2.1 Helmet

The use of the helmet is more and more spread between all the categories of skiers from the very beginner to the most skilled ones. For example, the data from annual report of National Ski Area Association (NSAA) [8] highlight that in the US area 73 % of all skiers and snowboarders wore helmets during the 2013–2014 ski season. Analysing the trend in a longer period one can observe that the helmet usage increased continuously from the 25 % of the 2002–2003 season till the most recent results expressed above. The usage of the helmet was enormously encouraged during the years, and also manufactures boosted a lot on safety and ergonomics (and for sure also on design).

Goal of the helmets (for general applications) is to prevent potentially dangerous impacts and injuries of the head, such as skull fractures or problems deriving from long-term brain damages. Design of helmets is changed during years to protect the skull to repeated impacts and severe concussions, with a high probability related to head accelerations and head injury criterion (HIC)[1] (see [9, 10]). Several studies show the benefit of the helmet usage in skiing for preventing the head/face and cervical spine injuries.[2] Thus, starting from children under 13 years old, where it was demonstrated that the helmet brings to a 43 % of reduction in head, face and

[1] It is a measurement of the likelihood of head injury arising from an impact, which takes into account both the acceleration of the head, measured in its centre of mass, and the time interval of the force application (see, for example, [11]).

[2] A different point of view is adopted by other authors, who tried to correlate the usage of the helmet with a more risk-taking behaviour in the skiing. Scott et al. [12] found no evidence of risk compensation among helmet wearers and probably the decision of wearing an helmet is part of a global risk reduction decision. This hypothesis clashes with the results of Ružić et al. [13] where the trend is different. Here the overall risk-tendency is higher in helmeted skiers than nonhelmeted ones. In particular the adult male population under the age of 35 exhibits the higher inclination to risk. Funnily, in helmeted female skiers riskier behaviour is not declining while the age is growing. This dichotomy between author's results is maybe dependent on the different data acquisition

neck injuries [14], up to other studies with a more widespread set of athletes (see, for example, [1, 15]), the beneficial effects of the helmet use are well known by now.

The three deformable bodies which absorb the energy during a head impact are the head, the object involved in the crash and the helmet lining [9]. The deformation of the helmet is fundamental to dissipate energy and thus to attenuate the dangerous consequences of the impact, and typical energy absorbing materials (diffused also in vehicle, ballistic armor, protective clothing, etc.) are polymer foams with microstructures of open and closed cells [16]. A widespread solution for recent helmets is the use of an expanded polystyrene (EPS) with an external thin plastic covering to prevent penetration by external objects. EPS is a lightweight material, good for energy absorption and it is also not expensive, but its irreversible deformation implies a performance deterioration for subsequent impacts (see also Swarén et al. [17] for a study about subsequent impacts effect on the helmet), and further the EPS reacts locally to external forces, leaving the force concentrated in a small area (and then with a high local pressure) [9]. One idea to improve the protective performances of an helmet is to incorporate, in a vinyl nitrile (VN) foam, a series of channel filled with a high viscosity fluid (see [18, 19] for previous prototypes with the VN foam surrounded by the fluid). Stewart et al. [9] conducted several tests on this kind of system. Performances with respect to the classic EPS material are better, in particular for multiple impacts: fluid channels help to better distribute the force (and so the pressure), obtaining a better stress distribution, less localized in a small area. Quantitatively, a reduction of about the 17 % in peak headform acceleration on the first impact and of about 50 % on the sixth impact is gained with respect to the EPS foam, using an aqueous solution containing 30 % glycerin in five 0.95 cm diameter channels [9].

4.2.2 Ski Suit

The suit covers the most part of the athlete's body and has consequently a leading role in aerodynamic performances determination (and for sure it protects the skier by crashes and …cold!). Focusing on professional competitions, the regulations about the suit completely changes between disciplines: for example, in Ski-Cross skin suits are not allowed, so it requires loose garments. Test on the aerodynamic behavior in this discipline has been proposed, for example, by Chua et al. [20] with an accurate analysis of the flapping garments using different textiles and by Oggiano and Saetran [21] who analysed the beneficial transition from laminar to turbulent regime, improved by the use of different garments on the same suit. Referring now specifically to the Alpine Ski, in Sect. 4.3.3 an analysis of aerodynamic loads and the influence of garments is provided.[3] Focusing now more on textiles, typically the

method: in Scott the amount of risk taking was self-reported, whereas in Ružić's study it was estimated from a series of items analysing the skiing style of skiers [13].

[3]One has anyway to bear in mind that a large part of the aerodynamic contribution is due to the body shape and posture, rather than to the garments.

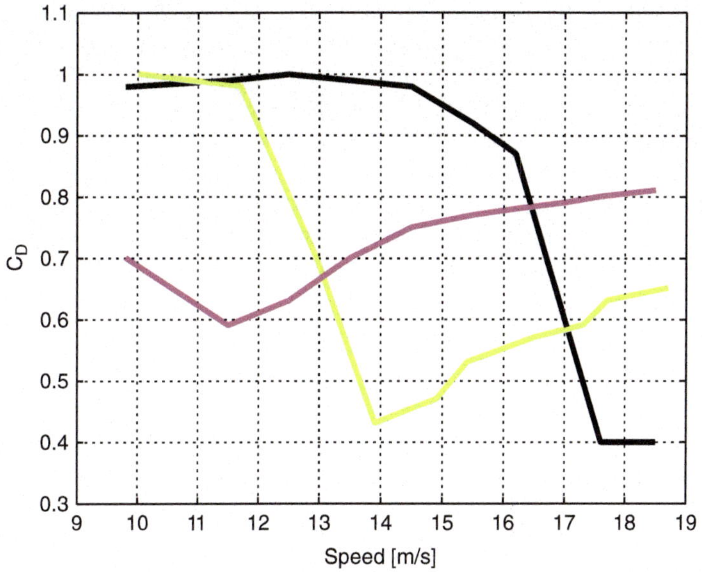

Fig. 4.2 Variation of drag coefficient C_D with velocity for three different fabrics. *Yellow*: rather thick fabric, fine texture. *Magenta*: rougher fabric, macro roughness pattern. *Black*: thin, very smooth fabric. Qualitatively adapted from [22]

ones used for alpine ski suits are composed by three layers: the external layer is protective, the inner one is absorptive, and in the middle an adhesive layer glues the other two together and reduces air permeability (the minimum allowed by regulations is $30\,\mathrm{l/m^2\,s}$ [22]). Brownlie et al. [23] provided data about the effects of wear, apparel fitting, air permeability and other factors on skier drag through a set of experiments conducted with full race suits: this approach is useful to get practical information, but cannot isolate the variable. Other tests were conducted on circular cylinders (often human limbs are modeled in this way, see, for example, [24, 25]): Bardal and Reid [22] tested different textiles both on cylinders and on a thin, symmetrical airfoil, to provide better indications about the more streamlined parts of the body. In Fig. 4.2 an example of the results in terms of drag coefficient C_D changing the velocity. The data (here qualitatively plotted) are obtained testing three different fabrics on a circular cylinder of diameter $D = 20\,\mathrm{cm}$ (note that this trends are dependent on the diameter of the cylinders, see [22]).

The fabric represented by the yellow line is a rather thick fabric with a smooth knit outer layer, giving it a fine texture: its behaviour resembles to the one of other typical sport fabrics [26]. The magenta line refers to a rougher fabric made with an outer layer featuring a macro roughness pattern designed to trip the boundary layer at low velocities: it could be used for slalom competitions (even though for higher velocities the rough surface induces a rapid growth of drag coefficient). Finally, the fabrics represented by the black line is the smoother and thinner one, guaranteeing a high transition speed and a low minimum C_D, good characteristics to make the fabric a possible candidate for use in downhill suits [22]. Also the effect of fabric stretching was considered in [22]. A realistic stretch for ski suits while wearing is about 25 %, and during tests it was raised until 42 % (probably excessive athlete's comfort and movement freedom). Raising the stretch, the rougher textiles experience flow transition at somewhat higher speeds (around 1m/s), while no relevant effect is seen for smoother fabrics.[4] Full suit wind tunnel drag measurements with athletes proved to be difficult with respect to repeatability and stability of measurements. In spite of the limited quantitative outcome of these measurements, significant drag reductions were achieved in good agreement with cylinder test results [22].

4.2.3 Boots

Ski boots have experienced several developments in different aspects: not only for performance enhancement, but also for ergonomics and injury prevention. The most part of the technological changes in boot technology have been introduced in the second half of the twentieth century, due to both the development of new skies and techniques (higher velocities, stresses and control) and the growth of materials like resins, polymers, etc. A first step was to pass from leather to stiffer materials, and Robert Lange was the first who tried to use a fibreglass reinforced epoxy resin in 1947, reaching the first ski-boot completely made of plastic, using acrylonitrile butadiene styrene (ABS) polymers in 1960 [27]. In the same years, Hans Martin of Henke Speedfit patented the levered buckles for the closure of the boot [28], while in 1965 the ABS was substituted by Adiprene, a thermoplastic polyurethane which allowed to overtake some problems of mechanical failures and to introduce the injection molding production process. Rosemount then introduced the first ski-boots completely made of composite materials and with an easier way to introduce the foot in it [29]. In 1966 a true mass production of plastic ski-boots started with Lange (with the so called *Overlap* design, made of two parts, the lower part called shell and the upper part called cuff) and with Nordica in Montebelluna, Italy, using a polyurethane used in aerospace applications (Desmopan) (see [30] and [31]). One of the last improvements in boot design was made by Eric Giese in 1979, who introduced an additional tongue to control the flex of the boot [32]: this construction was named *Flexon* (also known as *Cabrio* or *3-pieces* design), which remains one of the most common together with the *Overlap* design [31].

The choice of the right material and design is fundamental, as previously stated, for a good union between performances and ergonomics; one can summarize the most important characteristics of a ski-boot [31]:

- to have a precise control of the ski, through an efficient transmission of loads from the skier to the ski edge;
- quick connection (bindings) with the ski and fast release in case of tumble;

[4]The stretch of fabrics is also related to air permeability: results indicate that fabric air permeability itself has little if any correlation with drag for tight, form fitting fabrics.

- shock absorption;
- protection of foot and ankle during fall, in particular from overloads;
- ergonomics, comfort and optimal temperature/humidity conditions.

For a complete overview of this discussion and a definition of the most important involved parameters, please refer to [31].

Several studies exist about these arguments, with a particular attention at the safety requirements during the design of a boot (see, for example, [33, 34]). Starting from the problem of comfort of the boot, one of the most affecting factors are cold feet. Ski boots possess in general a good thermal insulation, but the sweat of the foot and the lack of moisture vapour transmission through the liner lead to moisture saturation inside it, facilitating the conductive heat loss. Hofer et al. [35, 36] proposed a study about the microclimate in the ski boot assessed during imitated skiing in a climate chamber and during field tests using four sensors in the liner between skin and sock. General considerations were obtained: the prevention of cooling of feet and toeing in particular is fundamental to avoid excessive temperature decrease. Ambient temperature and moisture affected strongly the foot temperature, while a high water content in socks reduces the isolation properties. A recommendation is the use of waterproof shells to stop snow and water penetration, and further a complete overnight dryer of the liner is suggested. Good absorption of liners and hydrophobic socks are a considerable suggestion to keep the foot dryer.

Ergonomically speaking one can, for example, look at the work of Pinter et al. [37], who proposed a study of the pressure distribution between a ski boot and a subject's foot as well as the lower leg. The tests showed that it is necessary to measure the pressure of a widespread area of the foot, instead of points, to ensure that there is a steady pressure over the whole area. One can also infer from the results which boot feels painful to the subject, as well as how the pressure distribution should look like within the boot. A sensor state during closing could help manufacturers to analyse how different materials and wall thicknesses of the shell or different shapes and materials of the liner influences the pressure distribution, but it could be useful also to speed up the decision-making process of a customer.

The boot was studied also from other points of view. Petrone et al. [38, 39] formalized the definitions of Flex Index (Nm) and of boot stiffness (Nm/°) to quantify the flexural behaviour of boots and their classification, useful to guide the users in the selection process. Further, they investigated the engineering parameters correlated to the enhanced performances with an innovative bootboard placed in the boot. It was instrumented both with strain gauges and markers (to goal was to measure displacements and deformations also through a motion capture technique). Some tests on stiffness highlighted a significant increase (20 %) of Tip-Heel Torsional stiffness of the closed boot, while lateral stiffness tests gave no evidence of any global lateral stiffening, therefore inducing to consider the experienced advantage in skiing as correlated more with the contribution of the reinforced ski-boot to the overall ski-shovel torsional stiffness during transition from edge to edge rather than to its contribution to lateral stiffness during the steady curve conduction [38]. The application of motion capture techniques to the structural analysis of such deformable structures can be seen as a powerful advance in the functional analysis of boots: the trend of local twist angles along the shell can highlight areas where stiffening (or softening) is appropriate. Further, the analysis of deformation of cross sections along the boot during loading cycles can give quantitative information about possible discomfort reasons at the medio/lateral malleoli.

Injuries and their prevention are not negligible aspects [40]: for example, Senner et al. [33, 34, 41–43] studied in depth the problem of safety and injury prevention of ski boots. The discussion is complex due to the fact that it is a mix of boot contribution (e.g., flexible rear spoiler), ski interventions (e.g., "nominal breaking point", a release function in the rear section of the ski) and the binding, and further it is an interdisciplinary discussion between different fields of study. Potential possible developments, so to better include also other criteria (e.g., effectiveness, efficiency, feasibility) different, but related to injury prevention, can come to a more large-scale effort coming from other sports. Further studies must be conducted on cadavers/simulations to better understand the failure conditions of structures (in the knee [34]) and to better understand the loading of these structures in specific skiing situations. In parallel, in vivo measurements of the joint angles and muscle activation during skiing to better understand the role that muscle forces may play in the loading or in the prevention of loading of the joint [34].

4.2.4 Skies

Up to 1930, the skies were basically constituted by wood, and in this year Rudolph Lettner invented the first steel edge, useful for an easier wear but also a significant help in turning on hard snow. The 1933 was the year of the introduction of laminated skies and the first useful release binding was invented in 1939 by Hjalmar Hvam. In the following years a lot of different materials were used to reinforce the ski structure: aluminum (1949), fibreglass (1952), polyethylene (1955), the first plastic fibreglass ski (1959). It was not until 1990 that the classical telemark geometry of the skies were replaced by deep-side cut shaped skies: the emerging carving skies were able to revitalize the whole skiing sport [44].

Nowadays skies are made by superimposed layers of different materials, like the example in Fig. 4.3 (see also [45] for a complete dissertation).

Usually, the bottom is composed by a layer of polyethylene enframed by steel edges. The structure composed by a wood laminate enclosed between two metal face layers determines the bending and torsional stiffness of the ski. One has further to consider that the thickness of the core is not constant along the ski, inducing a change and an adjustment of stiffness along the ski axis. Additionally layers of fibreglass laminates (often embedded in polyurethane) increase the torsional stiffness of a ski, and a superficial layer is used for decoration, without a relevant structural contribution [44]. All these layers are glued together thanks to a special

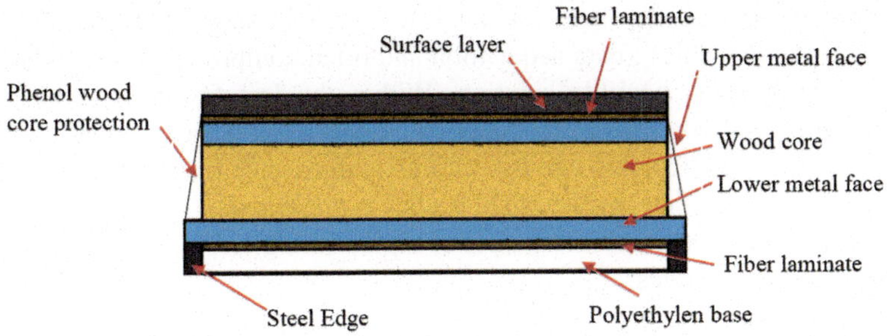

Fig. 4.3 Schematic of a modern laminate ski. Adapted from [44]

epoxy resin by press sizing. The press induces a deformation, and the ski starts to camber. If the ski is loaded in the middle section, the camber transfers pressure to the ski end and the shovel, which improves the snow grip of the ski edges at shovel and tail and facilities turning [44]. Finally few notes about the bindings. The two most relevant requirements are to hold the ski to the skier's boot and to easily allow the ski release in case of excessive stresses in order to prevent injuries. And for sure they have to preserve the interaction between skier, ski and snow. Usually the binding consists of separate toe and heel parts, which comprise the clamps holding the ski boot and the release mechanism. The connection between these two parts is usually flexible, such that the bending of the ski is not significantly interfered. Recently some ski manufacturers have developed skies, which are already equipped with a fixed binding system. This allows a better adaptation of the single components and thus a better tuning of the overall properties of the ski-binding-system [44].

4.3 Experimental Tests

Methods for analysing skiers' movements (kinetics and kinematics), ski-snow contact forces and aerodynamic forces in field conditions are the keys for introducing/evaluating changes in equipment design, investigating skiing techniques and improving training methods. Performance diagnostics are in fact of utter importance in modern sport.

Main challenges in developing measurement systems allowing the analysis of athletes' motion and/or the determination of ground and aerodynamic forces consist in:

- the measurement system has to fulfil high accuracy standards in order to detect the small but substantial differences between athletes trajectories;
- the measurement system should cause minimal interference with the athlete natural course of motion;
- the measurement system should allow the largest possible capture volumes in order to be able to analyse entire competitions and/or training runs.

In the past, it was not always possible to simultaneously achieve these partly contradictory aims during data acquisition and often compromises had to be made to arrive at the best possible solution. Within the last few years, changes in measurement methods have, especially in biomechanical field studies, made it possible to gain new insights into the various performance techniques [46]. In this section a review of biomechanical field studies is presented.

Technical literature on biomechanics of alpine skiing can be essentially divided into three main categories:

- qualitative studies of forces and resulting athlete motion sequences during skiing (1930–present);
- quantitative description of motion process in alpine skiing using biomechanical methods and systems for analysing skiing techniques (1955–present);
- optimization of course run line based on measurement of key variables for identifying skier movements (1980–present).

The first category contains studies concerned with qualitative descriptions of skiing such as recording of motion sequences of skilled athletes and the evaluation of forces during skiing [47, 48].

The first comprehensive description of motion process owing to the second category was achieved in the 1950s using a dynamograph in the form of a ski-mounted mechanical level gear, which allowed the measurement of ground reaction forces acting on the ski during a run. Since the 1970s, several studies have been carried out, which make use of biomechanical analysis in order to establish objective evaluation criteria for skiing technique of international elite ski racers based on actual measurements [49–51].

The third category is characterized by using biomechanical analyses and mathematical modelling in order to measure, predict and sometimes suggest technical improvements. Many studies were concerned with optimization of course run line and aerodynamic position [23, 52, 53]. Other studies have investigated how ski geometry affects run line [54, 55]. Several researches tried to correlate ski deflections (bending and torsion) and modes of vibration with skier performance. On the purpose of determining ski deflections while skiing, strain gauges were directly applied on the skies in [56] and [57], while in [58] a sensor beam was used. A sensor node including a tri-axial accelerometer and a GPS (Global Positioning System) was instead developed in [59].

In the following main methodologies developed for capturing 3D kinetics and kinematics of alpine skiers, ground reaction forces during field tests and aerodynamic forces by means of wind tunnel tests are described. Their application to analyse/evaluate skiers' performance is also presented.

4.3.1 *Measurement Systems for Capturing 3D Kinetics and Kinematics of Alpine Skiers*

In the sport of competitive alpine skiing there is a constant change of position of body, legs, arms and therefore of skier CoM (Centre of Mass) in all three dimensions. In order to analyse skier kinetics and kinematics with respect to performance an accurate three-dimensional (3D) reconstruction of the motion of his body segments (legs, arms, hips, chest, etc.) is needed. A variety of measurement methods were proposed to quantify skier kinetics and kinematics in field. Essentially they can be grouped in:

- optical systems (infrared-based or video-based 3D kinematics);
- GPS (Global Positioning System) or GNSS (Global Navigation Satellite System)[5] based systems;
- IMU (Inertial Measurement Unit) based systems.

Infrared-based systems are known as the gold standard for analysing 3D kinematics of human movement under laboratory conditions. However their application under field conditions (low temperature, wind, solar radiation, snow spraying, variable light conditions) is limited. Moreover an enormous amount of cameras (up to 24) is needed in order to analyse capture volume of one ski turn. In contrast, video-based 3D kinematic measurements have shown to be accurate and applicable under field conditions on the purpose of estimating movements of skier CoM and body segments [1, 3, 60–62]. In addition this method introduces only minimal interference with the athlete and enables capture volume of 1–2 giant slalom turns with only 5–6 panned, tilted and zoomed cameras. GPS/IMU-based systems (which can be worn by athletes during in field tests) allow analysing much wider sections (several turns) and even the entire run. Main drawbacks associated with these systems regard their higher interference with the athlete and their accuracy in estimating CoM position. To improve accuracy, GPS and IMUs data are often combined, using sensor fusion approach [63–66].

4.3.1.1 Video-Based 3D Kinematics Systems

Video-based measurements have been widely used to reconstruct 3D kinematics of skiers in field conditions [3, 60–62, 67]. The general procedure for 3D kinematics reconstruction is shown in Fig. 4.4 and consists of the following steps.

- Geodetic measurement. For analysing the 3D kinematics of a skier in field, large capture volumes are required. Consequently, to be able to track the skier and

[5]GNSS is the umbrella term for global navigation satellite systems, while the more widely term GPS is the name of the American global navigation satellite system. GLONASS is instead the name of Russian global navigation satellite system.

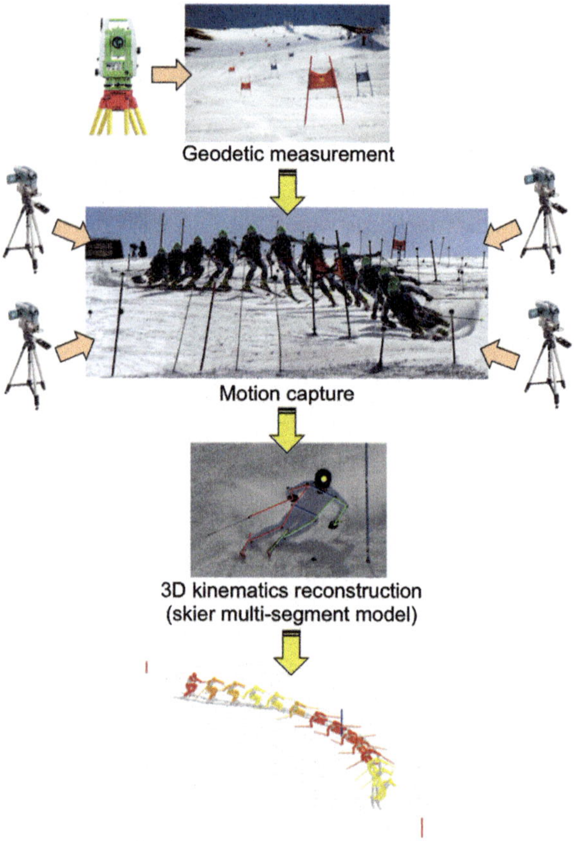

Fig. 4.4 Procedure for video-based 3D kinematics reconstruction

keep him in a sufficient size within the camera image, the use of panned, tilted and zoomed cameras and/or the use of multiple cameras with fixed position and image settings is required. In any case, the spatial coordinates of some reference points (markers) and video cameras are needed in order to keep the cameras calibrated at every instant. For that reason, before and after motion capture, the 3D space positions of all cameras and reference points have to be measured geodetically by theodolite (or tachymeter).

- Motion capture. The skier must be filmed by several cameras (the number odd cameras depends on the length of monitored course), time synchronized by a gen-lock signal. Camera positions should be selected so that:
  - a clear identification of anatomical markers must be ensured by at least two camera perspectives over the entire capture volume [3];
  - the optical axes of the cameras used for 3D reconstruction should ideally intersect rectangular over entire capture volume;
  - cameras should be placed as close to the analysed object as possible [3];

  - large panning angle should be avoided (max. $\approx 30°$) [3];
  - if monitored portion of the course is divided into subspaces, camera positions should guarantee an overlap within two and four recorded frames between subsequent subspaces [61].

- 3D kinematics reconstruction. Video capture of each camera must be time synchronized with corresponding sequences of other cameras and imported into a 3D video analysis software (e.g., Peack Motus or APAS), capable of transforming 2D pixel-coordinates into 3D data based on geodetically measured 3D space coordinates of reference points. 3D video analysis software is then used to reconstruct kinematics of skier movements, based on a rigid multi-segment model of the skier. In other words, trajectory and speed of reference points (or markers) of skier rigid multi-segment model are determined by the 3D video analysis software. Twelve [61] up to seventeen [3] rigid segment models are generally used on the purpose. Methods for calculating segment masses and centre of mass position based on anthropometric studies of human body can be used on the purpose [68, 69].
- Post-processing and data analysis. After reconstruction, trajectory and speed of CoM of each body segment are available for data analysis. Thus, parameters characterizing skier technique, line, energy and loading patterns can be calculated.

As an example, video-based 3D kinematic measurements were carried out during a men's WC (World Cup) slalom race in Kranjska Gora, Slovenia, involving eighteen male WC alpine slalom skiers [61]. Measurements were performed from slalom gate numbers 7–11 on a 56-gate course. Four complete slalom turns were recorded with approximately 11 m between the turning poles on an average slope inclination of 24°. The following experimental setup was proposed [61]. All ski runs were recorded using six generator-locked cameras covering three kinematical subspaces. Each subspace was covered with a pair of cameras (Fig. 4.5) having fixed position and image settings, placed higher for better visibility and located between 40 and 70 m away from the course setup to minimize errors involving lenses. The pixel size from the camcorders was consistently between 1 and 2 cm. The overlap between subspaces was between two and four recorded frames. The frame rate of the cameras was 50 Hz with a standard 720×576 pixel resolution. In addition, two panned, tilted and zoomed cameras were used to record the entire section from both sides. These recordings were only used to assist in confirming joint positions.

All subspaces were calibrated using geodetic measurements provided by a thoedolite (tachymeter) and eight 1.95 m long aluminium calibrating poles with reflecting markers. The calibrating poles were positioned on the racecourse, where they were filmed before the race. Two additional poles were placed and filmed to define the actual fall-line of the slope.

The kinematics of the skiers' movements was represented using a 12 rigid multi-segment model defined by 15 reference points/joints, where segment masses and the centre of mass for each segment were calculated according to [69]. APAS Ariel 3D kinematic software (Ariel Dynamics Inc.) was used to transform the $2 \times 2D$ data into 3D data. The raw data from each subspace were taken from APAS software.

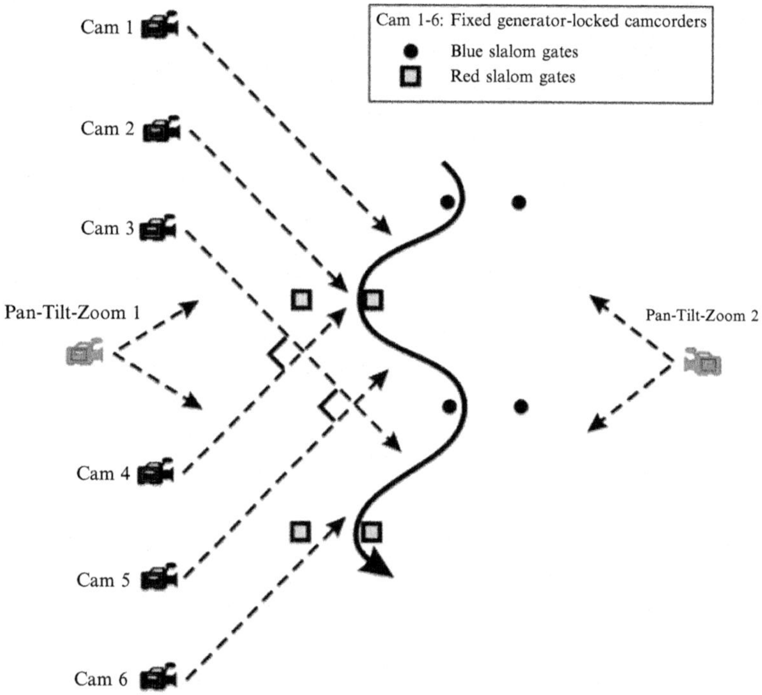

Fig. 4.5 Scheme of measurement setup proposed by Supej et al. [61]

The subspaces were joined together with specially designed routines. The centre of mass trajectory and its first differential were used to check for consistency between the data of different subspace overlap and near surrounding points.

To verify accuracy of the proposed measurement system, inside turning poles in the monitored section were used. These poles were first measured with the theodolite and compared with the 3D kinematics measurements. The standard deviation of the differences in location was lower than 2 cm.

A different setup was proposed in [3, 67] to measure 3D kinematics of a top WC athlete (world champion in giant slalom within the same year) performing a total of 12 runs on two different course settings. For the first six runs, the vertical gate distances were 26 m with an offset of 12 m. For another 6 runs, the offset was changed to 10 m. A total of 78 reference points, geodetic measured by theodolite, were used to calibrate a capture volume corridor of approximately $52 \times 12 \times 2\,\text{m}^3$ (Fig. 4.6). In this area, the skier was filmed with a system of five panned, tilted and zoomed cameras. Cameras were time synchronized by a gen-lock signal and their frame rate was 50 Hz. All runs were recorded in a manner in which the skier covered approximately two-thirds of the picture. In each frame of each camera, a 17 rigid segment model with 28 points on the skier, the skies and the ski poles, as well as the

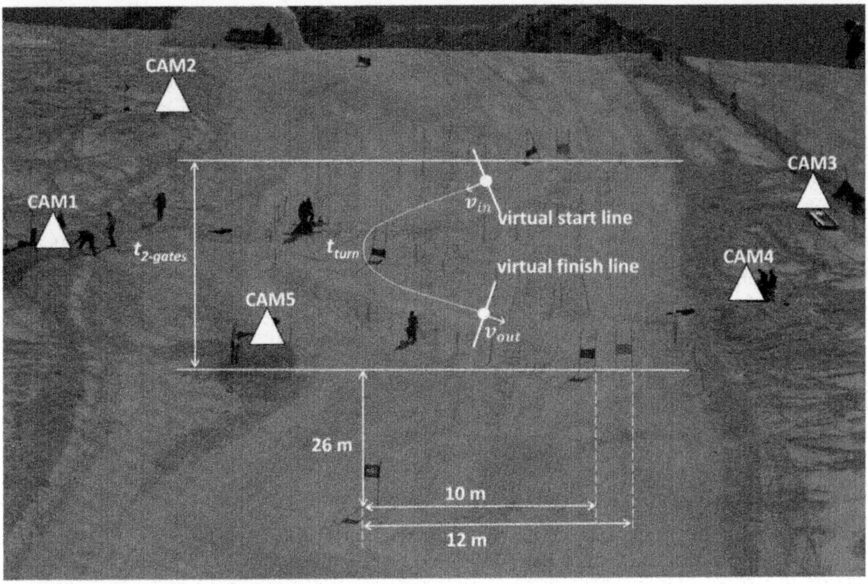

Fig. 4.6 Overview of the measurement setup proposed by Spörri [1]

three best visible reference points were manually digitized. The skier 3D position data were reconstructed using the software Peak Motus and a DLT (Direct Linear Transformation)-based panning algorithm [62].

The measurement error for reconstructed 3D positions of anatomical markers was estimated to be in the order of 1 cm [3].

4.3.1.2 GPS/IMU-Based Systems

Video-based 3D kinematic systems are limited in capture volume and need extensive processing time, which is unfortunate since the number of conditions to be investigated in alpine skiing is large. Therefore differential GNSS (dGNSS) [53, 70] or combinations of GNSS/GPS and inertial measurement systems [63–66, 71] have been applied to reconstruct skier CoM position, velocity and acceleration as well as segment kinematics in racing situations more efficiently. It must be however noticed that, contrarily to video-based systems, none of these wearable system-based methods has been validated against reference measurements which have been proven valid and have been extensively used [70].

An evaluation of position accuracy of available differential GNSS systems was provided in [70]. Specifically, position accuracy and time taken to compute differential GNSS solutions for five typical GNSS methods commonly used in sport science were investigated, in the case of alpine ski racing under three different satellite signal obstruction conditions. On the purpose, the GNSS methods were compared to an independent video-based photogrammetric reference system during one turn of a giant slalom course. Then, the most accurate and consistent GNSS method was defined as the reference method and was compared to the other methods over the larger capture volume of an entire giant slalom run. In addition, the time required to compute differential GNSS solutions was assessed for each method and condition. During the tests, six male European Cup or former WC skiers performed two runs on a typical giant slalom course, consisting of 12 gates. Each run lasted approximately 20 s, during which athletes skied in average at 17 m/s and reached accelerations of up to 4 m/s^2. Minimal turn radius was 16 m. The athletes' head trajectory was captured during the entire run using GNSS. Specifically, a GNSS antenna was mounted on the helmet of the athlete and was connected to a GPS/GLONASS dual frequency (L1/L2) receiver carried in a cushioned backpack and recording position signals at 50 Hz. Moreover two GNSS base stations were mounted close to the start of the course.

The trajectory of the GNSS antenna placed on the skier helmet was computed using the geodetic post-processing software Justin in double difference mode. The kinematic position solutions were computed for five different GNSS methods. Method A was a differential phase solution that included both GPS and GLONASS satellite signals and the signal frequencies L1 and L2. Method B was similar to method A, but using frequency L1 only. Method C was similar to method A except that GPS signals only were used. Method D was similar to C but using frequency L1 only. Method E was a non-differential code solution using only GPS code signals.

During turn 7 of the course the GNSS antenna was simultaneously captured by a video-based photogrammetric reference system. Similarly to what described in previous section, the reference system consisted of six panned, tilted and zoomed cameras, which captured the skier at 50 Hz. Camera images were time-synchronized using an electronic gen-lock signal. The GNSS antenna was manually digitized in the camera images. Reference points were mounted along the course for turns 6, 7 and 8 and their positions were determined using a theodolite in a local coordinate system (LCS). The reference points and the camera positions were used along with a DLT-based panning algorithm to reconstruct the GNSS antenna trajectory during turn 7.

In order to compare the GNSS-based trajectories with the reference trajectory, the GNSS coordinates in the global coordinate system were transformed into the local coordinate system (LCS).

The following were identified in the comparison of the five dGNSS methods with video-based system:

- the only GNSS method consistently yielding sub-decimeter position accuracy in typical alpine skiing conditions was Method A, i.e. a differential method using GPS and GLONASS satellite systems and both frequencies L1 and L2;
- more in particular, spatial accuracy was higher for method A, followed by C, B, D and E.

Comparing the five dGNSS methods during an entire giant slalom course emerged that:

- a certain amount of measurements with methods B–D resulted in differences in the order of meters consisting both of integer fixed ambiguity and real number float ambiguity solutions;
- in conditions with minimal satellite signal obstruction, valid results were also found if either the satellite system GLONASS or frequency L2 was dropped from the best configuration (Method A);
- all other methods failed to meet the accuracy requirements for alpine skiing even in conditions favourable for GNSS measurements;
- the methods with good positioning accuracy had also the shortest times to compute differential solutions. Specifically, time to fix integer ambiguities was shortest for method A followed by methods C, B and D.

It must be pointed out that GNSS/GPS data only allow the reconstruction of trajectory and velocity of the antenna, not of the complete 3D kinematics of the skier. It is however possible to estimate the trajectory and the velocity of the skier CoM and of the skies based on a single degree-of-freedom model reproducing the biomechanical phenomenon that skiers incline laterally in order to balance the radial inertial force during turns [53, 72]. Specifically, the skier can be modelled like an inverted pendulum attached to the GNSS antenna (Fig. 4.7), which is dynamically balanced at any given time. The neutral position of the pendulum occurs during straight skiing, when the skier radial acceleration is zero. During turns instead the pendulum is deflected from its neutral position. The intersection of the pendulum with the snow surface identifies the skies position. The CoM of the pendulum (which is coincident with the CoM of the skier) can be determined based on full body segment kinematic datasets. The deflection γ representing the skier lateral tilt can be calculated as a linear combination of the gravitational and the dGNSS antenna radial acceleration, $a_{r,\mathrm{GNSS}}$ (Fig. 4.7a) [72].

Alternatively, the skier inclination γ can be obtained from the balance equation describing the inverted pendulum model [53]:

$$\ddot{\gamma} = g\cos\gamma + \omega^2 R_{\mathrm{CoM}} \sin\gamma \tag{4.1}$$

where the skier mass m (including equipment) and the distance between the ground and the skier CoM (l_{CoM}) have been eliminated (Fig. 4.7b), ω is the angular turning velocity of the skier around the vector normal to the slope and R_{CoM} is the turning radius of the skier CoM.

The angular acceleration $\ddot{\gamma}$ can in fact be employed from previous time steps, the angular speed ω is known by measurements provided by GNSS, while the CoM turn radius is a known function of GNSS antenna trajectory and skier inclination angle.

To reconstruct complete 3D kinematics of the skier, GPS data must instead be combined with inertial measurements (provided by IMUs). An example can be found in [63], where Fusion Motion Capture (FMC) was used to reconstruct 3D kinematics of alpine skiers. In the proposed application, FMC consisted in developing a composite system able to fuse data from IMU, GPS, video, and pressure insoles. Specifically, to determine the kinematics of skier body segments,

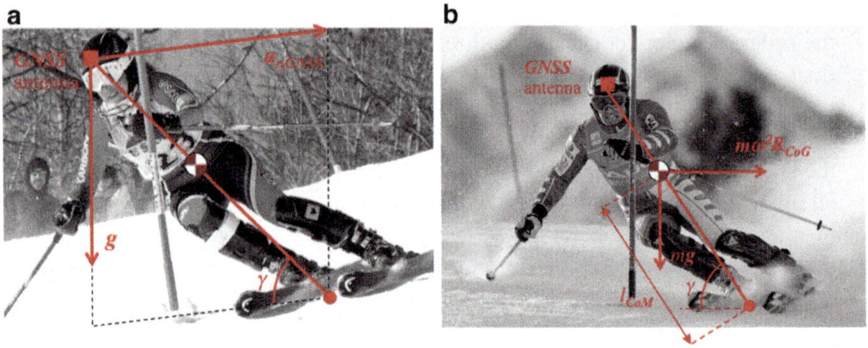

Fig. 4.7 Inverted pendulum model of the skier. (**a**) Gravity acceleration and GNSS antenna radial acceleration; (**b**) forces acting on the skier pendulum model

thirteen IMUs were attached to head, torso, pelvis, upper and lower arms, thighs, shanks, and ski boots of the athlete. IMUs were attached so that to minimize their interference with the athlete: a lycra bodysuit was built to contain connecting wires. The suit had gaps at the location of each IMU so that it could be attached directly to the athlete skin with the use of double-sided tape. Each IMU was fastened with a firmly fitting elastic strap attached with velcro to the suit.

All the IMUs (50 Hz sampling frequency) contained three gyroscopes, three accelerometers, three magnetometers and a thermometer in a 35-g box about the size of a matchbox.

A GPS (1 Hz output, dilution of precision $\pm$ 5 m) was attached to the athlete helmet and a local GPS base station was placed near the course.

A pressure measurement system (see Sect. 4.3.2.2) was used to determine plantar pressures and the ratio of loading between the skies.

Video from a hand-held digital camera, panned from a fixed position on the skiers on the left side of the course was used as an external reference and to confirm validity of the data.

The algorithm for fusing IMUs and GPS data developed in [63] to estimate position, velocity and orientation of skier body segments is shown in Fig. 4.8 (alternative approaches can be found in [64–66]).

The trajectory of the athlete helmet was calculated by fusing the GPS data at 1 Hz (both base station and helmet) with the data from the IMU attached to the helmet and sampled at 50 Hz. In this way, errors caused by drift of dynamic acceleration measurements and angular rates could be compensated.

The precise start and finish locations of the athlete and each gate position were surveyed using a theodolite and a GPS.

FMC was used to capture 3D kinematics of a 20-year-old male member of the New Zealand national team (body mass 78 kg), who completed five runs through a 10-gate and 300 m-long giant slalom training course at Mt Ruapehu Ski Area, North Island, New Zealand.

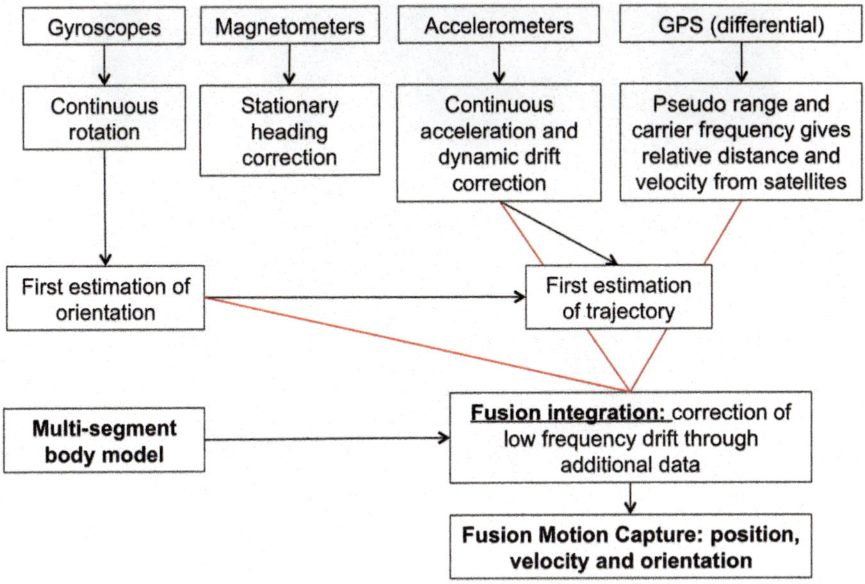

Fig. 4.8 Block diagram of sensor fusion integration algorithm for IMUs and GPS measurements

Although not confirmed by comparison with a reference measurement system, authors estimated an improvement in GPS accuracy from a dilution precision of ±5 m (meaning 50 % of the measurements will be within 5 m of the true value) to a maximum error of ±1.5 m over the race course, and a reduction of IMU orientation error from over 20° to less than 5°.

Moreover, comparing two giant slalom runs, analysis of skier kinetics and kinematics was able to point out the essential subtle differences explaining a variation lower than 1 % in time between the two runs (see Sect. 4.3.4).

4.3.2 *Measurement of Ground Forces*

Forces acting on the skier are shown in Fig. 4.9. Gravitational force, $\mathbf{F}_g$, is the main driving force for the skier. It accelerates the skier down the slope, giving the skier momentum (i.e. the product of skier mass and velocity).

Ground reaction force,[6] $\mathbf{F}_r$, arises as a result of snow resistance to compaction and shear under the applied load from the skier and it can slow down, speed up and turn the skier.

[6]In Fig. 4.9, the ground reaction force is applied in a point midway between skies. The real point of application depends on the force distribution between skies.

Fig. 4.9 Forces acting on the skier

Note that dynamic balance in slalom turns require that gravity must balance the vertical component of the ground reaction force. In addition, due to the torque created about the CoM by the horizontal and vertical reaction forces, they must dynamically balance to control a skier rotation about the CoM and avoid falls toward inside or outside of the turn [73].

Internal forces are produced by the skier muscles to control the motion by directing limbs and the body in a desired manner, manipulating the equipment and pushing against the snow [73].

Skier direction of travel is changed by the edging of the skies. The snow properties allow the skier to glide on it by being slippery. As the skier manipulates edging and inclines the CoM, the snow changes the skier momentum by resisting compression and making turning possible [73].

Aerodynamic drag, $\mathbf{F}_d$, acts opposite to the skier instantaneous velocity of the CoM and snow friction acts opposite to the instantaneous velocity of the skies. Air drag depends on several factors, among which the ones under the skier control are his posture and frontal area (see Sect. 4.3.3). Ski-snow friction encompasses the tribological properties of the skies on different snow conditions, and most importantly, precision when guiding the skies.

Ground reaction force, aerodynamic drag and gravitational force must balance, according to the Newton's second law, the inertial force acting on the skier, $\mathbf{F}_{\text{in}}$.

In this section, methods for measuring ground reaction forces are described, while Sect. 4.3.3 reports main approaches for evaluating aerodynamic drag.

Measurement of ground reaction force has been analysed in several studies. In particular, recent alpine skiing researches focused on three independent methods: estimation based on 3D kinematics analysis (either video-based or IMU/GPS-based)

[63, 72], force transducer placed between boot and binding [74] or between binding and ski [75–77] and plantar pressure measurement systems (or pressure insole systems) [46, 63, 76–78]. In addition, electromyographic (EMG) measurements were often used to monitor muscles activation in relation to ground reaction forces [46, 74].

4.3.2.1 Force Transducers

Ground reaction forces were assessed by several force transducers placed either between the ski and the binding [75, 76] or the binding and the boot [74]. Force transducers typically consist of two to four sensing cells and measure forces in all three dimensions.

The most common commercial force transducer applied in alpine skiing was developed by Kistler (Winterthur, Switzerland) and consists of three 3D dynamometers using piezoelectric sensors, and is able to measure 3D forces and torques.

Ad hoc designed transducers were instead developed in [74] and in [75]. More in details, the force transducer proposed in [74] is composed of two measuring plates, each containing two load cells. The measuring plates were mounted onto the ski boot sole.

The six component dynamometer proposed in [75] instead consists of three parts: the fixation to the boot, the fixation to the binding and the load cell. To measure the loads a mechanically uncoupled assembling of shear beams attached with gauges are mounted in load cell. Power supply and Bluetooth™ are integrated into the system. Main feature of this dynamometer is the possibility to be used with any kind of skiing equipment, being adjustable to different boot sizes and provided with wireless data transmission.

4.3.2.2 Pressure Measurement Systems

Plantar pressure measurement systems (or pressure insoles) provide electrical signal output (either voltage or current) that is proportional to the measured pressure, which is applied by the foot. Thus plantar pressure measurement systems do not directly measure ground reaction force, but the pressure normal to the sensing insole.

The required key specifications for a pressure sensor in terms of sensor performance include linearity, hysteresis, temperature sensitivity, sensing size and pressure range. Moreover, to be located in the insole (shoe sole) the sensor must be thin (about 2 mm), flexible and light (not heavier than 300 g to avoid affecting movements) [79].

The most common pressure sensors used on the purpose are capacitive sensors, resistive sensors, piezoelectric sensors and piezoresistive sensors [79]. Commercial products based on capacitive sensors are Pedar in-shoe systems (Novel, Germany), while resistive sensors are used in F-Scan in-shoe systems (Tekscan, USA).

4.3.2.3 3D Kinematics

Newton's second law is used to estimate ground reaction force, $\mathbf{F}_r$, [63, 72]:

$$\mathbf{F}_r = \mathbf{F}_{\text{in}} - \mathbf{F}_g - \mathbf{F}_d \tag{4.2}$$

where the inertial, $\mathbf{F}_{\text{in}}$, and the gravitational forces, $\mathbf{F}_g$, are calculated multiplying the mass of the skier (including equipment) by the skier CoM acceleration and the gravitational acceleration vectors, respectively. Acceleration of skier CoM can be determined either using video-based or IMU/GPS-based methods to reconstruct the skier 3D kinematics. Aerodynamic drag, F_d, can instead be calculated according to (see Sect. 4.3.3):

$$F_d = \frac{1}{2}\rho C_d A v_G^2 \tag{4.3}$$

being ρ the air density and v_G the skier CoM velocity determined by means of 3D kinematics reconstruction. The drag area $A_d = C_d A$ can instead be estimated according to the models proposed in [80] or in [81] (see Sect. 4.3.3).

4.3.2.4 Comparison of Ground Reaction Force Measuring Methods

Estimation of ground reaction forces based on 3D kinematics analysis (especially in the case of video-based methods) is time consuming (at least in the case of video-based methods) and its accuracy was never validated against reference measurements. Moreover 3D kinematics analysis alone does not allow the evaluation of force distribution between inside and outside ski.

Comparing force transducers and pressure measurement systems, they present advantages and disadvantages. Main advantages of pressure insoles are:

- their tiny dimensions make pressure insoles easy to use in field studies since they do not substantially change the ski equipment and their interference with athletes is minimal. In contrast, force transducers are heavier (about 0.5–1.0 kg) and their installation to skiing equipment changes their characteristics substantially as they change the stiffness of the skies and standing high (force transducers height is in the order of 0.4 mm) [78]. Force transducers thus may affect negatively performance of skiers and consequently a kinetic analysis may not be possible during high performance skiing [76];
- athletes can use their own skies, bindings and boots. On the other hand, being force transducers mounted between boots and bindings or between skies and bindings, athletes generally cannot use their own skies, bindings and boots.

On the other hand, main advantages of force transducers are:

- their capability to measure 3D forces and moments, while pressure insoles only measure pressures normal to the insoles themselves [76];

- measured forces are very reliable and determined values are considered as standard. Accuracy of force measurements provided by pressure insoles is still an open point of discussion [76, 77].

A comparison of force measurements provided by force transducers and pressure insoles for alpine skiing was given in [76]. Ten expert and ten intermediate level skiers performed ten double turns with different techniques (carving and parallel ski techniques, see Sect. 4.3.4.1) on both a steep (23°) and a flat (15°) slope twice. All athletes skied with both force transducers installed between skies and bindings and pressure insoles placed inside the boots, simultaneously. Specifically, two force transducers produced by Kistler (Winterthur, Switzerland) were, respectively, mounted under the toe binding and the heel binding of each ski, while pressure insoles produced by Pedar (Pedar, Novel, Munich, Germany) were located inside the boots of the skier. All skiers were allowed to use their own boots. Comparison included only the vertical component of the ground reaction force of the left leg. It was noted that [76], generally, pressure insoles underestimate both mean and peak values of the vertical component of ground reaction force, as measured by pressure transducers. Underestimation of vertical ground reaction force was found dependant on the turn phase (loading or unloading phase): during loading phases, the relative mean difference ranged from 22 to 35 %, while during unloading phase, it ranged from 13 to 39 %.

It was additionally noted that differences between the two measurement systems are also affected by skier ability, skiing technique and terrain slope.

4.3.3 Aerodynamic Forces

Aerodynamics in alpine skiing is basically the pressure interaction between the athlete and his equipment with the surrounding air. Aerodynamic force is generally divided into two components: the drag force $\mathbf{F}_d$ and the lift force $\mathbf{F}_l$ (Fig. 4.10).

Aerodynamic drag $\mathbf{F}_d$ is defined as the projection of the aerodynamic force along the direction of the relative wind. This means that if the relative wind is aligned with the athlete, drag opposes the direction of motion (i.e. the direction of the skier velocity vector $\mathbf{v}_G$). The drag force depends on three main parameters: the couple athlete/equipment frontal area A (defined as the surface area of the couple athlete/equipment projected on the plane perpendicular to the direction of motion), the drag coefficient C_d depending on the shape and the surface roughness of the couple skier/equipment and the relative velocity between air and skier v. Aerodynamic drag is thus expressed as:

$$F_d = \frac{1}{2}\rho C_d A v^2 \tag{4.4}$$

Fig. 4.10 Aerodynamic force components: drag, $\mathbf{F}_d$ and lift, $\mathbf{F}_l$

The relative velocity between air and body is usually considered coincident with the skier velocity v_G neglecting the wind speed contribution.

Lift force $\mathbf{F}_l$ is the component of the aerodynamic force acting normal to the drag component. As drag, it depends on three main parameters: the couple athlete/equipment frontal area A, the lift coefficient C_l depending on the shape and the surface roughness of the couple athlete/equipment and relative velocity between air and skier v. Aerodynamic lift is thus expressed as:

$$F_l = \frac{1}{2}\rho C_l A v^2 \tag{4.5}$$

Aerodynamic lift is generally neglected in alpine skiing, except for downhill when jumps are encountered.

Aerodynamic force in alpine skiing has been largely investigated both through wind tunnel and field tests [23, 52, 53, 80–82].

In [23], the results of an extensive wind tunnel experimental campaign involving the Canadian Men's and Women's Alpine Ski Teams were reported. Aim of the tests was to assess influence of skiers' position (or posture), frontal area and suit permeability on aerodynamic drag.

Tests were conducted at the General Motors Wind Tunnel (having a 5.5 m × 10.4 m test section) and at the National Research Council of Canada Wind Tunnel (having a 1.94 m × 2.7 m × 5.2 m test section).

During the tests, the skiers were placed in either a simulated downhill tuck or a more upright position on the wind tunnel six-components dynamometric balance. Specifically six positions were analysed: tight tuck, mid to high tuck, high knees, pre-jump/arms-in, in-air/arms-out and standing.

All skiers wore their normal racing equipment (helmet, goggles, gloves, back protectors, boots, underwear and ski poles) during each test session. New and worn downhill and giant slalom suits having different sizes were used. Suit permeability measurements were conducted with a Steinel porosity meter (SteinelElektronik GmbH & Co., Germany).

Measurements were carried out at four air velocities of 80, 90, 95 and 100 km/h.

A digital frontal view photograph of each skier was taken with a 10 cm × 10 cm grid reference board in the background and a reference post in the foreground. The frontal area of the skier in each photograph was then measured by digital planimetry.

It was observed that a skier who opens up by spreading his arms and becomes more upright causes a large increase in the frontal area and so in the drag force. For example, a skier who passes from pre-jump/arms-in to in-air/arms-out position would experience a 41 % increase in F_d. For this reason skiers are coached to keep their arms close to the body during a jump, thereby significantly reducing their resistance to air.

Drag area values (AC_d) ranging from 0.17 to 0.23 m^2 for full tuck position and ranging from 0.25 to 0.33 m^2 for upright position were found, consistently with the results reported in [52]. Drag coefficient instead was found to be in the range 0.46–0.6, although in [52] it was suggested that C_d is independent of body position in straight line skiing at racing velocities higher than 75 m/s.

Drag force variations induced by different suits were identified to be in the order of 5–7 %. The following criteria were moreover defined to select suits:

- downhill suits generate much less drag than giant slalom suits;
- adding giant slalom protective padding under a downhill suit invariably leads to an increase of the frontal area and so of the drag force;
- if the skier wears an undersized suit, the suit may overstretch and create a higher drag force;
- if a skier wears an oversized suit, the suit may be baggy and induce increased drag force;
- worn suits, having lost elasticity, become too porous and cause more aerodynamic drag than new suits;
- overlaying suits increases the frontal area of the skier and creates air pockets between the suit layers, increasing the drag force.

Wind tunnel tests, such as the above-mentioned ones, promoted the development of empirical models for aerodynamic force evaluation accounting for skiers' posture and anthropometric characteristics.

As an example, in [80], the drag area, C_dA, was estimated based on postural parameters (length and direction of skier body segments):

$$C_dA = a_1(L_1 \sin\theta_1 + L_2 \sin\theta_2 + L_3 \sin\theta_3) - a_2 + a_3(|\sin\xi_1| + |\sin\xi_2|) \quad (4.6)$$

where L_1, $L2$, and L_3 are, respectively, the lengths of legs, thigh and chest, while angles θ_i represent their inclination with respect to the snow surface in the sagittal plane; angles ξ_1 and ξ_2 are instead the orientations of the arms in the horizontal plane. Note that angles θ_i and ξ_i can be determined by reconstructing kinematics of body segments using video-based or GPS/IMU-based methods (Sect. 4.3.1). Finally a_1 and a_2 are fitting coefficient experimentally identified in [80] during wind tunnel tests on several skiers assuming different skiing postures.

A similar model was proposed in [81]. According to this model, the drag area, C_dA, is estimated as a function of the body length, U_{pH}, the skier frontal area A and the skier instantaneous height and width, H and W:

$$C_d\, A = b_1 + b_2\, U_{pH} + b_3\, A + b_4\, H + b_5\, W \tag{4.7}$$

Even in this case, A, H and W can be determined by reconstructing kinematics of body segments, which allow to estimate the skier silhouette [72]. Fitting coefficients b_i have instead to be experimentally identified through wind tunnel tests.

4.3.4 *Evaluation and Analysis of Skiing Technique and Performance*

Measurement systems described in previous sections promoted:

- the investigation of skiing technique with regard to the radical changes in the shape and materials of slalom skies;
- the development of objective methods for classifying alpine skiing performance.

4.3.4.1 Comparison of Traditional Parallel Skiing and Carving Techniques

Introduction of carving skies strongly changed skiing technique [83]. Carving skies are essentially more strongly waisted and markedly shorter than conventional skies. Additionally, a binding plate is mounted between the ski and binding whereby the standing height of the skier is increased by 1–2 cm. In carving, the "curved turn" is important. Using this technique, steering takes place along the ski edges without any lateral skid component. The navigable curve radius during carved turns is a function of the following variables: ski waist, on-edge angle and ski flexion (Fig. 4.11). The more strongly waisted the ski and the greater the on-edge angle, the more strongly the ski must flex to maintain contact with the slope along the total length of the edge. The curve cut into the snow under full contact with the slope is designated the turn radius [46].

A comparison between conventional parallel turn technique and new carving turn technique is provided in [46]. On this purpose, the same athletes performed a series of parallel and carved turns with conventional and carving skies, under otherwise identical boundary conditions. Test skiers had the task of making six runs each with eight turns per run with differently waisted skies. Video-based measurements were

Fig. 4.11 Side cut (ski waist), on-edge angle and ski flexion of carving skies [46]

used to reconstruct 3D kinematics of skiers. Ground reaction forces and pressure distribution in the ski boot were measured by two pressure insoles. Simultaneously, it was possible to observe activities of gluteus maximus, vastus medialis, vastus lateralis, rectus femoralis, biceps femoralis, tibialis anterius and peroneus longus muscles.

Summarizing the following were identified in the comparison of the traditional parallel and the new carving turn [46]:

- in the traditional parallel turn, during the steering phases, the predominant load is on the outer ski, whereas intensive co-loading of the inner leg is found in all turning phases with the carving technique;
- the carving technique has a relatively short second steering phase (steering the skies out of the fall line) and a comparatively long initiation phase;
- the turn initiation phase of the carving technique is in structure very similar to that of the traditional parallel turn. Up-unweighting in combination with a changing edge is seen with both techniques;
- turns with carving skies demand better sagittal balance as well as an improved edge steering ability to remain centrally positioned over the ski;
- the more waisted and more flexible skies, together with the greater on-edge angles during the steering phase, make the much smaller turning radii with the new carving technique possible.

4.3.4.2 Evaluation of Alpine Skiing Performance

The difference among top-level WC skiers is often hundredths of a second in a race. However, it has been shown that time differences in shorter sections along the course can vary by 10 % among the fastest skiers. Consequently, even at the highest level, skiers can significantly improve their sectional performance. A variety of studies have been performed to investigate the physical mechanisms that affect skiing performance:

- friction between the skies and the snow [84];
- the body posture effect on air drag [23, 52];
- the racing technique effect on ground reaction forces [46, 83];
- the impact of skier movements in changing the point of application on the gliding times [85];
- velocity maximization by altering the route down the slope [86].

A number of questions however remain about the fundamental understanding of a well-performed ski turn.

Several attempts to develop a decisive and pertinent method for classifying alpine skiing performance on a short section (one or a few turns) were proposed in the literature, based on data provided by previously described measurement systems (Sect. 4.3.1). Among the others, the following parameters were introduced to explain time differences, although a direct relationship between them and performance over short sections has not been completely demonstrated:

- section entrance velocity: intuitively higher entrance velocities may lead to a shorter section time. For consecutive sections, this would mean that a skier should aim for high section exit velocities in order to increase the performance of the following sections. However there also might be a kind of "velocity barrier" above which the athlete needs to control speed to avoid mistakes;
- path length: it is a common belief among coaches that a shorter path length may result in a shorter sector time. However, a shorter path length requires smaller turn radii and thus sharper trajectories, which may therefore lead to a loss of speed;
- skier CoM turn radii and average skies' turn radii: previous considerations on the path followed by the skier suggest that these parameters may identify excessively sharp trajectories leading to a loss of speed;
- energy dissipation: quality of a turn might be estimated based on energy losses due to snow friction and air drag. Assuming that main driving force in alpine ski racing is potential energy and that energy dissipations are mainly due ski-snow friction and aerodynamic drag, the rate of energy dissipation can be estimated based on the difference in total mechanical energy (kinetic and potential energy) per mass and entrance velocity:

$$\frac{\Delta(e_{\text{mech}})}{v_{\text{in}}} = -\frac{\frac{\Delta(v^2/2)}{\Delta z} + g}{v_{\text{in}}} \tag{4.8}$$

  where v represents the skier CoM velocity, v_{in} the entrance velocity, z the altitude, g the gravitational acceleration and Δ the differential;
- ground reaction force: an alternative way to evaluate energy dissipation may be provided by ground reaction forces during turns.

On the purpose of investigating whether these parameters can be used as predictors of performance in slalom ski racing and the relationships among them, in [61] 18 male WC alpine skiers were compared during WC slalom race in Kranjska

Gora. Video-based 3D kinematics measurements were used to monitor four slalom gates (gate numbers 7–11) on a 56-gate course. The experimental setup described in [61] and here reported in Sect. 4.4 was used.

The following considerations were drawn [61]:

- $\Delta(e_{mech})$ and speed of skiers are significantly different between elite and other skiers, while there are not significant difference in turn radii and ground reaction forces;
- section time has the largest negative correlation with $\frac{\Delta(e_{mech})}{v_{in}}$ and the second largest to v_{in};
- exit velocity v_{out} has the largest correlation to $\frac{\Delta(e_{mech})}{v_{in}}$;
- consequently, a high entrance velocity v_{in} is essential to achieve a short time section and thus a high v_{out}. Since the largest correlation of v_{out} is to $\frac{\Delta(e_{mech})}{v_{in}}$, this latter parameter can be considered a good estimator of sectional performance. It must be however noticed that it does not account for skier orientation at entrance and exit of the section, which strongly affect performance and possible mistakes;
- entrance velocity v_{in} has very large negative correlation to $v_{out}-v_{in}$. This suggests that skiers may have a "velocity barrier" above which they make mistakes;
- there is a positive relationship between turn radii and $\Delta(e_{mech})$, this meaning that smaller turn radii are associated with highest energy losses;
- smallest turn radii and highest ground reaction forces should be reduced in order to decrease energy dissipation; thus choosing a smooth round line between the gates may lead to better performance than skiing a more direct line from gate to gate;
- section time and skier velocity are thus not sufficient to estimate the quality of performance in short section.

Results partially in contrast with the findings of Supej et al. [61] were presented in [67]. Even in this case video-based 3D kinematics measurements were used to investigate performance of a skier performing a single turn course. Specifically, the experimental setup reported in Sect. 4.4 [67] was used during the tests, which were carried out by a top WC athlete performing a total of 12 runs on two different curse settings: vertical distance of gates was 26 m with an offset of 12 m and 10 m, respectively.

In contrast with [61], it was found that, comparing the fastest and slowest run, there was no difference regarding $\frac{\Delta(e_{mech})}{v_{in}}$ turn average, while the run time differed of 3.6 %. According to these results, it is surprisingly possible to reach a higher performance despite the same energy dissipation throughout the turn and so the index $\frac{\Delta(e_{mech})}{v_{in}}$ alone cannot predict turn performance in every case. Possible explanations for this discrepancy could be:

- in [67] a single athlete was used during the tests;
- in single turn analysis a mistake close to the end has only a small impact on the monitored section;

- in single turn analysis high exit velocity, as well as skier position/orientation at the end of the turn has marginal influence on the analysed section, but may be important for the following sections.

Thus the use of $\frac{\Delta(e_{\text{mech}})}{v_{\text{in}}}$ as a performance index might only apply to longer sections.

Coherently with [61], it was found that the turn radius of skier CoM was constantly larger throughout the turn in the fastest trial, although the angle between the instant direction of the skier CoM and the fall-line was smaller, implying that a higher rate of potential energy was transformed into kinetic energy. It was eventually observed that the fastest turn was initiated and terminated higher regarding the vertical position on the slope plane and was turning less out of the direction of the fall-line at the turn. Consequently, a higher percentage of the turn must be executed before passing the gate to improve performance.

Similar considerations with regard to the trajectory followed by the skier were obtained in [71] and in [63], where force vector analysis of two different runs performed by the same skier on a 10-gate giant slalom course was carried out in combination with 3D kinematics analysis.

Specifically, using the experimental setup described in [63] and here reported in Sect. 4.3.1.2, forces acting the skier were estimated through GPS/IMU-based 3D kinematics reconstruction (Sect. 4.3.1.2).

It was identified as key-point for reducing time, gate 6, which was placed before an increase of the terrain slope. In the fastest run, the skier made the apex of the turn well before the gate. Although the result was a longer CoM trajectory than in the slowest run, this allowed the athlete to ski straighter after the gate (i.e., reducing the angle between the skier CoM orientation and the fall-line) and accelerate out of gate 6, thus reducing global run time due to a higher momentum at the entrance of the steepest part of the course. In comparison, the apex of the slowest run was closer to the gate and the athlete was forced to make a very small radius turn after the gate. Force vector analysis offered a further explanation of this result. It in fact revealed that, in the fastest run, the skier was able to add pressure around the apex of the turn (the point during the turn at which the ski impulse has the greatest effect on changing skier momentum) and this led to an acceleration during the turn. On the contrary, in the slowest run, the smaller turn radius at the curve exit resulted in skidding and braking forces reducing the skier velocity.

4.4 Mathematical Models for Alpine Skiing

4.4.1 Aims of Models for Alpine Skiing

Modelling an alpine skier while racing still represents a challenging task: describing the motion of a human body continuously changing its position in 3D space while interacting with snow through a pair of skies is probably one of the most tricky problems that a sport engineer has to face.

Ski-snow interaction by itself represents a very complex matter (see Sect. 2.2): skies are flexible components of rather complex geometry which interact with snow, a non-uniform "material" whose response changes according to a number of parameters (primarily temperature and humidity). Several friction mechanisms may arise not only in different sections of a track, but even along the ski itself as function of pressure distribution and temperature profile.

Last but not least, the dynamics of human body plays the most important role on the performance; its description requires several degrees of freedom which can be combined in almost infinite ways. A realistic mathematical reproduction of skier's motion and reactions is extremely difficult to obtain but is essential to get consistent results.

Despite all these factors and limitations that still have to be properly addressed, several models have been proposed in technical literature. Mathematical models present in fact many advantages that make them attractive:

- exhaustive sensitivity analysis can be carried out with less time and costs with respect to experimental tests and with no risk of injury for the athletes;
- numerical simulations are deterministic, while experimental tests are subjected to a number of variables (weather, snow characteristics, athletes' fatigue and errors);
- each single parameter of a model can be controlled and changed individually, allowing to clearly put into evidence its role;

Models presented in literature are characterized by different levels of complexity, ranging from particle models [87], rigid-body models [88], to multi-body models including the skier-skies system moving on snow while practising slalom, downhill skiing or even acrobatics [89].

Some models are mainly focused on ski-snow contact: in this case the skier is usually schematized in a quite simple way (a point mass or a rigid bar) since the main interest is on the comprehension of the influence of skies geometry and flexibility on the performance. In other words, some models are developed to improve and optimize skiing equipment.

Sakata and Tsukiyama [90] investigated the effect of shoe centre position on curving performance modelling the skier with a point mass and considering the ski as a non-homogeneous orthotropic plate. Federolf et al. [91, 92] developed a finite-element model to describe the contact between ski and snow and analyse the influence of edging angle, snow hardness and force acting on the binding on the radius of a carved turn.

Other models are instead focused on the athlete and try to reproduce his behaviour and reactions while performing slalom, jumps, downhill skiing or even acrobatics [89]. Optimizations of trajectories or of body position are possible outcomes of these researches whose results could be conveniently used to improve skiing technique and performance. Models of the skier can also be used to estimate forces and moments transmitted through body articulations, especially knees, and understand why specific reactions or body positions are more likely to cause injuries. Results of the research provide support for athlete's training, this time with the goal of minimizing injury risk.

With few exceptions [87, 88], models of the skier are based on multi-body approach. The high computational costs associated with this kind of modelling often push toward rather simple models for the ski-snow contact, especially when 3D dynamics is involved. Realism of the skier's motion is usually obtained driving the joints of the model on the basis of data recorded during experimental tests through motion capturing techniques. This means that, in most cases, only inverse dynamics is solved to estimate loads at the contact interface or in body articulations. Anyway, attempts to implement logics for controlling body position on the basis of skier's status as well as results of forward dynamics simulations were presented [93, 94].

Rudakov et al. [88] developed an algorithm for the optimization of the trajectory in slalom modelling the skier as a point mass linked to the ski through a rigid bar. Chen and Qi [89] used a 2D multi-body model to analyse the effect of body weight and posture on skier's velocity, acceleration and joints forces. Nachbauer et al. [93] developed a 2D multi-body model for the analysis of the dynamics of jumps in downhill skiing; their work included also a musculo-skeletal model and evidenced relations between athlete's position at landing with the peak of force on the anterior cruciate ligament (ACL). A 3D multi-body model was used by Kawai et al. [95] to characterize the control action of a skier (i.e. edging angles, inward displacement of centre of mass and edging forces) in parallel turn. In the work carried out by Filippi Oberegger et al. the 3D dynamics of the skier in carved turns was investigated [94].

The aforementioned models will be described in the following more in detail. Models will be divided into two groups according to their focus on skiing equipment or on skier's dynamics.

4.4.2 *Ski Modelling*

4.4.2.1 Prediction of the Effect of Bindings Position on Ski Turn

Research presented by Sakata [90] is focused on the analysis of the longitudinal position of the bindings (i.e. of the shoe centre) on the dynamics of the skier-ski system while turning.

The skier-ski system is modelled according to the schematic of Fig. 4.12. The picture shows a side view of the ski: the position of the shoe centre is marked with **C** and it is the origin of a reference frame characterized by a longitudinal axis ξ a vertical axis ζ, directed downward and a lateral axis η. The skier is simply schematized with a mass $\mathbf{M_2}$ concentrated in $\mathbf{G_2}$. Position of $\mathbf{G_2}$ can change with respect to the reference frame due to the skier's motion. The ski has a mass $\mathbf{M_1}$ and its centre of mass (CoM) is in $\mathbf{G_1}$. The CoM of the skier-ski system is represented with the point **G**.

The ski is modelled as a non-homogeneous orthotropic plate whose flexural rigidities vary along the ξ direction. Flexural rigidity along ξ direction is implemented considering a bell-shaped function with a maximum value around 11,000 Nm/rad at the ski centre. Variation of flexural rigidity along η direction is taken into account in a similar way, considering a maximum value around

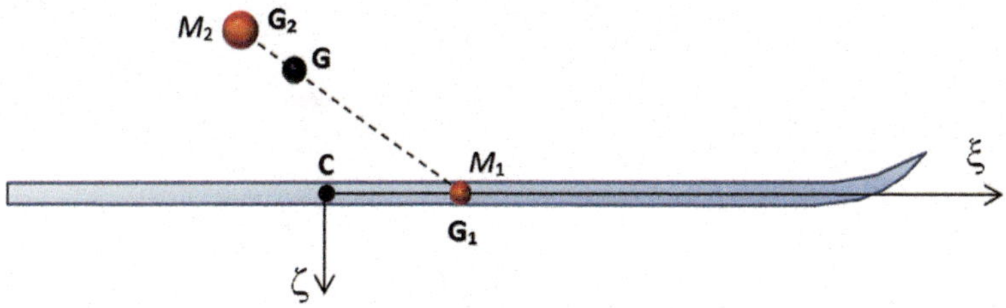

Fig. 4.12 *Side view* of the skier-ski system (adapted from [90])

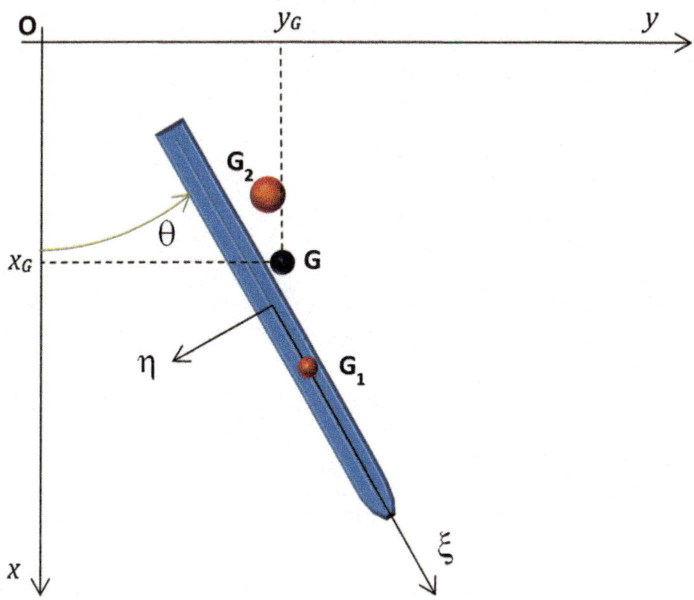

Fig. 4.13 *Top view* of the skier-ski system (adapted from [90])

7500 Nm/rad at the ski centre. Data are referred to a slalom ski 2.012 m long. The snow surface is schematized as a layer of linear springs with constant k_0 assumed equal to 1.0×10^7 N/m^2.

The ski-skier system is assumed to move in a horizontal plane as shown in Fig. 4.13. The position of the centre of mass of the system can be therefore identified with the coordinates x and y, while the orientation of the ski with respect to x-axis is given by θ.

Equations of motion of the skier-ski system are given by:

$$\begin{cases} (M_1 + M_2)\ddot{x_G} = f_x \\ (M_1 + M_2)\ddot{y_G} = f_y \\ I_G \ddot{\psi} = m_z \end{cases} \tag{4.9}$$

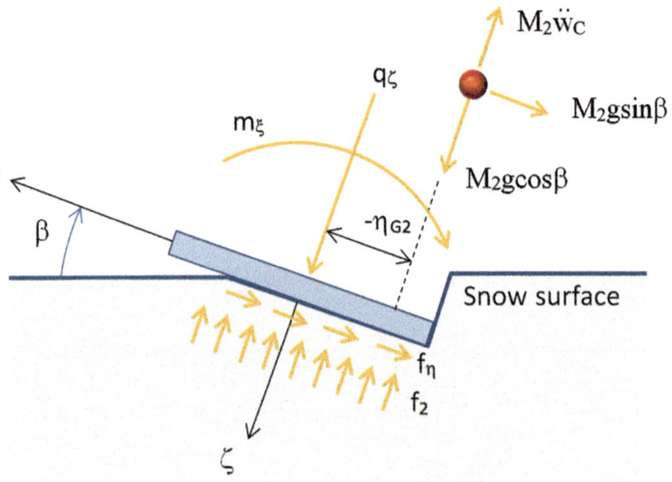

Fig. 4.14 *Front view* of the skier-ski system (adapted from [90])

I_G is the moment of inertia of the skier-ski system around a vertical axis passing through point **G**. In (4.9), f_x and f_y represent the components of the ski-snow contact force directed along x and y direction, respectively; m_z is the moment generated along a vertical axis by the contact forces with respect to **G**.

Lateral contact forces required for curve negotiation are generated by the edging action of the skier. Referring to Fig. 4.14, the edging action is reproduced considering a lateral displacement η_{G_2} of the skier's centre of gravity in the $\xi-\eta-\zeta$ reference. While cornering the skier applies to the ski a force q_ζ and a moment m_ξ, respectively, along the vertical and longitudinal axis of the ski. As a consequence of the edging action, a rotation β of the ski around its longitudinal axis (the *edging angle*) is generated. Gravity force and the inertia force $M_2\ddot{w}_c$ acting on the skier due to rigid displacement of the ski are considered. As far as the ski is concerned, f_2 and f_η represent the distributed contact forces, respectively, in ζ and η direction. Though not evidenced in Fig. 4.14, distributed forces due to gravity and to vibration of the ski are included in the modelling. The mass per unit area of the ski is assumed to vary along longitudinal direction ranging from 8 kg/m^2 at the ski edges to 20 kg/m^2 at the ski centre.

Several simulations were carried out by the authors of [90], assuming a slalom skier moving at 10 m/s and entering a turn gradually increasing his lateral displacement η_{G_2} from 0 to a regime value. Simultaneous shift of the skier's centre of gravity along longitudinal direction ξ_{G_2} was considered. Simulations were repeated changing the position of the shoe centre along the longitudinal axis of the ski. In particular, being l_c the distance of shoe centre from the ski tail and L the ski length, the ratio l_c/L was investigated in the range 39–45 %. The numerical analysis led to the following conclusions:

- Longitudinal position of the shoe centre affects the response of the skier-ski system during a turn, in terms of rotation speed around a vertical axis. Shifting the shoe centre forward allows faster rotation speeds. The opposite effect is obtained when the shoe centre is shifted backward.
- Longitudinal position of the shoe centre has instead little influence on the trajectory of the centre of mass of the skier-ski system.
- Different positions of shoe centre affect the angle between the longitudinal axis of the ski and the tangent to the trajectory of the centre of mass of the system **G**. The shift in longitudinal direction (ξ_{G_2}) of the skier's centre of mass can be adjusted so that the ski is tangent to the trajectory of **G**. When the shoe centre is shifted forward, this result will be achieved if the skier moves forward too. When the shoe centre is shifted backward, the skier has to move his centre of mass toward the ski tail to obtain the same result. Variations of ξ_{G_2} reported in [90] are anyway in the order of few millimetres.

4.4.2.2 Prediction of Carved Turns Radius Through a Finite Element Model

A finite element model was adopted by Federolf et al. [91, 92] to describe the interaction between ski and snow during carved turns. Aim of the work [92] was to investigate the effect of parameters like edging angle, snow hardness and load on ski binding on the actual turn radius.

The turn radius R_c with a carving ski can be estimated through Howe's model [96]:

$$R_c = \frac{L^2}{8\dfrac{\mathrm{SC}}{\cos\theta} + d\tan\theta} \tag{4.10}$$

where L is the ski length, θ the edging angle, d the penetration depth, and SC the ski sidecut which can be obtained as:

$$\mathrm{SC} = \frac{w_1 - 2w_2 + w_3}{4} \tag{4.11}$$

In Eq. (4.11), w_1 is the width of the ski at the shovel, w_2 at the ski waist and w_3 at the ski end. Results of Eq. (4.10) are shown in Fig. 4.15 where the turn radius is reported as function of the edging angle for three vales of FIS-radius R_{SC}. This last can be estimated as follows:

$$R_{\mathrm{SC}} = \frac{L^2}{8\cdot\mathrm{SC}} \tag{4.12}$$

The proposed formulation is based on an ideal carved turn, where no skidding occurs; Eq. (4.10) requires only few geometrical parameters to compute the turn

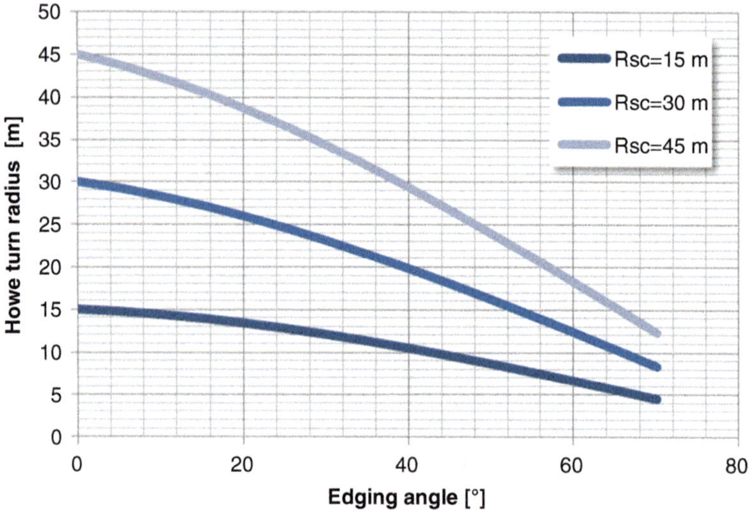

Fig. 4.15 Turn radius according to Howe's formulation for different SC radius; $d = 3\,\text{mm}$

radius and this makes its application very easy. At the same time, this constitutes a shortcoming of the approach since, dependence of the turn radius on physical parameters like ski flexibility, force exerted on the binding or snow hardness, cannot be evidenced. Experimental tests show a general agreement with the prediction of (4.10) for edging angles up to 40°. Beyond this limit experimental turn radii are generally larger than those obtained with Eq. (4.10); major differences arise with increasing penetration depth.

The model presented in [91, 92] can be divided into a ski model and a binding model. The ski model consists of a series of shell elements based on Kirchoff plate theory. Presence of thin adhesive layers and of non-isotropic materials like wood-core and fibre glass composite is considered. The binding is instead schematized through a series of 8-node volume elements.

Boundary conditions are introduced to take into account forces transmitted to the binding by the skier, gravity and inertial effects on ski and ski-snow interactions. In particular, experimental tests were carried out to collect data for boundary conditions. An experienced skier performed several carved turns while forces and moments transferred to the skies were recorded through dynamometric bindings. In addition, motion of skier and skies was captured with a video system allowing identification of edging angle and system accelerations.

The interaction between ski and snow is described through an analytical model which takes into account snow plasticity. During the loading phase, the pressure p developed by the snow surface with penetration depth d is given by:

$$\begin{cases} p(d) = 2A(\beta)\cdot d + B & d > 0 \\ p(d) = 0 & d \leq 0 \end{cases} \tag{4.13}$$

According to experimental tests [97], A should be in the range 2–20 kPa/mm for edging angles $\beta > 40°$, suitable values for B are between 0 and 400 kPa. When the unloading phase is considered snow pressure becomes:

$$\begin{cases} p(d) = C \cdot (d - d_{\max}) + p_{\max} & d_{\max} \geq d > d_0 \\ p(d) = 0 & d \leq d_0 \end{cases} \tag{4.14}$$

The value of C is related with elastic recovery of snow and is between 0.15 and 1.15 MPa/mm [98]. In Eq. (4.14), $d_{\max}$ is the maximum penetration depth reached in the loading phase and $p_{\max}$ the corresponding pressure; d_0 represents the penetration depth where contact pressure zeroes during unloading phase ($d_0 = d_m - p_m/C$).

Outputs of the model have been validated with exhaustive comparisons against experimental data. At first, the snow model was validated reproducing through finite elements the measuring set-up used in [97] where a penetrating plate was pressed into the snow with different edging angles. Test results allowed identification of data for Eq. (4.13) and were adequately reproduced with the model. As second step, quasi-static simulations were carried out considering only the loading phase of a carving ski pressed into the snow. The deformed shape of the ski was compared to the one determined in static experiments where a ski was loaded and pressed into the snow with various edging angles. Comparison with the turn radius of an actual carved turn was eventually carried out considering the data collected during the aforementioned outdoor tests with the experienced skier. Satisfying results were obtained when the presence of the groove created by the front part of the ski in the snow was considered.

The analysis carried out after model validation led to the following results.

- Increasing the edging angle reduces the turn radius but numerical results differ from the output of Howe's model [96] for edging angles higher than 40°. As qualitatively shown in Fig. 4.16, beyond this value the turn radius reaches a sort of asymptote instead of falling off. According to the analysis, dependence of turn radius on edging angle can be conveniently approximated with a Gaussian function. The phenomenon is explained considering that rear section of the ski slides in the groove formed by the front part of the ski itself. This reduces the bending of the ski with respect to the purely geometrical model proposed in [96] which, therefore, underestimates the effective turn radius.
- Soft snow allows smaller turn radii with respect to hard snow when small edging angles are considered; this effect is associated with a deeper penetration of the ski into the snow. For the same reason, increasing edging angles produces wider grooves in soft snow thus anticipating the asymptote region. As a consequence, compared to soft snow, hard snow allows smaller turn radii with higher edging angles.
- Increasing normal load on binding decreases the turn radius. A heavier skier or higher speeds thus allow smaller turn radii. Anyway, the effect of normal load is quite small: depending on the edging angle the decrease of turn radius is in the order 0.5–1.2 m/kN. This results means that variations of normal contact force due unevenness of snow surface do not significantly undermine the skier's stability during carved turns.

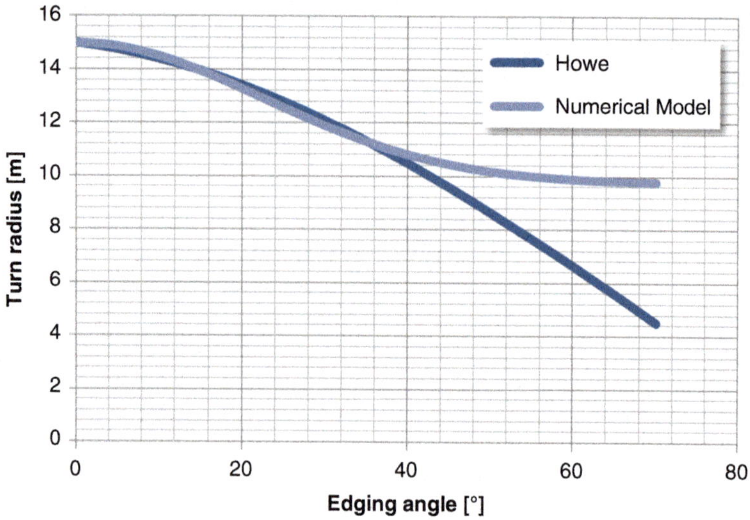

Fig. 4.16 Comparison between turn radius obtained through Howe's formulation and numerical model; $d = 3$ mm (adapted from [92])

4.4.3 Skier Modelling

4.4.3.1 Point Mass Model: Trajectory Optimization in Slalom

The ability to follow an optimal path in slalom or downhill is essential to minimize final time. According to snow conditions, track geometry and response of skiing equipment, a skier has to identify the best trajectory, try to follow it and be ready to set a different path if something goes wrong. During a race, the athlete has only fractions of second to perform these tasks and, therefore, training and experience are indispensable to develop the required abilities.

A mathematical model can provide information for identification of optimal path; in particular, according to different track geometries, snow conditions, skiing equipments and athlete's characteristics, optimal trajectories can be identified. Results of numerical model can be profitably used to support the training of athletes.

A rather simple method for identifying the optimal trajectory in slalom is proposed by Rudakov et al. [88]. In this work the skier is modelled as a point mass rigidly linked to the skies through a weightless rigid bar.

The slalom track is characterized by two parameters: considering Fig. 4.17, L represents the track spatial "period", i.e. the longitudinal distance between two poles of the same side, while a is the distance from the line of maximal descent.

The slope of the slalom track is assumed to vary only along the direction of the track midline with the following relation:

$$\alpha(l) = \alpha_0 - l/Rs \tag{4.15}$$

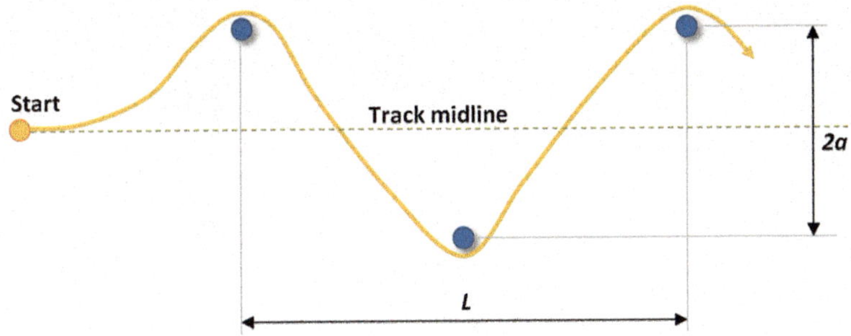

Fig. 4.17 Slalom track used for optimization in [87]

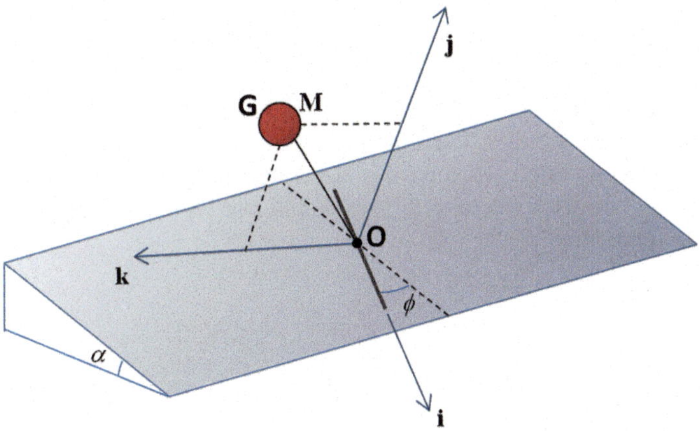

Fig. 4.18 Identification of skier-ski system orientation with respect to track (adapted from [88])

In (4.15), l is the curved abscissa of the slalom track taken along its midline and α_0 is the initial slope. R_s is the radius of the cylindrical surface representing the alpine slope.

As shown in Fig. 4.18, the whole mass of the skier-skies system M is concentrated in the CoM **G** joined through the weightless rigid bar $\overline{GO}$ to one single ski. This last is assumed to reproduce the response of the pair of skies. The ski is in contact with the snow and its longitudinal axis, identified with the unit vector **i**, forms the angle ϕ with the line of maximal descent. Always referring to Fig. 4.18, the unit vector **j** identifies the direction normal to the contact surface in **O**. A binormal unit vector **k** is introduced to identify the lateral direction.

$$\mathbf{k} = \mathbf{i} \times \mathbf{j} \tag{4.16}$$

While cornering, the rigid bar $\overline{GO}$ rotates around the longitudinal axis of the ski reproducing the behaviour of the skier who moves his centre of gravity inward to achieve balance. Figure 4.19 presents a front view of the system: the rigid bar $\overline{GO}$ forms an angle γ with the axis normal to the contact surface which can be obtained as:

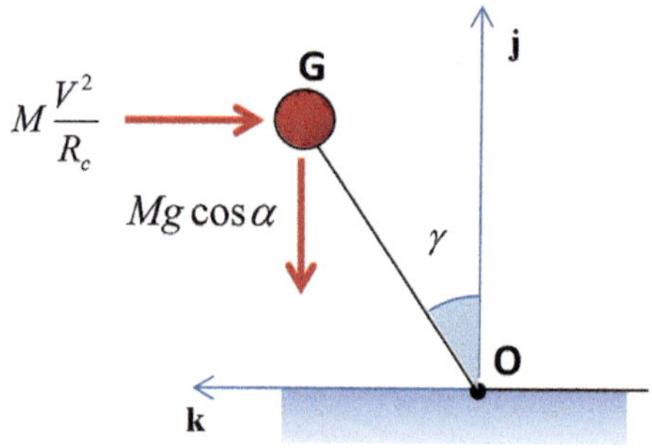

Fig. 4.19 *Front view* of the skier-ski system γ (adapted from [88])

$$\gamma = \arctan\left(\frac{V^2}{R_c g \cos\alpha}\right) \tag{4.17}$$

in (4.17) V is the modulus of skier speed (whose direction is identified by **i**), g is the acceleration of gravity, and R_c the turn radius.

The system is described with three degrees of freedom in the space and, therefore, three equations of motion along directions **i**, **j**, **k** can be written. Equation along **j** and **k** allows computing the contact reactions forces:

$$\begin{cases} N_j = -\dfrac{MV^2\cos^2\phi}{R_s} + Mg\cos\alpha \\ N_k = \dfrac{MV^2}{R_c} - Mg\sin\alpha\sin\phi \end{cases} \tag{4.18}$$

The last equation of motion, along **i** direction, can be used to compute the speed profile:

$$M\frac{dV}{dt} = Mg\sin\alpha\cos\phi - \frac{1}{2}\rho A_d V^2 - \mu\sqrt{N_1^2 + N_2^2} \tag{4.19}$$

In (4.19) μ is the friction coefficient between ski and snow. In [88] dependence of μ from γ is considered with the simple relation $\mu = 0.05 + K\tan\gamma$, with $K = 0.24$. The aerodynamic drag is computed considering the air density ρ and the drag area A_d. The best trajectory is obtained through an optimization algorithm aiming at minimizing the final time. Trajectory of the centre of mass is assumed to be a sixth order polynomial whose coefficients are changed by the minimization algorithm. For each combination of coefficients, a trajectory is defined. At first step Eqs. (4.17) and (4.18) can be solved to determine the contact forces (and also to check their plausibility with respect to maximum forces allowed by ski-snow contact). The speed profile can then be computed by solving Eq. (4.19). In this way, each trajectory can be associated with a final time.

Authors of the research presented also results of the optimization procedure considering a slalom track with $L = 20$ m, $a = 4$ m and an initial slope $\alpha_0 = 32°$. Assuming $M \simeq 80$ kg, $A_d = 0.48\,\text{m}^2$ and a length of 0.5 m for the rigid beam $\overline{GO}$, the optimal turn radius of the skies around the pole resulted equal to 2.78 m, while the optimal turn entry angle ϕ_c is 50.2°.

4.4.3.2 2D Multibody Model: Analysing Skier Motion Vertical Plane

Model complexity can be drastically increased with respect to the model analysed in previous paragraph. More degrees of freedom can be added to describe the motion of the human body and to include ski deformation. In addition, more refined formulations for ski-snow interaction can be introduced. Adding complexity usually leads to more information and accuracy but, at the same time, implies higher computational efforts and more data to be identified for reliable results.

2D multi-body models describing the dynamics of the skier-ski system in vertical plane represent a good compromise between accuracy and simplicity. Two examples of this approach can be found in [89] and [93].

Aim of the work presented by Chen and Qi [89] is the analysis of the 2D dynamics of the skier while moving on uneven slopes, jumping, landing and performing acrobatics.

The proposed model uses 7 degrees of freedom to describe the motion of the skier-ski system in the vertical plane. The ski is introduced as a rigid bar; therefore, considering the schematic of Fig. 4.20, its position in the plane with respect to the absolute reference $x - y$ is identified by the position of a point (x_O, y_O) and a rotation (θ_0). Point **O**, representing the position of the ski centre can be conveniently

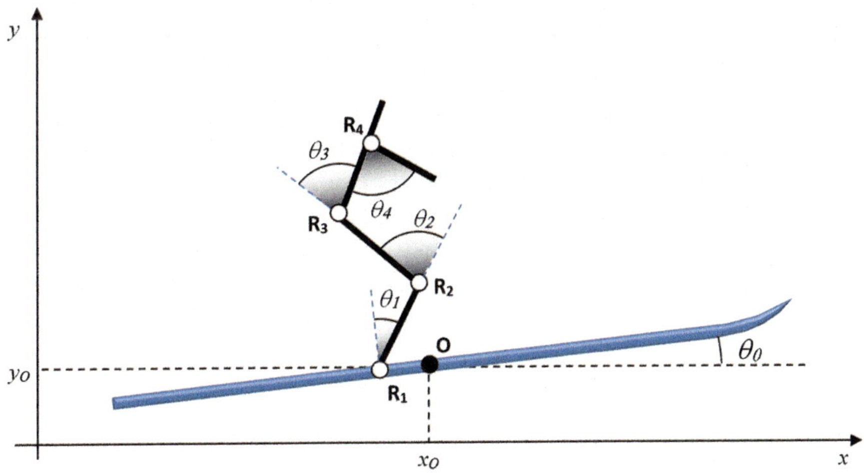

Fig. 4.20 Degrees of freedom of the model (adapted from [89])

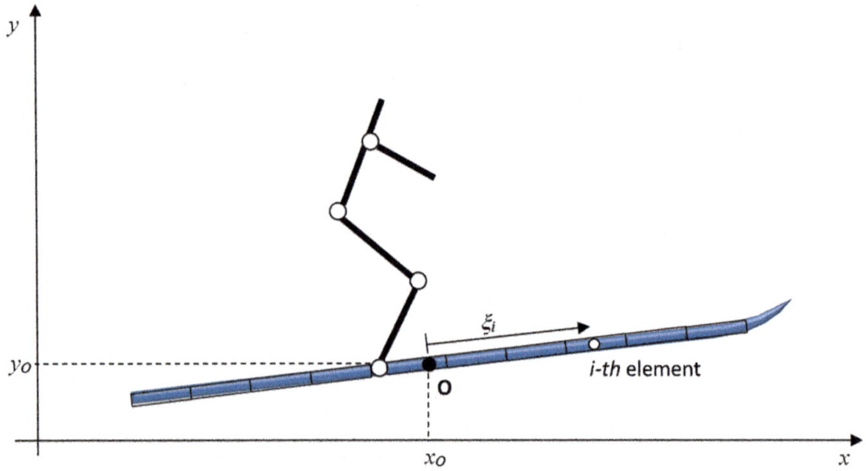

Fig. 4.21 Division of the ski in segments (adapted from [89])

chosen. The human body is schematized with four rigid bars, respectively, for shanks thighs, trunk, and arms. A rotational joint in $\mathbf{R_1}$ links the shanks with the ski; three additional rotational joints $\mathbf{R_2}$, $\mathbf{R_3}$, and $\mathbf{R_4}$, respectively, link the shanks with the thighs, the thighs with the trunk and the trunk with the arms. Four relative rotations $(\theta_1, \theta_2, \theta_3, \theta_4)$ between the rigid bars are then used to identify the position of each portion of human body.

Ski-snow interaction is introduced considering a local contact model where pressure p depends on both penetration depth and speed. As shown in Fig. 4.21, the ski is divided into several segments and the contact model is applied locally on each element. In particular the pressure acting on the ith element identified with the local coordinate ξ_i, is computed as:

$$p_i = A + B\delta_i + C\dot{\delta}_i \tag{4.20}$$

In Eq. (4.20) δ_i represents the penetration of the centre of the ith element into the snow while $\dot{\delta}_i$ is the penetration speed. Due to the assumption of rigid body, this simple model is not sufficient for estimating pressure distribution along the ski. Ski flexibility and variation of stiffness along the ski length determine how the load transmitted from the binding distributes over the contact surface. Usually the load is higher under the boot and spreads over the lower surface of the ski according to its flexural stiffness.

For this reason a weight function $E(\xi)$ is introduced, taking into account the distribution of flexural stiffness.

$$E(\xi) = \frac{1}{E_{\max}} \left(\frac{d_1}{x^2 + 0.5} - d_2 \right) \quad -0.5l \leq x \leq 0.5l \tag{4.21}$$

E_{max} represents the maximum flexural stiffness (measured at the ski centre) and l is the ski length; d_1 and d_2 are two parameters of the weight function. Once the weight function is introduced, the normal contact force N_i acting on the ith element can be computed:

$$N_i = E(\xi)A_i p_i \tag{4.22}$$

The contact force N_i is directed normally to the contact surface A_i. A simple Coulomb friction model is then employed to compute the tangential contact force. Forces acting on each element are then summed to obtain the total forces acting on the ski in x and y directions and the moment generated with respect to the ski centre.

As additional interesting feature, the model takes into account variations of drag forces associated with different body positions. Each element of the body is assumed cylindrical in shape and its drag force F_{dj} is computed considering the relative speed v_j of its centre of mass with respect to air and the projection of the front area A_j along the direction of relative speed.

$$F_{dj} = \frac{1}{2}\rho C_d A_j(v_j)v_j^2 \tag{4.23}$$

F_{dj} is obviously directed opposite to the relative speed v_j. Model equations can be written in vector form:

$$\mathbf{M}(\mathbf{q})\ddot{\mathbf{q}} = \mathbf{Q}(t, \mathbf{q}, \dot{\mathbf{q}}) \tag{4.24}$$

where $\mathbf{q}$ is the vector of the independent variables.

$$\mathbf{q} = [x_O \; y_O \; \alpha \; \theta_1 \; \theta_2 \; \theta_3 \; \theta_4] \tag{4.25}$$

Results of the proposed numerical model were compared with the output of a simplified model considering downhill skiing with constant slope. In this condition the contact model can be conveniently described also with a simple sliding joint. Speed profiles obtained with the two models were in good agreement.

The model was then used considering more complex scenarios with non-planar terrains. Jumps or even acrobatics during freestyle competitions were simulated using inverse dynamics. These simulations were performed driving the joints according to kinematic data previously recorded [99]. Skier's velocity and flying angle at take-off were compared to experimental measurements showing reasonable agreement.

Altogether, even if only 2D dynamics is considered, the proposed model shows how the effect of several parameters on performance and safety can be investigated with a rather simple approach. Simulations can be used to understand how body position influences speed profile, loads on articulations when landing after a jump or quality of acrobatics in freestyle. Such kind of model may thus provide additional information for improving training technique.

A step forward in this direction is presented in the work of Nachbauer et al. [93]. In this case a 2D multi-body model is developed with the specific goal of understanding the influence of jump landing kinematics on peaks of forces on ACL. Skies flexibility is considered and also muscolo-skeletal model is introduced for the actuation of the joints of the skier's model.

The skier is modelled considering seven rigid bodies, one representing head, arms and trunk and three bodies for each leg corresponding to thighs, shanks and foots. Each ski is divided into nine rigid segments connected with revolute joints. Linear spring-damper elements are used to link the segments, thus reproducing skies flexibility. Globally the skier-skies system is characterized with 25 degrees of freedom.

As mentioned before, a model for muscles actuation is included. Eight muscles for each lower extremity are considered (iliopsoas, glutei, hamstrings, rectus femoris, vasti, gastrocnemius, soleus and tibialis anterior) each modelled as a three-element Hill-type model. Model of contraction dynamics takes into account muscle fibre length, lengthening velocity, muscle length and activation, this last modelled as a first-order process.

Motion equations of the system are in the form:

$$\mathbf{M}(\mathbf{q})\ddot{\mathbf{q}} = \mathbf{F}(\mathbf{q}, \dot{\mathbf{q}}) + R(\mathbf{q})\mathbf{F_m} \tag{4.26}$$

where $\mathbf{M}(\mathbf{q})$ represents the system mass matrix function of the vector of the generalized coordinates $\mathbf{q}$. $\mathbf{F}$ is a vector of generalized forces (including gravity, drag force function of body position, inertial effects, contact forces) while $R(\mathbf{q})F_m$ is a vector of moments generated on the joints by the muscles forces $\mathbf{F_m}$ with the corresponding arms $R(\mathbf{q})$.

The contact between skies and snow is modelled considering the following formulation for the relation between normal contact force and penetration depth.

$$N_i = \frac{1}{2}k_p A_i \left(d_i + \sqrt{d_i^2 + d_0^2}\right)(1 + b_d \dot{d}_i) \tag{4.27}$$

In Eq. (4.27) N_i is the force acting normal to the snow surface and applied on the i-th segment of the skies characterized by an area A_i; d and $\dot{d}$ represent the penetration depth and penetration speed, respectively. Stiffness parameter k_p was set to 10^5 N/m^3 is while the damping parameter b_d was set to 0.75 s/m. A value of 0.01 m was assumed for the reference depth d_0.

The friction force T_i acting on the ith segment is instead computed as:

$$T_i = -\mu N_i \frac{v_i}{\sqrt{v_i^2 + v_0^2}} \tag{4.28}$$

where v_i is the sliding velocity parallel to the snow surface and v_0 a parameter set to 1 m/s. The friction coefficient μ is assumed equal to 0.1.

Several simulations were carried out considering a 2.11 m downhill ski whose visco-elastic properties were identified by means of experimental tests [100].

First of all, activation of muscles was optimized so that the model was capable of reproducing experimental data previously recorded. In particular, kinematic data of a downhill skier performing a jump were captured with a high speed camera. Three cost functions J_1, J_2 and J_3, whose sum was minimized, were introduced.

$$J_1 = \frac{1}{N_d T} \int_0^T \sum_{i=1}^{N_d} \left(\frac{q_i(t) - q_{i,\text{data}}(t)}{\sigma_i} \right)^2 dt \tag{4.29}$$

$$J_2 = \frac{w_m}{N_m T} \int_0^T \sum_{i=1}^{N_m} u_i^2(t) dt \tag{4.30}$$

$$J_3 = \frac{w_s}{2N_s T} \int_0^T \sum_{N_d+1}^{N_d+N_s} \left(q_i^2(t) + \dot{q}_i^2(t)\right) dt \tag{4.31}$$

The first function J_1 accounts for the error between model outputs $q_i(t)$ and experimental data $q_{i,\text{data}}(t)$ during all the jump, which lasts T seconds. N_d is the number of degrees of freedom of the skier model and σ_i is a scaling factor. The second function J_2 is introduced to optimize muscle redundancy: N_m is the number of muscles, $u_i(t)$ represents the neural excitation of the ith muscle, while and w_m is a weight factor. The last function J_3 weighted with w_s considers the movement (vibrations) of the skies during the jump; N_s is the number of degrees of freedom of the skies segments. Minimization of costs function thus allow to reduce discrepancies between skier model kinematics and experimental data, to minimize the activation energy of muscles and to avoid excessive and unrealistic vibrations of skies.

The sum of cost functions was then minimized assuming $w_m = 10$ and $w_s = 0.01$ obtaining a reference simulation. Comparison with experimental data showed satisfying results, evidencing minimal differences in the joint angles ($\leq 6.7°$).

A series of numerical simulations was then performed perturbing the landing position of the athlete prior to ground contact. In particular, following a Monte Carlo approach, random numbers were added to the degrees of freedom of the skier (with respect to the reference simulation). A Gaussian distribution was assumed for the perturbation, considering data for standard deviations published in previous analyses [41]. For example, considering the knee rotation, a standard deviation of 8.7° was assumed.

With these conditions, an optimal control problem based on the costs functions J_1, J_2 and J_3 was solved. This time J_1 was referred only on the final time, so that the skier model attempted to recover the correct position at landing. In parallel, in J_2, w_m was set to 1 to increase weight in tracking the correct position.

Aim of these simulations was to investigate the effect of perturbed landing positions on the peak force on ACL. Numerical results evidenced how increase of backward lean, hip flexion, knee extension and ankle dorsiflexion are all associated with higher peaks of ACL force. Increasing the distance between skier's centre of mass and base of support leads to higher injury risk. These conclusions are in good agreement with those resulting from the analysis of the dynamics of injuries caused by a jump.

Altogether, results of the work can provide useful information for training the athletes: induce forward rotation of the body during takeoff, keep tuck position during flight. In addition, also indications for proper track design can be derived: straight and long inruns, low flight trajectories and low speeds should be preferred to reduce injury risk.

4.4.3.3 3D Multi-Body Models

In several winter sports, the athlete has to perform complex movements in 3D space continuously changing his position. Though interesting and useful in several scenarios, the approach analysed in previous paragraph does not apply in these conditions. For example, investigating the dynamics of skier while performing turns on a slope requires a 3D multi-body model. This dramatically increases the number of degrees of freedom of the model leading to significant computational cost.

Obtaining realistic motion and reactions of the skier represents the most difficult part of the approach. Due to the high number of degrees of freedom, this problem is typically solved using kinematic data captured in outdoor tests, usually with high-speed cameras. This means that several degrees of freedom of the model, or even all, are driven and inverse dynamics is computed allowing estimation of forces and moments acting on skies and body's articulations. The approach may also be used to investigate the effect of different skier's kinematics on speed profile and on contact forces, providing useful data for athletes' training.

At present, due to computational costs, these models are coupled with simplified models for the skies and for ski-snow contact. For the same reason, and for the issues related with obtaining a realistic motion of the skier, it is still very difficult to perform analyses in forward dynamics trying to optimize the skier's posture for speed maximization or injuries minimization.

The research presented by Kawai et al. [95], is focused on characterizing the skier's movements and forces transferred through the bindings during parallel turn. In particular, the so-called *ski control* is identified by the displacement of the skier's centre of mass, the edging angles and the edging forces. The goal is reached combining 3D-CAD drawings and multi-body simulations.

A skier's model made up of 15 body parts (head, neck, trunk, upper arms, lower arms, hands, thighs, shanks and foots) was implemented. The model was drawn using 3D-CAD system. Each part was considered rigid and its inertial properties were estimated using data taken from [101]. The model did not include ski, boots and ski poles. Most of relative displacements and rotations between body parts were constrained, while others (like elbow rotation, trunk-thighs rotations, knees rotation, edging angles) were not.

As second step, videos referred to a skier performing parallel turns were examined and divided into frames with a time interval of 0.1 s. The position of the skier in each frame was then carefully reproduced with the 3D-CAD model. The sequence of movements was played back and visually compared to videos till the motion of the model resulted natural and smooth.

The sequence of movements was eventually applied to the skier multi-body model. Motion of the non-constrained joints was driven to reproduce a sequence of two opposite parallel turns. No ski-snow contact model was introduced, since the motion of the skier was pre-determined and an inverse dynamics problem was solved.

The kinematic analysis allowed to determine the trajectory of the skier's CoM; lateral and vertical displacement of this point were of particular interest since they characterize the ski control during a turn. Referring to a local reference frame centred on the skier's centre of mass at the beginning of the simulation (just before entering the first turn), maximum lateral and vertical displacement were in the order of 0.2 and 0.5 m, respectively. Longitudinal displacement was instead negligible.

As far as accelerations are concerned, peaks of lateral acceleration were recorded in the central section of a turn and, considering the simulation performed in the work, the maximum value was around 7–8 m/s^2. Though the vertical displacement was significantly lower than the lateral one, peaks of vertical acceleration higher than 5 m/s^2 were estimated. This is explained considering that the period of the vertical motion, characterized by the *down-unweighting* and the *up-weighting* phases, is about half the one of lateral motion.

An additional output of the kinematic analysis is represented by edging angles which constitute another element of the ski control in turns. Edging angles up to 60° were reached.

The dynamic analysis led to prediction of edging forces (i.e. the forces transferred from the skier to the skies through bindings). These forces result from the combined effect of gravity and inertia forces acting on different parts of the body. Outputs of the model were successfully compared with experimental results, as outdoor tests used to capture the motion of the skier were performed with load cells allowing measurement of edging forces. Considering that a skier weighting 62 kg was used for the simulations, a maximum lateral force between 500–600 kN was reached during the central part of the turn. Values of vertical force, associated with down-unweighting and up-weighting actions, varied from 300 kN, reached while entering the turn, to 900 kN developed in the central part of the turn.

In the work presented by Filippi Oberegger et al. [94], the motion of the multi-body model of the skier is only partially driven on the basis of kinematic data; some of the degrees of freedom are actuated to satisfy kinematic constraints and one equilibrium condition. In addition, the speed profile of the skier is computed in forward dynamics.

The model of the skier-ski system is made up of seven rigid bodies: two for the skies, two for the shanks, two for the thighs and one for the trunk. This last incorporates the effect of arms, head and equipment like helmet and poles. Knees and ankles are modelled as rotational joints, while the hips are introduced as spherical joints.

The trajectory of the binding of right ski $\mathbf{P}_{\mathbf{rsk}}$ is constrained, meaning that the coordinates of its path on the snow surface are known, but the speed profile has to be determined through direct dynamics. Trajectory is identified on the basis of experimental data where a skier performed a series of carved turns. During outdoor tests, the motion of the skier was captured with three synchronized cameras working at 50 fps. Videos were then post-processed [102] to obtain 3D coordinates of markers on the skier and skies. In the same experimental campaign, two force plates were used to measure forces and moments on the right ski [103].

Coming back to the model, the longitudinal axis of the right ski forms an angle ψ with the tangent to the trajectory of its centre; the angle accounts for the effect of lateral skidding which could not be avoided during experimental tests. In addition a rotation β (the edging angle) along the longitudinal axis is applied. Computation of β is carried out considering the equilibrium of the whole skier during a turn. In particular, the skier has to lean the CoM inward to balance the effect of centrifugal force. This condition is reached when the resultant force $\mathbf{F}_\mathbf{r}$ in a plane normal with respect to the direction of motion points towards the base of support (i.e. between the skies).

The edging angle is assumed equal to the angle between vector $\mathbf{F}_\mathbf{r}$ and the unit vector $\mathbf{n}_\mathbf{h}$ normal to the contact surface in $\mathbf{P}_{\mathbf{rsk}}$. Vector $\mathbf{F}_\mathbf{r}$ can be computed as:

$$\mathbf{F}_\mathbf{r} = m\mathbf{g} \pm m\frac{v_G^2}{R}\mathbf{n}_\mathbf{t} \tag{4.32}$$

In Eq. (4.32), m denotes the total mass of the system, $\mathbf{g}$ the gravity acceleration, R the curve radius and v_G the speed of the centre of mass. The unit vector $\mathbf{n}_\mathbf{t}$ is tangent to the snow surface in $\mathbf{P}_{\mathbf{rsk}}$, normal to $\mathbf{n}_\mathbf{h}$ and directed leftward. Therefore, the edging angle becomes:

$$\beta = -\arctan\left(\frac{\mathbf{F}_\mathbf{r}\cdot\mathbf{n}_\mathbf{t}}{\mathbf{F}_\mathbf{r}\cdot\mathbf{n}_\mathbf{h}}\right) \tag{4.33}$$

where "·" indicates a scalar product between two vectors. According to the conventions, β is positive for a right turn.

As the ankle is modelled with a rotational joint, the edging angle β corresponds also to the lean angle of the right shank.

Time histories of rotations of right knee and of right and left thighs are driven on the basis of experimental data, while rotations of left knee and ankle are computed during numerical integration so that two kinematic conditions are satisfied.

Once the position of the left knee is defined, rotation of this joint determines the distance of the ski boot from the ground. According to the first kinematic condition, rotation of the left knee (and thus of the left shank) should be set so that the ski boot is in contact with the snow surface. This condition is obtained by zeroing the distance between ski boot and snow surface which is locally approximated with the osculating plane.

As second kinematic condition, the left ankle is rotated until the longitudinal axis of the ski is tangent to the osculating plane.

Considering now the contact model, since the trajectory on the snow surface is predetermined, two components of the contact forces on the skies can be directly computed as reactions forces. The sole component that needs to be modelled is the longitudinal frictional force $\mathbf{T_i}$ that influences the speed profile.

$$\mathbf{T_i} = -\mu|\mathbf{N_i}|\frac{\mathbf{v_i}}{|\mathbf{v_i}|} \tag{4.34}$$

The friction force acting on the ith ski is simply given by the friction coefficient μ times the normal contact force $\mathbf{N_i}$. In Eq. (4.34) $\mathbf{v_i}$ is the speed of the contact point of the ith ski. Variation of friction coefficient with speed is taken into account with the following linear model.

$$\mu = \mu_0 + \mu_1|\mathbf{v_i}| \tag{4.35}$$

Motion equations of the system are derived assuming 6 degrees of freedom for each body and adding a set of algebraic equations representing the effect of joints. Globally 42 differential motion equations can be obtained and combined with 41 algebraic constrain equations. The system has in fact only 1 real degree of freedom represented by the position of the right ski along the predetermined trajectory. Free coordinates are represented by the positions of the centre of mass of each body, and by Euler's parameters used to describe the spatial orientation. Lagrange parameters are introduced for computing reaction forces and moments.

The model was used to simulate a sequence of opposite turns (half right turn, complete left turn, complete right turn) performed by the skier during the aforementioned experimental tests. The same sequence was at first analysed using inverse dynamics (ID) to obtain reference values for kinematic quantities and also for reaction forces and moments on hip knee and ankle joints [104]. Obviously, also binding forces measured through load cells were used for direct comparison.

As far as skier speed is concerned, a proper tuning of the friction coefficient in Eq. (4.35) allowed to get almost the same result for inverse dynamics (ID) and forward dynamics (FD) simulations. The speed profile presented a maximum of 45 km/h in the central part of the left turn. The almost perfect agreement between simulations means that the skier is following the same trajectory with the same speed profile. Despite this condition, the inward lean angle computed according to Eqs. (4.32) and (4.33) resulted a little bit larger, in the order of few degrees, with respect to ID simulation. Small adjustments seem required to improve the model of lean angle estimation. During the central part of both left and right turns, lean angles around 50° were reached.

Results of FD simulation were consistent with measurements of reaction force on the right ski obtained through the force plates. Maximum values around 1400 N were developed during the left turn. In addition, very good agreement was found when comparing forces on left and right knees obtained with ID and FD simulations.

Altogether, the approach followed in this work tries to reproduce a realistic motion of a skier not only on the basis of motion capture technique but also implementing relations between the movements of the skier and track geometry trajectory, and speed. Simple relations were applied to determine the shank lean angle on one leg, the knee and ankle angles on the other leg. Though small refinements are needed, this work represents an interesting step towards a model able solve a full forward dynamics aiming at optimization of skier's posture.

References

1. J. Spörri, Helmet use and risk of head injuries in Alpine Skiers and snow-boarders. Ph.D. thesis, University of Salzburg, 2012
2. J. Allen, Mathias Zdarsky: the father of Alpine Skiing. Skiing Herit. J. **20**, 8–14 (2008)
3. J. Spörri, Biomechanical aspects of performance enhancement and injury prevention in alpine ski racing. Ph.D. thesis, University of Salzburg, 2012
4. International Ski Federation (FIS), History of skiing (2015), http://history.fis-ski.com/
5. H. Polednik, *Das Glück in Schnee* (Amalthea, Vienna, 1991)
6. J. Suttner, *Die Grossen Skisport-Stars* (Copress, München, 1992)
7. International Ski Federation (FIS), Document library (2015), http://www.fis-ski.com/inside-fis/document-library/alpine-skiing/#deeplink=rules
8. D. Byrd, Helmet usage and safety fact sheet. Tech. rep. National Ski Area Association (2014)
9. D. Stewart et al., Evaluating the performance of helmet linings incorporating fluid channels. J. ASTM Int. **7**, 10 (2010)
10. R.M. Greenwald, G.T. Gwin, J.J. Crisco, Head impact severity measures for evaluating mild traumatic brain injury risk exposure. Neurosurgery **62**, 789–798 (2008)
11. A. Patel, T. Goswami, Comparison of intracranial pressure by lateral and frontal impacts -validation of computational model, in *Injury and Skeletal Biomechanics* (InTech, Rijeka, 2012)
12. M.D. Scott, D.B. Buller, P.A. Andersen, Testing the risk compensation hypothesis for safety helmets in alpine skiing and snowboarding. Inj. Prev. **13**, 173–177 (2007)
13. L. Ružić, A. Tudor, Risk-taking behavior in skiing among helmet wearers and nonwearers. Wilderness Environ. Med. **22**, 291–296 (2011)
14. A.J. Macnab et al., Effect of helmet wear on the incidence of head/face and cervical spine injuries in young skiers and snowboarders. Inj. Prev. **8**, 324–327 (2002)
15. B.E. Hagel et al., Effectiveness of helmets in skiers and snowboarders: case-control and case crossover study. Br. Med. J. **330**, 281 (2005)
16. N.J. Mills, in *Polymer Foams Handbook: Engineering and Biomechanical Applications and Design Guide*, ed. by M.A. Butterworth (Heinemann, Burlington, 2007)
17. M. Swarén, H.-C. Holmberg, A. Eriksson, Repeated low impacts in alpine ski helmets. Sports Technol. **6**, 43–52 (2013)
18. L.R. Young, N. Chan, J. Ruchelsman, Incompressible fluid-lined protective helmet, 60/792, 287, 2006
19. A. Claire, B. Vue, Helmet liners with fluid filled channels. Tech. rep. MIT Course 16.621 (MIT, Cambridge, 2007)
20. J.J.C. Chua, F.K. Fuss, O. Troynikov, Aerodynamics of loose sports garments, in *5th Asia-Pacific Congress on Sports Technology (APCST)*, 2011
21. L. Oggiano, L. Saetran, A low drag suit for ski-cross competitions, in *8th Conference of the International Sports Engineering Association (ISEA)*, 2010

22. L.M. Bardal, R. Reid, Testing of fabrics for use in alpine ski competition suits, in *9th Conference of the International Sports Engineering Association (ISEA)*, 2012
23. L. Brownlie et al., Factors affecting the aerodynamic drag of alpine skiers. Proc. Eng. **2**, 2375–2380 (2010)
24. J.R. Shanebrook, R.D. Jaszczak, Aerodynamic drag analysis of runners. Med. Sci. Sports **8**, 43–47 (1976)
25. M.M. Zdravkovich, Conceptual overview of laminar and turbulent flows past smooth and rough circular cylinders. J. Wind Eng. Ind. Aerodyn. **33**, 53–62 (1990)
26. L. Brownlie et al., Streamlining the time trial apparel of cyclists: the Nike Swift Spin project. Sports Technol. **2**, 53–60 (2009)
27. M. Lund, The empire that exploded: Bob Lange and the plastic boot. Ski. Herit. J. **13**(3), 13–23 (2001)
28. W. Lert, The finish line. Ski. Herit. J. **8**(3), 40–45 (1996)
29. S. Masia, The short happy eras of the American ski boot. Ski. Herit. J. **14**(2), 35–38 (2002)
30. S. Masia, The selling of skiing. Ski. Herit. J. **17**(4) 36–40 (2005)
31. M. Colonna, M. Nicotra, M. Moncalero, Materials, designs and standards used in ski-boots for Alpine Skiing. Sports **1**, 78–113 (2013)
32. S. Masia, The rise and fall of the knee-high boot. Ski. Herit. J. **6**, 16–17 (2003)
33. V. Senner, F.I. Michel, S. Lehner, *Ski Equipment-Related Measures to Reduce Knee Injuries* (bfu – Swiss Council for Accident Prevention, Berne, 2013)
34. V. Senner et al., Technical possibilities for optimising the ski-binding-boot functional unit to reduce knee injuries in recreational alpine skiing. Sports Eng. **16**, 211–228 (2013)
35. P. Hofer et al., Temperature, relative humidity and water absorption in ski boots, in *5th Asia-Pacific Congress on Sports Technology (APCST)*, 2011
36. P. Hofer et al., Microclimate in ski boots -temperature, relative humidity, and water absorption. Appl. Ergon. **45**, 515–520 (2014)
37. M. Pinter, M. Eckelt, H. Schretter, Evaluation of ski boot fitting characteristics by means of different pressure distribution measurements, in *8th Conference of the International Sports Engineering Association (ISEA)*, 2010
38. N. Petrone et al., Analysis of the structural behavior of an innovative reinforced ski boot, in *8th Conference of the International Sports Engineering Association (ISEA)*, 2010
39. N. Petrone, G. Marcolin, F.A. Panizzolo, The effect of boot stiffness on field and laboratory flexural behavior of alpine ski boots. Sports Eng. **16**, 265–280 (2013)
40. M.S. Koehle, R. Lloyd-Smith, J.E. Taunton, Alpine ski injuries and their prevention. Sports Med. **32**(12), 785–793 (2002)
41. M. Barone, V. Senner, P. Schaff, ACL injury mechanism in alpine skiing: analysis of an accidental ACL rupture, in *Skiing Trauma and Safety*, ed. by R.J. Johnson (American Society for Testing and Materials, West Conshohocken, 1999)
42. V. Senner, S. Lehner, Skiing equipment: what is done towards more safety, performance and ergonomics? in *The Impact of Technology on Sport II* (Taylor & Francis, London, 2008), pp. 803–811
43. V. Senner, S. Lehner, H. Bohm, Equipment development and research for more performance and safety, in *Science and Skiing IV* (Meyer & Meyer Sport, Maidenhead, 2009), pp. 110–133
44. P.A. Federolf, Finite element simulation of a carving snow ski. Ph.D. thesis, ETH, 2005
45. E. Bernardo, Studio dei materiali impiegati nella costruzione degli sci. MA thesis, Universitá degli Studi di Padova, 2011
46. E. Muller, H. Schwameder, Biomechanical aspects of new techniques in alpine skiing and ski-jumping. J. Sports Sci. **21**, 679–692 (2003)
47. J. Howe, *Skiing Mechanics* (Poudre Canyon Pr, Laporte, 1983)
48. D.A. Lind, S.P. Sanders, *The Physics of Skiing. Skiing at the Triple Point* (Springer, New York, 2003)
49. E. Müller, Analysis of the biomechanical characteristics of different swinging techniques in alpine skiing. J. Sports Sci. **12**, 261–278 (1994)

50. B.M. Nigg, Load of the locomotor system during skiing: a biomechanical perspective. Sci. Skiing **1997**, 27–35 (1994)
51. C. Raschner et al., Carving turns versus traditional parallel turns – a comparative biomechanical analysis, in *Proceedings of 2nd International Congress on Skiing and Science* (2001), pp. 56–57
52. P. Kaps, W. Nachbauer, M. Mössner, Determination of kinetic friction and drag area in alpine skiing. Ski Trauma Skiing Saf. **10**, 165–177 (1996)
53. M. Supej et al., Aerodynamic drag is not the major determinant of performance during giant slalom skiing at the elite level. Scand. J. Med. Sci Sports **23**(1), 38–47 (2012)
54. F. Casolo et al., A composite model for the simulation of skiing techniques, in *Science and Skiing* (Chapman and Hall, Cambridge, 1996), pp. 116–130
55. J. Margane et al., A mechanical apparatus executing turns on carver skis, in *Proceedings of the 16th ISBS Symposium* (1998), pp. 285–289
56. R. Adelsberger et al., On bending characteristics of skis in use. Proc. Eng. **72**, 362–367 (2014)
57. M. Fauve et al., Measurement of dynamical ski behaviour during alpine skiing, in *Science and Skiing* (Meyer & Meyer Sport, Maidenhead, 2007), pp. 195–206
58. T. Yoneyama, M. Kitake, K. Osada, Investigation on the ski-snow interaction in a carved turn based on the actual measurement. Proc. Eng. **2**, 2901–2906 (2010)
59. M. Bassetti et al., Sensor nodes for the dynamic assessment of alpine skis, in *Topics in Modal Analysis II-Conference Proceedings of the Society for Experimental Mechanics Series*, vol. 6 (Springer, New York, 2012), pp. 471–479
60. R. Reid, A kinematic and kinetic study of alpine skiing technique in slalom. Ph.D. thesis, Norwegian School of Sport Sciences, 2010
61. M. Supej, R. Kipp, H.C. Holmberg, Mechanical parameters as predictors of performance in alpine World Cup slalom racing. Scand. J. Med. Sci. Sports **21**, e72–e81 (2011)
62. M. Mössner, P. Kaps, W. Nachbauer, A method for obtaining 3-D data in alpine skiing using pan and tilt cameras with zoom lenses. Ski Trauma Skiing Saf. **10**, 261–278 (1996)
63. M. Brodie, A. Walmsley, W. Page, Fusion motion capture: a prototype system using IMUs and GPS for biomechanical analysis of alpine ski racing. Sports Technol. **1**, 17–28 (2008)
64. K. Hirose, H. Doki, A proposal for motion analysis method of skiing turn by measurement of orientation and gliding trajectory. Proc. Eng. **13**, 17–22 (2011)
65. A. Kondo, H. Doki, K. Hirose, An attempt of new motion measurement method for alpine ski turns using inertial sensors. Proc. Eng. **34**, 421–426 (2012)
66. A. Kondo, H. Doki, K. Hirose, Motion analysis and joint angle measurment of skier gliding on the actual snow filed using inertial sensors. Proc. Eng. **60**, 307–312 (2013)
67. J. Spörri et al., Turn characteristics of a top world class athlete in giant slalom: a case study assessing current performance prediction concepts. J. Sports Sci. **7**, 647–659 (2012)
68. C.E. Clauser, J.T. Mc Conville, J.W. Young, Volume and centre of mass segments of the human body. Tech. rep. (Wright-Patterson Airforce Base, Dayton, 1969)
69. W.T. Dempster, Space requirements of the seated operator. Tech. rep. (Wright-Patterson Airforce Base, Dayton, 1955)
70. M. Gilgien et al., The effect of different global navigation satellite system methods on positioning accuracy in elite alpine skiing. Sensors **14**, 18433–18453 (2014)
71. M. Brodie, A. Walmsley, W. Page, Force vector analysis of ski racing technique using fusion motion capture, in *Proceedings of the XXV International Symposium on Biomechanics in Sports* (2007), pp. 71–74
72. M. Gilgien et al., Determination of external forces in alpine skiing using a differential global navigation satellite system. Sensors **13**, 9821–9835 (2013)
73. R. LeMaster, *The Skier's Edge* (Human kinetic, Champaign, 1999)
74. O. Radenovic, B. Nemec, V. Medved, A new biomechanical measurement and testing method for turns in alpine skiing, in *IFMBE Proceedings of MEDICON 2001* (2001), pp. 624–627
75. A. Kiefmann et al., A new six component dynamometer for measuring ground reaction forces in alpine skiing, in *The Engineering of Sport 6* (Springer, New York, 2006), pp. 87–92

76. K. Nakazato, P. Scheiber, E. Müller, A comparison of ground reaction forces determined by portable force-plate and pressure-insole systems in alpine skiing. J. Sports Sci. Med. **10**, 754–762 (2011)
77. G. Stricker et al., Determination of forces in alpine skiing and snowboarding: validation of a mobile data acquisition system. Eur. J. Sport Sci. **10**(1), 31–41 (2010)
78. M. Lappi, The kinetic characteristics in competitive slalom skiing. Ph.D. thesis, Norwegian School of Sport Sciences, 2009
79. A.H.A. Razak et al., Foot plantar pressure measurement system: a review. Sensors **12**, 9884–9912 (2012)
80. C. Barelle, M. Tavernier, Experimental model of the aerodynamic drag coefficient inalpine skiing. J. Appl. Biomech. **20**, 167–176 (2004)
81. F. Meyer, D. Le Pelley, F. Borrani, Aerodynamic drag modeling of alpine skiers performing giant slalom turns. Med. Sci. Sports Exerc. **44**, 1109–1115 (2011)
82. K. Watanabe, T. Ohtsuki, Postural changes and aerodynamic forces in alpine skiing. Ergonomics **20**, 121–131 (1977)
83. M. Supej, O. Kugovnik, B. Nemec, New advances in racing slalom technique. Kinesiologia Slovenica **8**(1), 25–29 (2002)
84. S.C. Colbeck, A review of the friction of snow skis. J. Sports Sci. **12**, 285–295 (1994)
85. P. Federolf et al., Impact of skier actions on the gliding times in alpine skiing. Scand. J. Med. Sci. Sports **18**, 790–797 (2008)
86. R. Von Hertzen, U. Holmlund, M.A. Ranta, On the velocity maximization in downhill skiing. J. Biomech. **30**, 525–529 (1997)
87. R. Rudakov et al., The slalom path optimization on the curvilinear slope. Russ. J. Biomech. **7**, 2–56 (2003)
88. R. Rudakov et al., Optimisation of the skier's trajectory in special slalom. Proc. Eng. **2**, 3179–3182 (2010)
89. L. Chen, Z.-H. Qi, A 2-dimensional multi rigid bodies skiing model. Multibody Syst. Dyn. **21**, 1–98 (2009)
90. T. Sakata, M. Tsukiyama, Effect of position of shoe centre on ski turn. JSME Int. J. **42**(4), 922–929 (1999)
91. P. Federolf et al., Finite element simulation of the ski-snow interaction of an alpine ski in a carved turn. Sports Eng. **12**, 123–133 (2010)
92. P. Federolf et al., Parameter study using a finite element simulation of a carving Alpine ski to investigate the turn radius and its dependence on edging angle, load and snow properties. Sports Eng. **12**, 135–141 (2010)
93. D. Heinrich, J. van den Bogert, W. Nachbauer, Relationship between jump landing kinematic and peak ACL force during a jump in downhill skiing: a simulation study. Scand. J. Med. Sci. Sports **24**, 180–187 (2013)
94. U. Filippi Oberegger et al., Simulation of turns with a 3D skier model. Procedia **2**, 3171–3177 (2010)
95. S. Kawai, K. Yamaguchi, T. Sakata, Ski control model for parallel turn using multibody system. JSME Int. J. **47**(4), 1095–1100 (2004)
96. J. Howe, *The New Skiing Mechanics* (McIntire, Waterford, 2001)
97. P. Federolf et al., Deformation of snow during a carved ski turn. Cold Reg. Sci. Technol. **46**(1), 69–77 (2006)
98. P. Federolf, Finite element simulation of a carving snow ski. Ph.D. thesis, ETH Zürich, 2005
99. Q.W. Hao et al., Kinematic study on biff movement of freestyle aerial in poor training by Ji Xiao-ou. J. Shenyang Inst. Phys. Educ. **22**, 18–20 (2003)
100. F. Bruck, P. Lugner, H. Schretter, A dynamic model for the performance of carving skism, in *Skiing Trauma and Safety*, ed. by R. Jhonson, M. Lamont, J. Shelaly (American Society for Testing and Materials, West Conshohocken, 2003)
101. M. Miura, Y. Ikegami, S. Matsui, Calculation of the compound C.G. by using coordinate measurement (in Japanese). Sci. Phys. Educ. **24**(8), 517–522 (1974)

102. W. Nachbauer et al., A video technique for obtaining 3D coordinate in alpine skiing. J. Appl. Biomech. **12**, 104–115 (1996)
103. K. Schindelwig et al., Prediction of moments at the knee joint for carving and parallel turns technique, in *Book of Abstracts, 17th Congress of the ISB, Calgary* (1999), p. 667
104. U. Filippi Oberegger et al., Reaction forces and moments in carved turns. J. ASTM Int. **7**(9), 187–205 (2010)

Chapter 5
Cross-Country Ski

Peter Carlsson, M. Ainegren, M. Tinnsten, D. Sundström, B. Esping, A. Koptioug, and M. Bäckström

5.1 Cross-Country Skiing and Roller Skiing

Cross-country skiing, biathlon and ski orienteering are competitive sports with practitioners who are mostly from countries in the northern hemisphere. The competition season is during the time when the ground is covered with snow, which roughly extends from mid-November to late March. During the rest time of the year, which is a long preparatory period of training for the skiers before the competition season, the skiers use roller skis for dryland training with the aim of imitating skiing on snow. Furthermore, over the last few decades, fairly specific indoor testing methods for cross-country skiers have become possible due to the development of treadmills that allow roller skiing using classical and freestyle techniques.

Outdoor experiments on snow, using cross-country skiers, biathletes and ski-orienteers, are difficult to standardize due to changes in factors that influence the grip and glide (static and dynamic friction coefficients) of the skis and the skier, and thus the skiing technique and energy expenditure. Such factors are air and snow temperature, humidity and snow and wind conditions. Also, it is difficult to control the intended speed and to find a track profile with proper, relatively constant, inclination for the specific core-technique during the time required to retrieve stable energy expenditure.

Moreover, cross-country skis are constructed in different lengths and with cambers of different heights and stiffness, enabling a skier to choose ski characteristics to suit his or her relative body length, mass and technical skiing skill. Thus, using

P. Carlsson (✉) • M. Ainegren • M. Tinnsten • D. Sundström • B. Esping • A. Koptioug • M. Bäckström
Department of Quality Technology and Management, Mechanical Engineering and Mathematics, Mid Sweden University, SE-831 25 Östersund, Sweden
e-mail: Peter.Carlsson@miun.se

F. Braghin et al. (eds.), *The Engineering Approach to Winter Sports*,
DOI 10.1007/978-1-4939-3020-3_5

the same pair of skis for all subjects means badly matched skis for the individual, while individually matched skis can be difficult to standardize as regards grip and glide. In addition, experience has shown that ski manufacturers have difficulties with the reproducibility of the skis' characteristics, such as the skis' grip and glide, even when attempting to produce them to the same specifications. Therefore, much of the current sports research into the physiology and biomechanics of cross-country skiing is conducted indoors on treadmills and using roller skis, due to the opportunity to use a wide range of advanced equipment for different types of analyses and for benefitting from comparisons that use relatively stationary and reproducible conditions. Even though roller skiing on a treadmill never will be 100 % similar to skiing on snow, it can be more or less similar depending on the equipment used by the skier and the skiing conditions, i.e. the similarity between cross-country skis and snow versus roller skis and treadmill conditions. Thus, in parallel with physiological and biomechanical research, the development of equipment and testing methods are important in order to minimize errors and to increase the specificity, validity, and reliability of the testing method and thereby the relevance of the research conducted using the athletes.

One important thing regarding roller skiing on a treadmill is mimicking the friction coefficients that exist when skiing with cross-country skis on snow. Variations in grip (static friction coefficient, μ_S), when skiing in the classical style, diagonal stride and kick double poling techniques, affect the ability to apply propulsive force from the legs, while variations in the rolling resistance coefficient (μ_R) mean that the need for propulsive force varies in overcoming the roller skis' rolling resistance. In terms of maximal oxygen uptake, results from experiments show that there is no statistical difference when skiing with different friction coefficients [1, 2]. However, what will be changed in a graded maximal test is the time from start to when the maximal oxygen uptake and voluntary exhaustion occurs. Further, when roller skiing at submaximal intensity around race pace, on a given speed and inclination but with different μ_S and μ_R, the recruitment of skeletal muscle cells in the upper and lower body will be slightly different, affecting the skiing technique, peripheral and central circulation and the submaximal oxygen uptake (skiing economy). Furthermore, the differences in performance-predicting factors among elite athletes are quite small, thus emphasizing the importance of standardizing the roller skis' friction coefficients used in the experiments on cross-country skiers, biathletes, and ski-orienteers. Last but not least, it is doubtful from a training point of view that it is useful to train a skiing technique on roller skis with friction coefficients which deviate significantly from the ones used for skiing on snow.

The following pages of this chapter will examine research assessing the importance of controlling the roller skis' friction coefficients and provide examples of innovative equipment and methods that can be used to measure and standardize these variables.

5.1.1 Mechanical Experiments on Gliding Friction and Rolling Resistance

Using roller skis in research experiments results in a need to control their rolling resistance coefficient (μ_R). One reason for controlling the roller skis' μ_R is to mimic the coefficients that exist when skiing on snow. However, cross-country skis' dynamic (gliding) friction coefficient varies due to different weather and snow conditions, types of ski base material, topography, and preparation with various types of glide waxes. Thus, a range of 0.02–0.18 has been reported from different studies in the subject area [3–5]. However, a coefficient of 0.02–0.06 applies in most circumstances when it comes to skiing on groomed trails. Another important reason for controlling the roller skis' μ_R is to secure good reproducibility for this specific testing method, in order to make accurate comparisons and conclusions regarding the results of the treadmill experiments. However, μ_R has not always been checked in scientific experiments using roller skis, particularly not in older published experiments.

In presented research, the conventional way of calculating μ_R is the ratio from measurements of the total resisting force to the total normal force

$$\mu_R(N_{\mathrm{TOTAL}}, \alpha) = \frac{F_r + F_f}{N_r + N_f} = \frac{S - mg \sin \alpha}{mg \cos \alpha} \tag{5.1}$$

where angle α is the inclination of travel (treadmill), S is the tangential force registered in a load cell, m is the total mass of the roller ski and the load applied on the roller ski, g is the acceleration due to gravity, N is normal force on the wheels, F is the rolling resistance and index r and f are the rear and forward positions of the forces. The individual normal forces can be calculated if the position of the center of mass, the horizontal distances between the wheels and where S is connected to the roller ski (or the skier), and the vertical distance between the ground surface (treadmill rubber mat) and S are known. However, it is the total resisting force rather than the individual values that has been of interest in presented studies on roller skis' μ_R.

Only a few authors have studied the μ_R of roller skis. The first method described in published studies was based on force measurements that were carried out using a skier wearing a backpack filled with varying mass. The skier was instructed to distribute the mass evenly on both roller skis whilst rolling on the treadmill [6]. However, the data presented when using this method showed some varying results and no reliability testing for the method was presented [6–8]. A similar method, which investigated roller blades' rolling resistance on an outdoor surface, showed a variability of 20 % [9].

Using this method, [6] observed that the coefficient of roller skis' rolling resistance was not dependent on speed, but that it increased with increasing body mass. However, in 1994 and 1995 Hoffman and co-workers found that body mass did not affect μ_R, but that μ_R was related to speed [7, 11]. Millet et al. [8], on the other hand, found that μ_R was not dependent on speed for low-resistance roller skis but that it was dependent on speed for high-resistance roller skis.

In 1990, [6] wrote that they allowed the roller skis to become warm prior to making force measurements, but they did not describe the amount of time that was needed, nor any temperature registrations, and neither did they describe how great the differences in μ_R were between the cooler and the warmer roller skis. If μ_R is temperature dependent, this could be of great importance when comparing physiological results, since the roller skis might have different initial temperatures depending on different previous use.

The question of whether a warm-up time is needed prior to experiments was addressed by Ainegren et al. [10] who found out that the time needed to retrieve a stabilized μ_R and temperature (measured close to the rear wheel bolt) was 30 min of continuous rolling. This study clearly showed that a proper warm-up period for roller skis must precede testing, otherwise the results of different physiological tests cannot be compared correctly. The authors raised the idea that warming up the roller skis on the treadmill could be replaced by controlled warming in a low-temperature oven. Based on the weight of the skier, the roller skis could be heated to the appropriate temperature and ready to use at once. This method of preheating the roller skis was applied in subsequent research [2, 10].

The second method, in chronological order, for measuring the roller skis' μ_R was presented by Ainegren et al. [10] who used specially designed equipment for roller ski rolling resistance measurements (RRMS) with heavy lead plates to simulate different body masses, see Fig. 5.1. The reproducibility of the rolling resistance measurement system was tested and the reproducibility study showed no difference between the two sets of paired measurements with either a classical or a freestyle roller ski ($p < 0.05$).

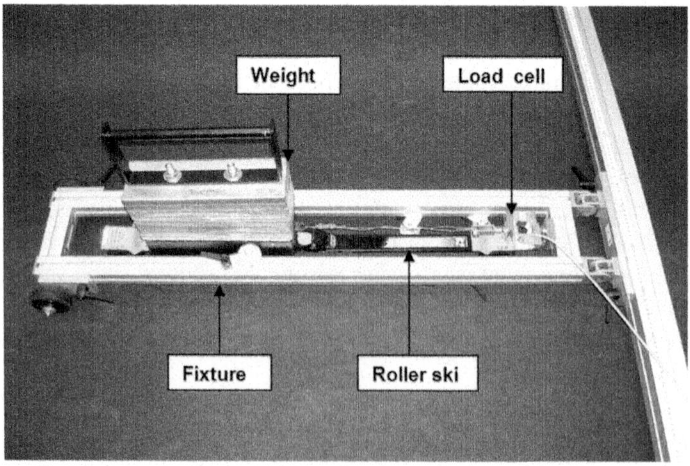

Fig. 5.1 Roller ski with load of lead plates and the equipment for rolling resistance measurements

In contrast to earlier studies [6–8, 11], this study showed a negative correlation between normal force and μ_R, while different treadmill speeds (8–28 km h^{-1}) only resulted in non-significant changes to μ_R [10]. However, vibrations in the treadmill and therefore in the RRMS fixture were observed, causing the roller ski to sway against the stabilizing wheels on each side of the lead plates and the roller ski. Later unpublished experiments indicate that this swaying reduces the force S measured in the load cell, especially with an increasing number of lead plates and thereby a higher vertical position of center of gravity and a greater mass-moment of inertia. Therefore, these vibrations caused some impact on the part of the study in the presented paper, which investigated the roller skis' μ_R as a function of different normal forces. However, it had probably little impact on the results of the reproducibility study, i.e. this bias affected the paired measurements of both series equally.

The error bars (±SD) in [10] give an indication of the variation in μ_R among the six classical and six freestyle roller skis of the same model from the same manufacturer. The range in μ_R was of a magnitude up to 0.0007 and 0.0010 (28 % and 34 %) for the freestyle and classical roller skis, respectively.

To remedy this in upcoming experiments, Ainegren and co-workers [1] developed a rolling resistance regulating function. Two roller bearings, whose pressure on the rubber wheel was regulated using a spring-loaded screw, regulated the rolling resistance of the individual roller ski. With this function applied to one of the roller ski's wheels it became possible to adjust and standardize μ_R between different test-occasions and to mimic coefficients that exist outdoors when skiing on snow.

Finally, a third method for measuring roller skis' μ_R was recently presented by Ainegren et al. [12], see Fig. 5.2. This method is based on the basic design of a small treadmill, equipped with beams, roller ski stabilizing lateral supports, consisting of ball bearing wheels, and a screwed bar for applying different loads normal to the surface in the ski binding at the ski boot fix point. Since this rolling resistance measurement system uses only one force sensor, measuring forces around three axes, a simple and superior method is obtained for measuring forces in the various directions. Unlike the RRMS system that used lead plates, this system has fewer vibrations in the treadmill and no lead plates swaying from side to side. Also, it is much more ergonomic to work with. So far, very few results from experiments with the system have been presented, but a reproducibility study showed good reliability [12].

5.1.2 Physiological Effects of Different Rolling Resistance Coefficients

There are few studies which have investigated the biomechanical and physiological responses to different μ_R. The roller skis' μ_R were measured on a treadmill belt, using the methods described above, while in some of the studies the biomechanical and physiological measurements were made outdoors, in other environments and on other surfaces, i.e. on an asphalt oval [8] and on an asphalt roadway [13].

a b

Fig. 5.2 The figures show a roller ski rolling resistance measurement system designed by Ainegren et al. [12]

Hoffman et al. [13] compared poling forces and metabolic demands of roller skiing in the freestyle technique called V2 alternate (gear 4) on submaximal intensities using roller skis with 100 % difference in μ_R (μ_R: 0.012 vs. 0.024–0.026). The study showed that poling time was similar but recovery time and cycle length was shorter for the skiers when using roller skis with higher μ_R ($p < 0.01$). Also, peak and average poling force, cycle rate, oxygen uptake, heart rate, blood lactate and ratings of perceived exertion were higher when using roller skis with the higher μ_R ($p < 0.01$). Further, [13] compared the physiological effects of roller skiing in the V1 (gear 3) freestyle technique at submaximal intensities with 100 % difference in μ_R (μ_R: 0.011 vs. 0.021). This study retrieved a 17 % difference in external power, 13 % in oxygen uptake and 5 % difference in heart rate ($p < 0.05$). Further, to elicit the same oxygen uptake, the lower resistance roller skis required a 0.4 m/s (15 %) faster speed. Finally, [2] investigated the physiological effects of ~50 % change in roller skis' μ_R during treadmill roller skiing using classical (μ_R: 0.020 vs. 0.029) and freestyle roller skis (μ_R: 0.017 vs. 0.009) on the two sub-techniques diagonal stride and gear 3. This study showed that roller skiing on submaximal steady rate exercises (~55 % and ~75 % of maximal oxygen uptake), changed external power by 8–12 %, oxygen uptake 5–9 %, heart rate 3–8 %, and blood lactate 14–46 % ($p < 0.05$), while there were non-significant or small changes to cycle rate, cycle length and ratings of perceived exertion. Incremental maximal tests showed that time to exhaustion was significantly changed by 12–24 % for the two techniques and this occurred without a change in maximal oxygen uptake, power, heart rate, and blood lactate.

5.1.2.1 Summary

In summary, the overall results from mechanical and physiological experiments on μ_R clearly emphasize the importance of controlling the μ_R of each pair of roller skis used in experiments on skiers. The published research from mechanical experiments shows that μ_R varies greatly among different pairs of roller skis, but that μ_R is about the same size as the friction coefficient for cross-country skis gliding on snow. Thus, the reproducibility for roller skis' rolling resistance seems to be a larger problem to deal with in experiments with human roller skiing than that of an invalid rolling resistance coefficient. Further, some differences in results exist due to the influence on μ_R of different normal forces and speed. It is uncertain whether this is caused by deficiencies in the methods being used or if it is because different roller skis react differently to different speeds and normal forces. A more comprehensive study that includes many roller ski manufacturers could clear this up. Moreover, the rolling resistance measurements were carried out using static normal force, in contrast to the more dynamic normal force acting on the wheels during human roller skiing. In the studies it was assumed that a particular value of μ_R, established with the static normal force, likewise existed during human roller skiing. Another source of uncertainty is the side forces on the wheels that occur when the roller skis are edged during freestyle roller skiing. How these forces affect rolling resistance was not examined in the studies. Both differences in construction and the side forces will probably have some influence on the rolling resistance. Altogether, this also shows the importance of using a method that quantifies μ_R with high validity and reliability.

5.1.3 *Classical Style Cross-Country Skis and Roller Skis*

Cross-country skis that are intended for use in competitions and for training in the classical style have a surface on the snow that can be divided into three zones [14]. The middle zone is waxed with grip wax (grip zone), while the front and rear zones are waxed with glide wax (glide zones). The grip zone's task is to achieve sufficient friction between the ski and the snow to enable propulsive force that comes from the kick force from the legs, in the diagonal stride, herringbone and kick double poling techniques, during the time when the ski surface is stationary on the snow (static friction). The glide zone's task is the opposite, to minimize friction, which is of interest during the time when the ski is gliding on the snow surface (dynamic friction). There is therefore a conflict and a compromise between the desire for as low friction as possible when the ski (skier) is gliding on the snow and a high friction when the ski is stationary in order to be able to apply propulsive force against the snow surface.

In order to mitigate the grip zone's negative impact on the glide, cross-country skis have a concave camber of a certain height and stiffness, so that as much as possible of the grip zone is not in contact with the snow when the ski glides against it in the gliding phase [14]. The amount of kick zone that does not have snow contact in the gliding phase mainly depends on how stiff and high the ski's camber is, according to the skier's body mass and the amount of grip wax that is applied. Cross-country skis' camber is thus constructed with different heights and stiffnesses, to enable skiers to choose them according to their body mass and technical skiing skills.

The classical style roller skis on the market have a design where one of the two wheels (one wheel at the front and one at the back) has a ratchet that allows a grip on the surface (static friction) during a leg kick (for example; PRO-SKI C2, Sterners, Dala-Järna, Sweden; Swenor Fiberglass, Sarpsborg, Norway; Marwe Classic, Hyvinkään Kumi Oy, Finland). Since this ratcheted mechanism is not dependent on a load being applied to the roller ski, in practice this type of construction provides a high static friction coefficient (μ_S) between the ratcheted wheel and the surface, independently of the skier's body mass and technical skiing skills. This is a great contrast to on-snow skiing on groomed trails, where proper technique is essential to obtain a good grip on a par with what is possible, i.e. in the vicinity of grip waxed cross country skis' μ_S.

Even if ratcheted wheel roller skis provide the opportunity to apply a relatively higher tangential force in comparison with grip-waxed cross-country skis, and thereby a higher propulsive force, until recently it was not known whether technically skilled and aware cross-country skiers were (mis)using this opportunity. A possible awareness of the problem that faulty learning on roller skis may cause with their on-snow skiing technique can make the skiers decide not to use all the potential μ_S.

Therefore, until recently, the size of the difference in μ_S between ratcheted wheel roller skis and grip-waxed cross-country skis and how the difference in μ_S affects the physiology and biomechanics of cross-country skiers was unknown. The following pages present the results of very few published research papers in this area.

5.1.4 Mechanical Experiments on Static Friction Coefficients

Classical style cross-country skis' static friction coefficients (μ_S), defined as the ratio between the tangential and normal forces acting on the ski when it is stationary on the snow, just before the ski starts gliding, have been studied by very few researchers. Ekström [14] used a force plate system attached to the skis and reported a μ_S of ~0.25 as a starting friction coefficient, before the skis started gliding, in his experiments on snow.

The μ_S used during human skiing on snow has been studied using a force plate system attached to the skis [14, 15] and by using a long force platform system mounted under the snow [16–18]. The advantages and disadvantages of the two methods are discussed by Komi and by Smith [15, 19]. As a result, μ_S of 0.1–0.2 have been reported, estimated from tangential and normal forces of 0.1–0.2 and 1–3 times bodyweight, respectively [14, 15, 17, 18].

To be able to quantify roller skis' μ_S, [20] used a fixture with a function for applying different loads normal to the surface and generating tangential traction on the roller ski. The fixture was equipped with either a bottom plate with an overlying rubber mat of the same type as used on many treadmills or a bottom plate with an asphalt surface. A reproducibility study showed an error of about 5 % for this method.

Within the studied range of vertical loads (500–1500 N), ratcheted wheel roller skis from three different manufacturers showed results of μ_S of ~0.81 for the treadmill rubber mat and ~1.05 and ~0.92 for dry and wet asphalt surfaces, respectively [20]. This was about 5–8 times the values reported from on-snow skiing with grip waxed cross-country skis [14, 16–18].

In [20] a new type of roller ski with a camber and adjustable grip function was presented (Camber-Ski). The functionality for this was applied to the forward wheel of the roller ski. When a load that was sufficient to press down the camber was exerted, the forward wheel established contact with a ratcheted spool situated above the forward wheel. The degree of grip was therefore dependent on the stiffness of the roller ski's camber, which was easily adjusted via a spring-loaded screw.

The results for the Camber-Skis showed that μ_S, with this type of construction, can be varied within a range of no grip at all up to the level of the tested ratcheted wheel roller skis [20].

5.1.5 Physiological Effects of Different Static Friction Coefficients

In order to be able to study μ_S used during roller skiing, [1] supplemented ratcheted wheel roller skis and the roller skis with a camber with a force plate measurement system located between the ski binding and the roller ski. The system measured forces in three directions, using strain gauges. Two of the coordinates, the tangential and normal forces, were analyzed in order to be able to calculate μ_S during the leg push-off phases when using roller skis with different μ_S, established in the fixture described above.

The μ_S used during roller skiing at a submaximal steady rate around race pace, using the ratcheted wheel roller skis, was about 3–4 times higher than what has been reported from on-snow skiing [1, 16–18] and $2\frac{1}{2}$ times higher than the starting friction coefficient on snow reported by Ekström [14]. Thus, the skiers clearly made use of the advantage it gives to be able to apply a greater propulsive force from the legs without slipping backwards. However, the μ_S established in the fixture for the ratcheted wheel roller skis was thus not entirely used in practical skiing, i.e. the skiers had a margin to when the roller skis would slip backwards.

Moreover, roller skiing with the Camber-Ski, with a used μ_S similar to that reported for on-snow skiing, resulted in a ~14 % higher oxygen consumption compared to roller skiing with the ratcheted wheel roller skis [1]. The explanation given for the higher oxygen consumption for a given speed and slope was that it costs energy to push down a camber to retrieve a certain grip, which is not necessary when roller skiing with the ratcheted wheel roller skis. An additional factor in the total higher energy demand was that a decreased ability to apply propulsive force from the legs, due to a lower μ_S, needs to be compensated with a corresponding increase in propulsive force coming from the ski-poles. This transition, from less leg to more upper body work, seems to be less efficient when skiing with the diagonal stride and kick double poling techniques. Further, in [1] the poor intra class correlation for oxygen consumption (skiing economy) and time to exhaustion, respectively, showed that skiers with the best economy and performance time, with roller skis with a high μ_S, did not have the best economy and performance time when using roller skis with a μ_S more similar to on-snow skiing. Thus, considering that skiing economy and efficiency is considered to be so important in the performance times for skiers, and it is apparently influenced by different μ_S, the authors raised the question of whether the detection of a skier's economy and efficiency is valid when using roller skis with a μ_S several times higher than the μ_S of skiing on snow.

In [1] there were also graded maximal tests performed by the skiers using the different μ_S. Similarly to the results using roller skis with different μ_R, maximal oxygen uptake was not influenced by different μ_S, while time to exhaustion was changed dramatically. Thus, roller skiing with the lower μ_S shortened the time to exhaustion by $\sim$30 %.

5.1.5.1 Summary

In summary, the results from mechanical and physiological experiments on μ_S clearly show a difference in μ_S between cross-country skis and ratcheted wheel roller skis and between the μ_S used when skiing with cross-country skis compared to μ_S used when skiing with ratcheted wheel roller skis. This difference in μ_S ultimately affects the power and energy demand for the skiers. To resolve this difference, roller skis of the same type as the Camber-Ski can be used for physiological and biomechanical experiments and for dryland training.

5.1.6 The Multifunctional Roller Ski

At the Science and Nordic Skiing II conference in Vuokatti, Ainegren and co-workers presented an instrumented roller ski for laboratory experiments (the multifunctional roller ski) [21]. The roller ski's base was manufactured in a titanium alloy, using additive manufacturing (3-D printing), and prepared with a socket for the easy application of a force plate measurement system, see Fig. 5.3.

In the paper, the authors describe in detail the process of calibration and control of the force plate system. The roller ski was also equipped with the adjustable grip and rolling resistance functions described above. Without the adjustable grip function

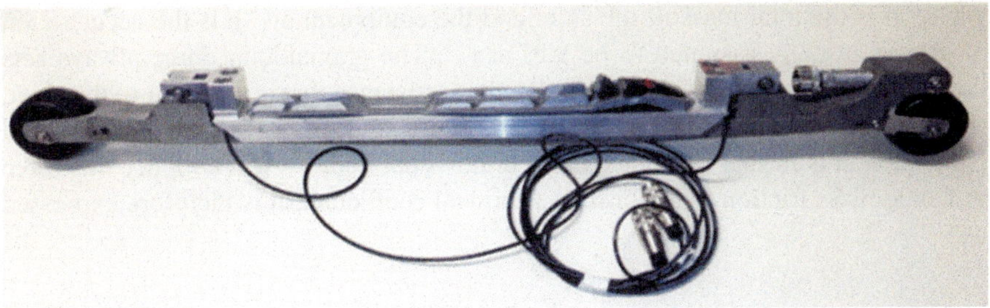

Fig. 5.3 The classical style roller ski designed for laboratory experiments [21]

and with narrower wheels found on freestyle roller skis on the market, but with the rolling resistance function and the force plate system, a roller ski for laboratory use in the free style technique is obtained. Because the multifunctional roller ski was designed for indoor use only, on treadmills, it was designed with a lower ground clearance and thereby a lower center of gravity in comparison with the roller skis currently on the market. This makes the multifunctional roller ski more stable and provides better "ground feel" compared to the roller skis currently available on the open market.

5.2 Describing the Locomotion of Cross-Country Skiing

The ski track in cross-country skiing has a groomed snow surface, which may be specially prepared for its intended use for either classical or freestyle skiing. In cross-country skiing competitions, the skier is obliged to keep to the course set up by the organizers. The outline of this course is regulated by the International Ski Federation (FIS) in the homologation manual [22]. One main principle in this set of guidelines is that the course should have a terrain distribution that is approximately one-third uphill, one-third downhill, and one-third undulating terrain. Furthermore, climbs are defined as A, B, or C climbs depending on their height difference and their gradient. The smallest climbs are 4 m in height difference, while the largest climbs should not exceed 80 m. The gradient of each climb varies from about 6 % to more than 18 %, depending on the type of climb, while climbs less than 4 m or flatter than 6 % are considered to be undulating terrain. From this, one can conclude that cross-country skiing is performed on courses with a great amount of elevation compared to other locomotive sports, such as track cycling and speed skating. Therefore, the action of gravity will have a considerable impact on the locomotion of the skier. The gravitational force is expressed as:

$$F_G = mg \tag{5.2}$$

where m is the total mass of the skier and the equipment and g is the acceleration of gravity, usually assumed to be $9.81\,\mathrm{m\,s^{-2}}$. The gravitational force always acts downwards in the vertical direction. Another resistance that is apparent to the skier, is the glide friction developed between the skis and the ski track. This physical phenomenon is further described in section 5.4 but here we will simplify this force as CoulombŠs friction with constant frictional coefficient; it is therefore expressed as:

$$F_\mu = \mu_K N \tag{5.3}$$

where μ_K is the kinetic friction coefficient and N is the normal force between the skis and the ski track. This expression may be rewritten as

$$F_\mu = \mu_K m\left(g\cos\alpha + \frac{v^2}{R}\right) \tag{5.4}$$

where α is the angle of inclination, v is the speed and R is the curvature radius in the vertical plane of the course. The second term within the brackets in Eq. (5.4) is termed the normal acceleration and derives from the curvilinear contour of the course. The kinetic friction coefficient is reported to vary between 0.02 and 0.10 [23]. The frictional force is always directed in the opposite direction to the movement.

While the skier travels along the ski track, nearly all of the skier's surface area is in contact with the ambient air. At speed, the gas molecules of air will have to move around the body to let it through. This motion will not be totally unrestricted because of the inertia of these molecules, but is more or less pressure driven at high speeds. The force exerted by this pressure is termed the aerodynamic drag. The aerodynamic drag is often expressed as:

$$F_D = \frac{1}{2} C_D A \rho v^2 \tag{5.5}$$

where C_D is the drag coefficient dependent on the form of the body, A is the projected frontal area, and ρ is the air density. The direction of the drag force is directed in the reverse direction to locomotion if there is no ambient wind affecting the skier. With an ambient wind present, the direction of aerodynamic drag is decided by the sum of relative wind vectors. In a strong tailwind, the drag force may act to propel the skier. The product of C_D and A, termed the drag area, can be measured in a wind tunnel. For a cross-country skier, the drag area varies depending on the body's posture. In downhills, where speeds are high, the skier often assumes a posture with a low drag area to decrease the aerodynamic drag. According to [24] the drag area may be reduced by as much as 58 % by assuming a semi-squatting posture rather than an upright posture.

In most circumstances, the skier has to overcome retarding forces by generating a propulsive force in the direction of travel. The propulsive force may be expressed as:

$$F_S = \frac{\dot{E}\eta}{v} \tag{5.6}$$

where $\dot{E}$ is the rate of energy expenditure and η is the mechanical efficiency of the skier. The energy expenditure is the sum of both aerobic and anaerobic energy expenditure and the mechanical efficiency is highly dependent on the sub-technique being used and the skier's technical skills. In downhills skiers are often passive, simply standing on their skis, thus producing no propulsive force. According to Newton's second law, the sum of all external forces acting on the skier will perform the acceleration of inertia associated with the skier. If the sum of all forces equals zero, no acceleration will take place. However, the skier may nevertheless travel at a considerable speed and produce a significant propulsive force to maintain that constant speed. With Newton's second law, the motion equation of the skier, in the direction of motion, is expressed as:

$$a = \frac{\dot{E}\eta}{vm} - g\sin\alpha - \mu_K\left(g\cos\alpha + \frac{v^2}{R}\right) - \frac{C_DA\rho}{2m}v^2 \tag{5.7}$$

where a is the acceleration in the direction of movement. One version of this equation of motion for the cross-country skier was introduced by Moxnes and Hausken [25]. Furthermore, they presented some mathematical solutions to this equation, based on constant propulsive power and speed dependent relationships for the propulsive force. In their study, Moxnes and Hausken showed how skiers with different body masses are affected by different inclinations and glide friction circumstances. All forces acting on the skier and the geometry of the ski track can be seen in the free-body diagram in Fig. 5.4.

5.2.1 Numerical Simulation and Optimization of the Pacing Strategy in Cross-Country Skiing

Further development in utilizing the motion equation is presented by Carlsson et al. [26], who introduced a numerical approach to simulate cross-country skiing via a numerical solution of the motion equation. This approach enables the assessment of continuously changing incline on courses and real-world course profiles. Therefore, Carlsson et al. [26] investigated the effect of combined uphills and downhills, as well as different friction coefficients and wind directions. To determine whether one or the other skier is favored, all these parameters have to be considered. However, generally a smaller skier is favored on uphills and the larger skier is favored on downhills and on the flat. This has to do with the mass scaling of the projected frontal area, which directly influences the aerodynamic drag, which in turn is the dominant resistance at high speeds. The smaller skier has a relatively larger frontal area with respect to body mass and, therefore, the aerodynamic drag has a greater

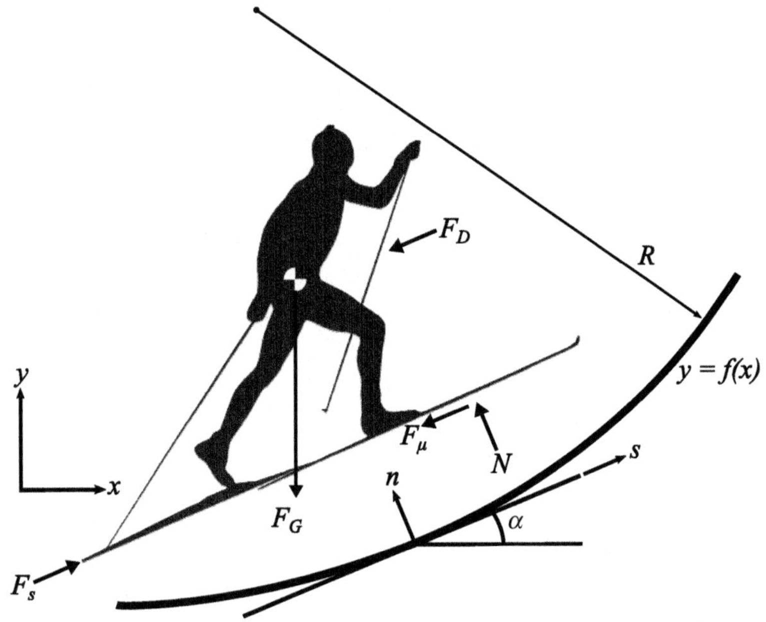

Fig. 5.4 Free-body diagram of a cross-country skier with global and natural coordinates of motion. The direction of movement is termed s and the normal direction to movement is termed n. The course profile equation is represented by $y = f(x)$ where y is the dependent variable and x is the independent variable, representing the altitude and horizontal position, respectively

effect than on a larger skier. Moreover, the smaller skier is favored in tailwinds while the opposite is true for the larger skier, who benefits more from a headwind than the smaller skier. This is also due to the scaling effects of projected frontal area.

Different snow conditions also have different effects on skiers with differing masses. The small skier is favored in slow snow conditions where the friction between skis and snow is high. This is because the low speed creates a lower aerodynamic drag which, according to the same scaling reasoning as above, is beneficial for the smaller skier. Of course, there are individual differences not mentioned here, such as the preparation of skis; this may have great importance for the friction between skis and snow, and thus also to performance. It is self-evident that even the smaller skier benefits from wearing well-prepared skis with low friction on the snow. For uphill terrain, the smaller skier is generally favored compared to the larger skier. This has nothing to do with aerodynamic drag as this resistance is marginal at the low speeds assumed in uphill terrain, but rather the gravitational force exerted on the skier. For skiers in the same category, elite for instance, the smaller skier tends to have a higher oxygen uptake to body mass ratio. This enables a higher rate of energy expenditure and thus higher power output to body mass ratio, which is beneficial when the major restriction derives from gravity. Moxnes et al. [27] further investigated the locomotion of cross-country skiing by using a similar simulation model. They fitted their model to experimental data and compared simulations to the experimental trial. Furthermore, sensitivity analyses showed the effect of varying each parameter, which is helpful to both athletes and coaches. According to the results of Moxnes et al. [27], the locomotive power and the body mass have the greatest effect on the finishing time for the specific course simulated in their study.

The rate of energy expenditure in the skier's bioenergetic system is constrained so that the athlete avoids exhaustion. Therefore, the power output generated through the skis and poles is also restricted. Because of the restriction on power output, the skier is forced to vary speed when travelling a hilly course. The way the skier varies speed along the course is termed the pacing strategy. Sundström et al. [28] introduced a numerical model that was able to optimize the pacing strategy so that the time between start and finish was minimized. This approach relies on the solution of the motion equation in Eq. (5.7) and a model that constrains the bioenergetics of the athlete. Sundström et al. [28] utilized a simple average power output constraint that does not distinguish between power produced from different substrates. However, more sophisticated bioenergetic models, such as the Margaria-Morton model [29], have been used for similar pacing strategy optimizations in road cycling [30]. Although not thoroughly validated, this more sophisticated bioenergetic model may be more realistic and unarguably further restricts the variation of power output compared to the average power constraint.

According to the study by Sundström et al. [28], the advantage of utilizing a optimized pacing strategy on a specific sprint course is 6.5 % in finishing time compared to a more or less even power output distribution. Judging by the above reasoning regarding bioenergetic models, the real advantage would be smaller. However, according to [30], numerical results with the M-M model still showed time gains of 3.6 % compared with a constant power strategy on a hilly course in road cycling. This is a significant improvement even in elite level cross-country skiing, where the winning margin is frequently below 1 %.

5.3 Optimization of an Early Skating Ski

This section mainly refers to: [31–35].

This part shows how very old and traditional constructions can be improved through the use of modern technology. In this case the old construction is a cross-country ski. The oldest such ski to have been found is approximately 5000 years old and was discovered in a swamp in Västerbotten in northern Sweden. According to early ideas, the goal of optimization was to obtain the best geometry for a skating ski, but these ideas are rather different today. However, the way of working shown here is general and, no matter the equipment or the purpose, optimization is a powerful design tool better equipment performance.

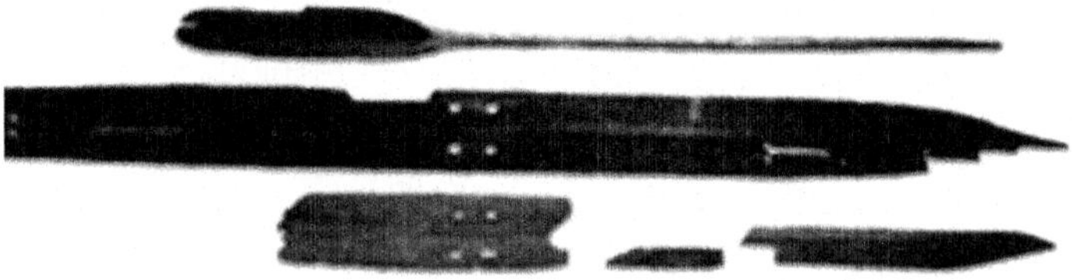

Fig. 5.5 The famous Kalvträsk skis with ski pole

Fig. 5.6 Early man on skies, from Rödöy engraving

5.3.1 Introduction

In countries with snow during the winter, different cultures developed different means for travel across open country. In northern America and Canada, snow shoeing was the most common method, while skis were frequently used in northern Scandinavia.

The earliest known ski is dated at 3500 BCE and is known as the "Kalvträskskidan." These skis were made of wood and were found in Västerbotten, Sweden. They were 2040 mm long and 155 mm wide (Fig. 5.5).

A rock engraving from the same period found in Rödöy, Norway, shows an early man skiing (Fig. 5.6).

Skiing is now mainly used for sport and exercise. Modern skiing with well-prepared tracks (smooth and hard) is the reason why cross-country ski racing is divided into two categories: classical and skating. These categories place different demands on the properties of the ski.

A ski for classic skiing requires a combination of good running, good grip, and low weight. Since the first two requirements are opposites, skis must be chosen for every skier individually. The problem here is to control the distribution of pressure during the different phases of the skiing cycle.

A skating ski doesn't need to have grip, only good running. Despite this, the geometry is the same as that of the classical ski, where the span in the middle of the ski is the grip zone. Grip in the span zone is, however, not relevant on a skating ski; according to Erik Ludvigsson[1] the requirements for this type of ski should instead be high stiffness combined with low weight. Ludvigsson has even proposed that the ski should be flat, i.e. without the span.

[1] Former head of Tegnässkidan AB, Granö, Sweden.

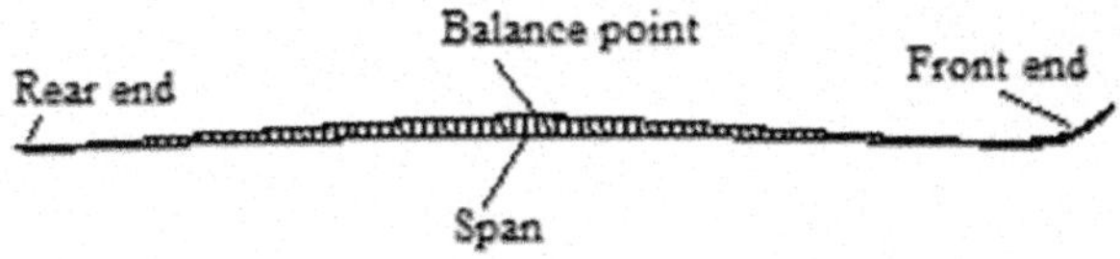

Fig. 5.7 Parts of a modern ski

Traditionally, skis were made of wood. Development of new materials began in the mid-1970s and today skis have an advanced composite structure, built up from carbon fiber, reinforced plastic, a honeycomb core, etc.

5.3.2 Problem Definition

5.3.2.1 Traditional Ski

The basic idea of a traditional ski is to reduce local pressure to the snow to make it possible to glide on the surface. This could be achieved with two flat boards, but in order to reduce the energy required and increase speed, even early skis had curved front ends. The skier's foot was placed at balance or c.g. point, i.e. the balance point. This means that the ski remains in balance even when the skier's foot is lifted. Effective skiing requires good glide forward and good grip backward. This problem has been solved in many different ways. An old technique, still in use, is to mount fur, bristle, or similar materials on the underside of the ski. Modern skiing with ski wax and well-prepared tracks has modified the geometry of the ski. Today's skis are manufactured with an elevated span in the middle, where the grip wax is placed. The glide wax is placed on the other parts. As long as the skier is standing on both skis, the span is not in contact with the snow surface. As soon as the skier stands on one ski, the span (and the grip wax) comes into contact with the surface (Fig. 5.7).

5.3.2.2 The Early Skating Ski

The early skating ski derived its geometry from the modern ski described above. This means that a traditional skating ski has a span despite having no function for the grip. Early experiences showed that a stiffer construction is better for the skating technique (Fig. 5.8).

5.3.2.3 The New Design of the Skating Ski

Erik Ludvigsson, together with Swedish Olympic skier Christer Majbäck, worked for many years on the development of a better skating ski. In their early results, they believed that a good skating ski should have the following properties:

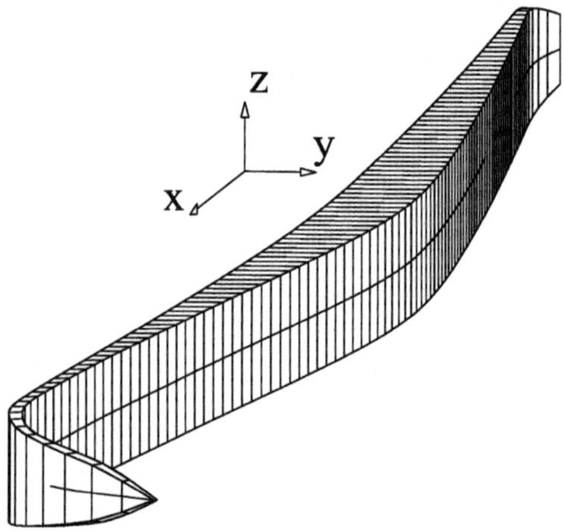

Fig. 5.8 3D view of the skating ski

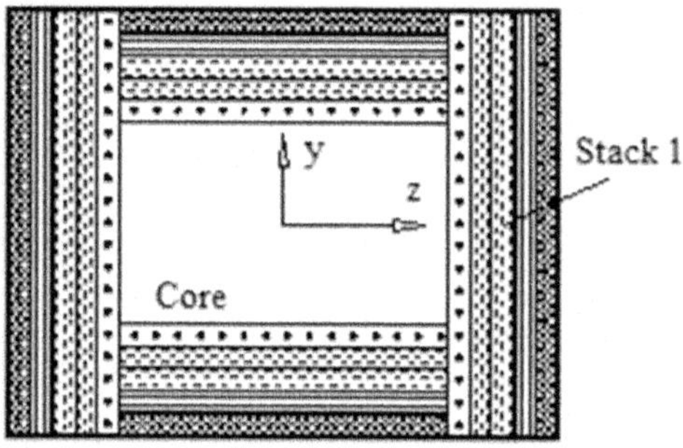

Fig. 5.9 Cross section of the ski

- High stiffness
- Initially flat (i.e., no span)
- Soft front end
- Low weight

The ski must fulfil the FIS rules for skis, and must have a geometry which doesn't disturb the skier's action. The prototypes were made using trial and error, an expensive and time-consuming method.

5.3.2.4 Material Configuration

The design problem was studied as an optimization problem. However, making a ski that is initially flat with a soft front end is not difficult, but the problem is to combine high stiffness with low weight.

The model was built up of a lightweight core surrounded with fiber layers oriented in specific directions.

A honeycomb material was selected for the core. The fiber material was assembled in stacks as shown in Fig. 5.9. The stacks were numbered c.c.w from

Table 5.1 Material properties in the core

Material no	E (MPa)	G (MPa)	ν	ρ (kg/m^3)
1	6.76	2.6	0.3	29

Table 5.2 Material properties in layers

Material no	E_1 (MPa)	E_2 (MPa)	G_{12} (MPA)	ν_{12}	ν_{12}	ρ (kg/m^3)
1	72,000	7200	3600	0.30	0.03	1700
2	82,000	6640	2540	0.21	0.017	1700
3	144,800	11,720	4482	0.21	0.017	1550

Table 5.3 Fiber orientation (all stacks)

	Layer 1	Layer 2	Layer 3	Layer 4	Layer 5
β	0	90	+45	−45	Isotropic

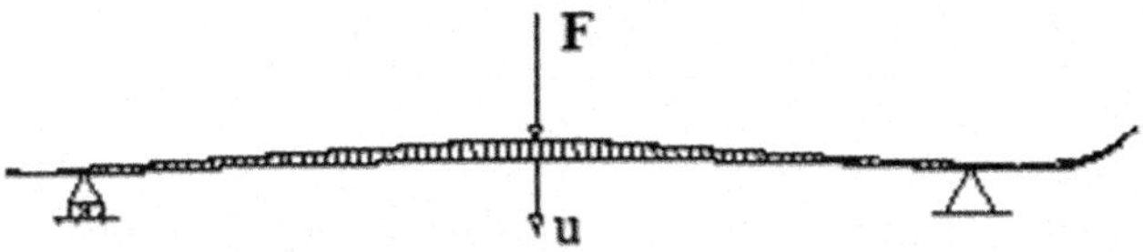

Fig. 5.10 Load case

1 to 4, which means that stacks 1 and 3 are the sides of the ski and stacks 2 and 4 are the top and bottom, respectively. Each stack consists of five layers of different materials, numbered from 1 to 5, starting from the core. In each of the first four layers the fibers have different orientations. The purpose of the fifth layer in stacks 1 to 3 is solely to give an attractive surface. In the fourth stack, the fifth layer is a running layer (Table 5.3).

The orientation of the fiber is given by an angle between the projected x-axis on the actual layer plane and the first direction in the fiber material. This angle is henceforth called β. The fiber orientation of each layer is the same in all stacks. For example, the fiber direction in layer 1 is zero degrees in all stacks (Table 5.3).

5.3.2.5 Load Case

The load was applied at the balance point of the ski, which was supported as shown in Fig. 5.10. The applied load F is given the value 200 N, u is the vertical displacement of the balance point.

5.3.2.6 Optimization Problem, Objective Function, and Variables

Optimization can be described as the act of obtaining the best result in the given circumstances. The origin of optimization methods can be traced back to the days of Newton, Lagrange, and Cauchy, but great progress in practical methods started with the development of modern, high-speed computers. In its broadest sense, optimization can be applied to solve almost any engineering problem.

In highly competitive contemporary world, it is not usually enough to simply develop an acceptable design; it is often necessary to design the best solution. Furthermore, increased concern for the environment raises questions like: Are we making the most effective use of our limited resources? Can we obtain a more economical design? Are we taking risks within acceptable limits? Mathematical optimization methods offer some of the tools necessary to answer such questions.

Some typical applications for optimization from different engineering disciplines are provided below.

- Design of aircraft and aerospace structures for minimum weight.
- Design of structures like buildings, bridges, etc.
- Design of chemical reactors.
- Design of pumps, turbines, and heat transfer equipment for maximum efficiency.
- Shortest route taken by a salesperson visiting various cities on one trip.
- Planning maintenance and the replacement of equipment to reduce operating costs.
- Design of electrical networks.

In this case, optimization is about a skating ski, and the optimization problem was formulated as:

$$\begin{aligned} &\text{Min } u(x, y, z) \\ &\text{Subject to } w(x, y, z) \leq \bar{w} \end{aligned} \tag{5.8}$$

where u is the deflection and w is the weight of the ski. The geometric variables were divided into two groups, outer geometry variables and layer thickness variables. The 12 variables for the outer geometry could vary in both the y- and z-directions, with respect to FIS regulations. The 36 layer thickness variables could vary between 0.01 and 5.0 mm.

5.3.2.7 Optimization

OASIS software for structural optimization was used in combination with the ALADDIN pre/postprocessor, see [32]. The system handles various kinds of design variables such as shape, thickness (also for composites), material orientation (fiber orientation), and material properties (material selection). Stresses, displacements, Eigen frequencies, weight and mass moment of inertia can be used as objective/constraint functions. The optimization problem was solved iteratively, and in each iteration an approximate subproblem was created and solved. The solutions to the sequence of subproblems usually converge within 10 iterations. The convergence speed, in terms of the number of iterations, is not dependent on the number of design variables or objective/constraint functions. However, the convergence is strongly dependent on the nonlinearity of these functions. Each subproblem was made convex and separable, using first-order derivatives, i.e. gradients, of the selected response functions. The computation cost for each iteration is approximately linear in the number of design variables and response functions.

5.3.3 *Results*

The deflection of a reference ski (one of Christer Majbäck's racing skis) loaded as in 5.3.2.5 was measured to 92 mm. The weight limit $\bar{w}$ was given the same value as the weight of the reference ski. With different types of material configurations, the results were as follows.

5.3.3.1 Material Configuration I

Material 2 in layers 1–4 and material 1 in layer 5 in all stacks (Table 5.4).
Deflection: $u = 17$ mm

5.3.3.2 Material Configuration II

Material 2 in layers 1–4 in stacks 1–3, material 3 in layers 1–4 in stack 4 and material 1 in layer 5 in all stacks (Table 5.5).
Deflection: $u = 12$ mm

Table 5.4 Layer thickness [mm] and fiber angle β [degrees] for different layers in different parts of the ski (see Table 5.3) where $\Delta 1$ means the x-value 0–500 [mm], $\Delta 2$ means 500–800 [mm], and $\Delta 3$ means 800–1800 [mm]

	Stack 1, 3			Stack 2			Stack 4		
β	$\Delta 1$	$\Delta 2$	$\Delta 3$	$\Delta 1$	$\Delta 2$	$\Delta 3$	$\Delta 1$	$\Delta 2$	$\Delta 3$
0	0.04	0.02	0.02	1.24	1.42	0.74	0.78	1.40	0.64
90	0	0	0	0	0	0	0	0	0
+45	0.18	0	0.01	0	0	0	0.02	0	0
−45	0.13	0	0.03	0	0	0	0	0	0

Table 5.5 Layer thickness [mm] and fiber angle β [degrees] for different layers in different parts of the ski (see Table 5.3) where $\Delta 1$ means the x-value 0–500 [mm], $\Delta 2$ means 500–800 [mm] and $\Delta 3$ means 800–1800 [mm]

	Stack 1,3			Stack 2			Stack 4		
β	$\Delta 1$	$\Delta 2$	$\Delta 3$	$\Delta 1$	$\Delta 2$	$\Delta 3$	$\Delta 1$	$\Delta 2$	$\Delta 3$
0	0.24	0.13	0.20	0.94	1.25	0.94	0.66	1.21	0.83
90	0	0	0	0	0	0	0	0	0
+45	0.13	0	0.01	0	0	0	0	0	0
−45	0.05	0	0.05	0	0	0	0	0	0

Table 5.6 Layer thickness [mm] and fiber angle β [degrees] for different layers in different parts of the ski (see Table 5.3) where $\Delta 1$ means the x-value 0–500 [mm], $\Delta 2$ means 500–800 [mm] and $\Delta 3$ means 800–1800 [mm]

	Stack 1,3			Stack 2			Stack 4		
β	$\Delta 1$	$\Delta 2$	$\Delta 3$	$\Delta 1$	$\Delta 2$	$\Delta 3$	$\Delta 1$	$\Delta 2$	$\Delta 3$
0	0.05	0.02	0.02	1.24	1.48	1.15	0.77	1.39	0.96
90	0	0	0	0	0	0	0	0	0
+45	0.13	0	0.03	0	0	0	0	0	0
−45	0.07	0	0.05	0	0	0	0	0	0

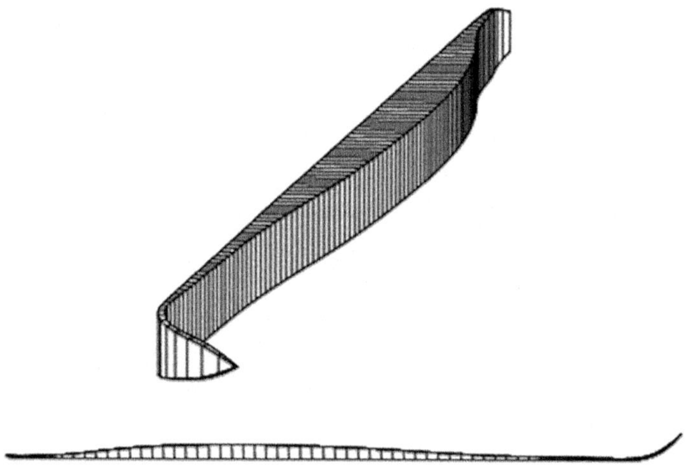

Fig. 5.11 Stylized figure of the new ski

5.3.3.3 Material Configuration III

Material 3 in layers 1–4 and material 1 in layer 5 in all stacks (Table 5.6).

Deflection: $u = 9\,\text{mm}$

The geometry variables received values according to Fig. 5.11.

5.3.4 Discussion

With no increase in weight, deflection was reduced from 92 to 9 mm with material configuration III. This means an increase in stiffness by approximately a factor of 10, and according to Christer Majbäck the overall action of the prototype was much better than that of the early skating ski.

The result shows that the layer thickness in stacks 1 and 3 (i.e., the sides of the ski) is so small, that is has no practical importance. Although the prototype was made without a fiber layer on the sides, the torsional stiffness was sufficiently high. The thickness of the layers in stacks 2 and 4 (the top and the bottom of the ski) was adapted to commercially available fiber materials.

In the model, the geometry was allowed to vary in y- and z-directions. While the optimum geometry shows that the ski should have a top with variable z-coordinates, manufacturing restraints still only permit constant thickness in the z-direction.

The evolution of skating skis went a different way to that anticipated when these calculations were carried out. The construction methods that are reviewed in these pages, however, are also applicable to other design requirements for skis; for example, optimization might be performed for the desired pressure distribution under the gliding surfaces, desired stiffness in the span, etc.

In general, any problem in which a certain parameter, or variable, must be chosen to satisfy constraints may be formulated and solved as an optimization problem. A proper problem formulation is essential for the success of the optimization process, but once a problem has been transcribed into the standard model they all look the same. This means that the same solution strategies can be applied to a great number of design problems.

5.4 Ski and Snowboard Friction

Frictional forces play a critical role in the ski and snowboard gliding. The ambition to decrease friction and improve gliding efficiency strongly impacts the construction and design of the corresponding equipment. Despite significant progress in research into the gliding friction with snow and ice many of the critical issues still remain unresolved. Even though the system may seem rather simple the processes and interactions involved are quite complex. Significant variations of the environmental and dynamic conditions in skiing and snowboarding also complicate the analysis and modeling of the gliding friction phenomenon.

The most common relation describing the dependence of the friction force upon the loading for the two planar objects commonly used in the analysis and modeling of gliding is (e.g., [14, 36–45]):

$$F_{\mathrm{Fr}} = \mu F_N \tag{5.9}$$

where F_{Fr} is the longitudinal (along the surface) friction force, μ is the friction coefficient, and F_N is the component of the loading force normal to the surfaces. But with more careful analysis certain complications start to arrive.

5.4.1 Static and Dynamic Friction

Noticeably, the relation between the friction force and loading cannot be truly described by a continuous function. In static conditions, when the surfaces do not yet slide one over another the system is in the equilibrium, and the force resisting sliding F_R is equal and opposite to the longitudinal pulling force F_L. At some point when the pulling force exceeds a threshold value F_{st} the sliding abruptly starts and the resisting force stabilizes (in the first approximation) at some lower value F_d. This is representing a characteristic stick-slip behavior with two different (static and dynamic) friction force values (e.g., [4, 36, 46–49]).

It is also known that in the dynamic (gliding) regime the dependence of the friction force upon the normal loading force is not linear. It has been experimentally proven that the friction of solid bodies with snow and ice that (for certain environmental conditions and for certain, not very large loading force values) may be approximated by the fitted equation(see, for example, [14, 40–44], see Fig. 5.12):

$$\mu = \mu_0 - \mu_x F_N \text{ or } F_{Fr} = \mu_0 F_N - \mu_x F_N^2 \tag{5.10}$$

In fact the nature of this nonlinearity may be much more complex, and possibly we just approximate it by the first terms of the polynomial expansion series:

$$F_{Fr} = \sum_{i=1}^{\infty} k_i F_N^i \tag{5.11}$$

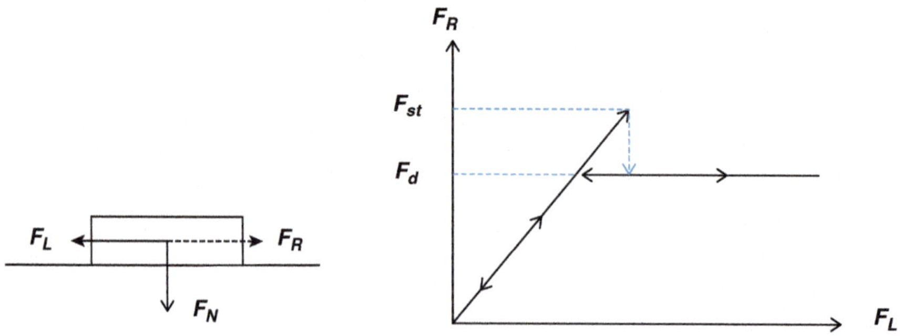

Fig. 5.12 Characteristic stick-slip behavior with two different (static and dynamic) friction force values

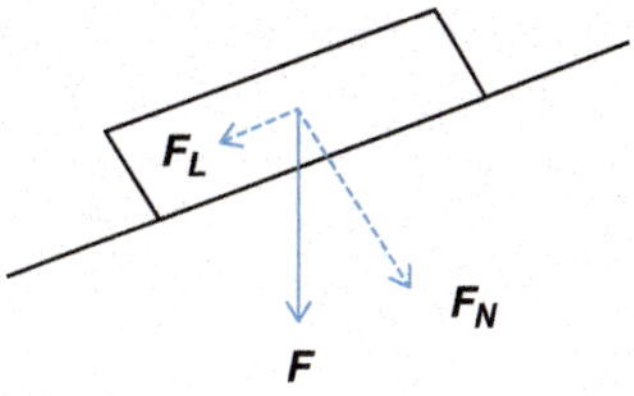

Fig. 5.13 Object sliding under its own weight over the inclined surface

Both linear [Eq. (5.9)] and second-order [Eq. (5.10)] approximations are quite useful, but as we will see in some cases they may not be good enough.

5.4.2 *Gliding Speed Effects*

The friction force between snow and solid objects shows clear dependence upon the gliding speed (see, for example, [4, 39, 42–44, 50–56]). The basic equation for the gliding of the object, sliding under its own weight over the inclined surface (in the case when one can neglect the air resistance, see Fig. 5.13) can be written as [36]:

$$F_L - F_{\mathrm{Fr}} = m\ddot{x}, \; F_L = mg \cdot \sin\alpha, \; F_{\mathrm{Fr}} = \mu \cdot mg \cdot \cos\alpha \tag{5.12}$$

where m is the mass of the slider, g is the free fall acceleration, α is the inclination angle, and x is the coordinate along the sliding surface. For the friction coefficient dependent on gliding speed this equation may be re-written as

$$\frac{1}{g}\ddot{x} + \mu(\dot{x})\cos\alpha - \sin\alpha = 0 \tag{5.13}$$

and the steady gliding condition will be given by $\dot{x} = \text{const}$ or $\ddot{x} = 0$, yielding a simple steady gliding condition

$$\mu(\dot{x}) = \tan\alpha \tag{5.14}$$

Interestingly, as we will see further, this simple equation does not have reasonable solutions for certain accepted dependencies of the friction coefficient upon the gliding speed.

The situation is complicated further by the impact of varying environmental parameters upon the gliding friction. For example, Fig. 5.14 presents the functional dependencies of the friction coefficient upon the gliding speed for different snow temperatures (analytical dependencies using the fit to experimental data are either taken from the literature or fitted to the data from the literature, [39, 40, 42, 43]).

Figure 5.15 presents the plots for the analytical solutions of Eq. (5.14), where the gliding speed dependencies of the friction coefficient is represented by the fitting equations, corresponding to the data from Fig. 5.14 (basing on the experimental data from [39, 40, 42, 43]).

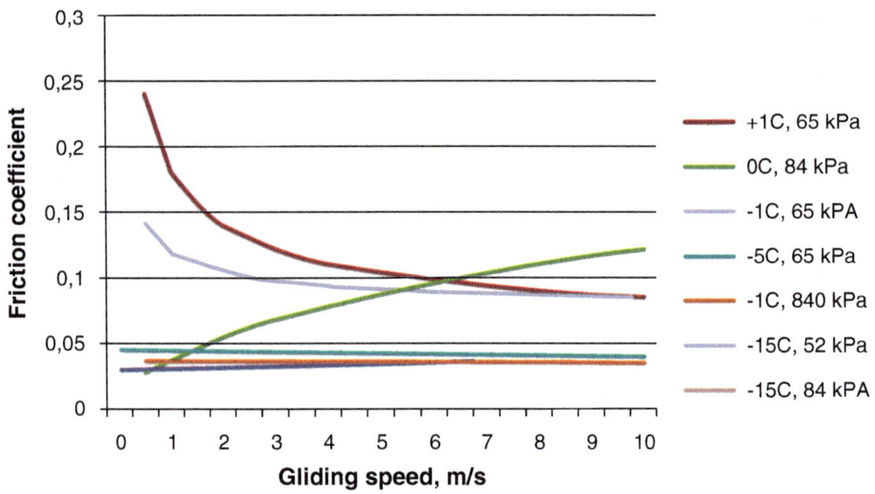

Fig. 5.14 Functional dependencies of the friction coefficient upon the gliding speed for different snow temperatures

It is easy to see that for the realistic slope angles (10–30 °C) the steady gliding speeds taken as the solutions for Eq. (5.14) are generally realistic for the temperatures below −5 °C. These values are some higher than the experimentally measured gliding speeds for the cross country skis (see, for example, [24, 40, 57, 58]). But in the discussed simple approach other factors like the air resistance were not taken into account. Also the fitting equations are based upon the experimental data for the gliding velocities not exceeding 10–20 m/s and projecting the solutions for the higher gliding speeds may be too ambiguous.

It is interesting to note that basing upon the experimentally measured dependence of the friction coefficient upon gliding velocity for the near-melting temperatures (from [39, 40, 42, 43]) solutions of Eq. (5.14) together with the analytical expressions fitting the gliding speed dependence of friction coefficient (from Fig. 5.14) yield steady gliding speeds as high as 50–100 m/s even for very small slope angles (Fig. 5.15). Of course these estimates are only valid for the system not affected by air resistance for example, when the gliding of the skis with large weight over them is tested. But still, our experience with the cross country skiing tells that free gliding at near melting temperatures is often much less effective than for the cold conditions, and steady gliding speeds should be lower, not higher.

One of the first suggestions is that possibly there exists another "rising" branch in the dependence of the friction coefficient on the gliding speed at low speed values. And indeed, experimental friction coefficient dependencies measured for the low gliding speeds (microcrystalline ice on glass and granite) do show such "rising branch" for the speeds below 1 m/s ([36], Fig. 14.7, [47]). Figure 5.16 illustrates the situation for the friction coefficient steadily growing with the gliding speed (similar to the experimentally measured dependence at 0 °C in Fig. 5.14), and for the case

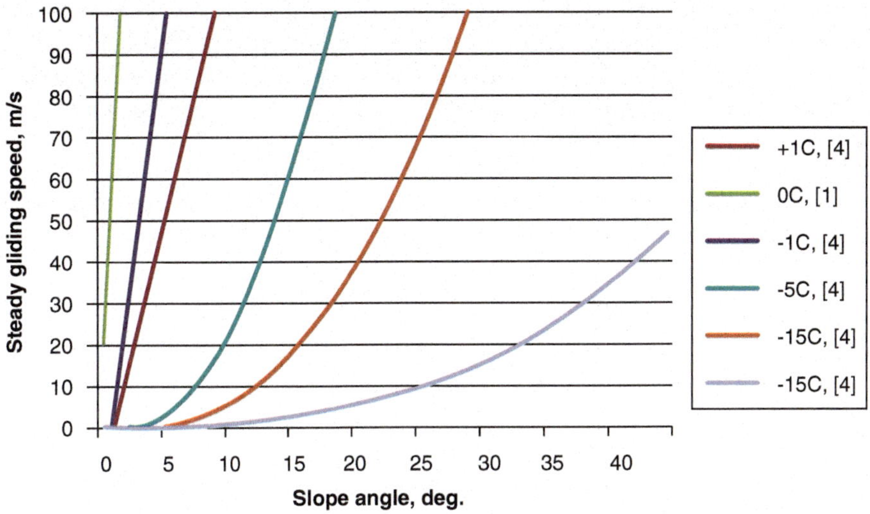

Fig. 5.15 Analytical solutions of Eq. (5.14) based upon functional dependences of friction coefficient (see Fig. 5.14)

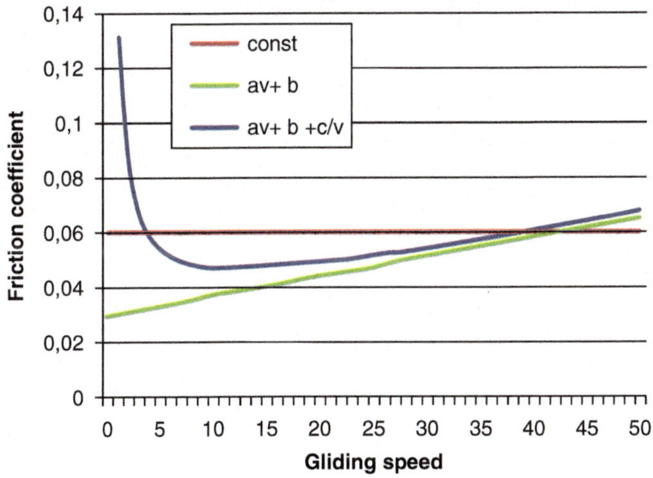

Fig. 5.16 Friction coefficient dependence on the gliding speed corrected for the low speed case

when additional "rising branch" of this dependence is present at low gliding speed values. In both of the cases there exists a solution at similarly high gliding speeds, but for the "corrected" dependence there exists another solution at low gliding speeds. The solutions of Eq. (5.14) in this case are given by the intersection of the curves $y = \mu(v)$ and $y = \text{const}$.

Unfortunately such solutions with the corrected friction coefficient dependencies yield the steady gliding speed values that are too low to be realistic. Cross country skiing experience indicates that steady gliding speeds at near melting temperatures are lower than for the cold conditions, but not as dramatically lower as such "corrected" model predicts. Also, none of the experimental dependencies of the friction coefficient on the gliding speed measured experimentally [36, 39] show any indications of presence of such "rising" branch for the speed values down to 0.5–1 m/s.

When air resistance is taken into account calculated steady gliding velocity values become some more realistic. To account for the air resistance forces (air drag) Eq. (5.12) should now be re-written as:

$$F_L - F_{\mathrm{Fr}} - F_D = m\ddot{x},\ F_L = mg \cdot \sin\alpha,\ F_{\mathrm{Fr}} = \mu \cdot mg \cdot \cos\alpha \tag{5.15}$$

and air drag force is

$$F_D = \frac{1}{2} C_D A \rho \dot{x}^2 \tag{5.16}$$

or using the notation $\dot{x} = v$

$$mg \cdot \sin\alpha = \mu(v) \cdot mg \cdot \cos\alpha + \frac{1}{2} C_D A \rho v^2 \tag{5.17}$$

Instead of solving Eq. (5.17) for all fitted analytical curves $\mu(v)$ [as it was done earlier for much simpler Eq. (5.14)] one can use simplified approach and use graphical way of solving (using corresponding plots). Figures 5.17 and 5.18 illustrate such solutions for the partially tuck position in the cases of the friction coefficient dependencies on gliding velocity fitted from experimental data (from [39, 40, 42, 43]). Normal loading force is 65 kg for 10° slope case and 69 kg for

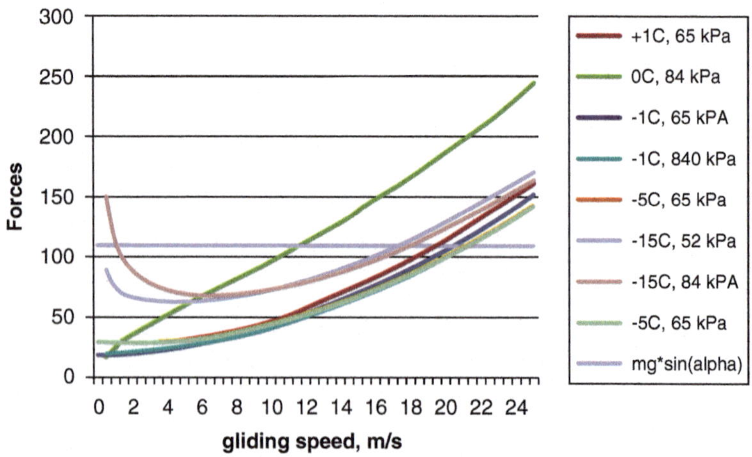

Fig. 5.17 Friction coefficient dependencies on gliding velocity fitted from experimental data (10° slope)

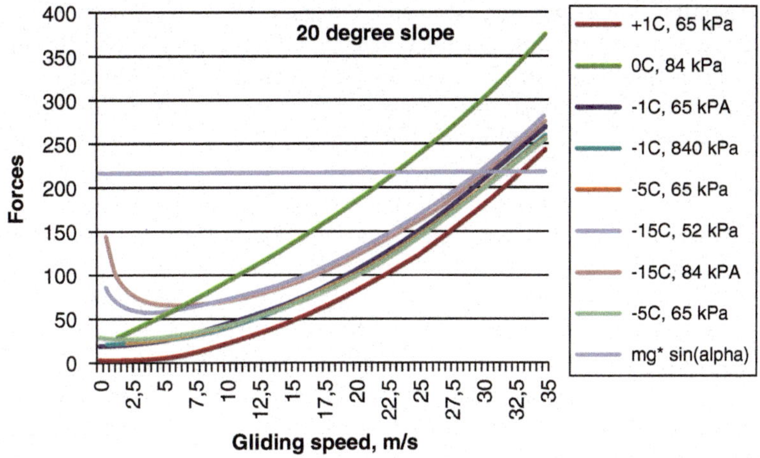

Fig. 5.18 Friction coefficient dependencies on gliding velocity fitted from experimental data (20° slope)

Table 5.7 Estimated gliding speed values

Temperature (°C)	Slope angle (°)	Upright position (m/s)	Partially-tuck position (m/s)
0	10	10	11
−5	10	15	20
−15	10	12	18
0	20	17	23
−5	20	22	32
−15	20	21	30

20° slope case. Table 5.7 summarizes the values for the steady gliding velocity estimated using this approach. Slope angle of 10° corresponds to about 18 % and of 20° to about 36 % slope inclination correspondingly.

Above estimates indicate that for used friction coefficient dependencies on gliding speed at low temperatures the air drag is responsible for about 30–40 % of the overall braking force (air drag plus friction) for the partially tuck position and about 60 % for the upright position of the skier gliding at 10 m/s on the slope when his weight produces for 65 kg normal loading. For the near-melting snow conditions according to this model air resistance produces more than 80 % of the breaking force on these mild slopes.

Thus if we assume that the braking forces in gliding are only due to the friction and air drag than according to the above estimates reasonable values of steady gliding velocity for near-melting snow conditions are due entirely to the presence of the air drag. This means that even smallest improvements in the skier position and streamlining of the clothing should dramatically improve cross country skier gliding efficiency at near melting temperatures. For example, terminal velocity (steady state value when the sum of the drag force and buoyancy equals the gravity) of the sky diver in the "belly down" position is estimated to be just above 50 m/s [24, 59]. And the current world record speed for the sky diving shown in 2013 is about 150 m/s [24]. Thus air resistance mechanisms though significant cannot be solely responsible for the braking forces leading to the lower steady gliding speeds at near melting temperatures.

The same time poor gliding of the cross country skis at near melting conditions is experienced by both elite athletes and amateur skiers. Thus one should possibly seek for other mechanisms strongly decreasing gliding efficiency at near melting snow conditions strongly contributing to the ski motion. One of the feasible effects may be due to energy loss into the non-elastic deformation of the snow caused by its compression in the direction normal to the ski track by the passing skis. Though significant amount of experimental data on the snow compression and its elastic properties both in the laboratory and in the field is available (see, for example, [42, 56, 60–62]), and some studies were made into the acoustic wave dumping in the snow at different environmental conditions [63, 64] and researchers were already interested in the non-elastic properties of the snow in relation to gliding [65], there are no models that can estimate how such effects can influence the ski and snowboard gliding.

Another effect can be caused by significantly larger snow displacement along the ski track at near-melting conditions. This may have multiple reasons. Firstly, due to the larger sinking of the skis into easier compressible (wet) snow a force component along the ski track caused by the angled profile of the front of the ski will cause the displacement of the snow in the direction of the gliding (micro-plowing effect). Secondly, there is a friction force between the ski running surface and the snow. In the case of near-melting conditions this force will cause non-elastic deformation of the snow in the direction of gliding. The snow displacement idea is not new. Early attempts to prove the concept that "better gliding skis produce less snow displacement" using accelerometers placed in the snow just under the ski track surface were not successful (though they have shown the presence of another interesting effect [57]). Again, significant amount of research into the gliding at near-melting conditions has been conducted and certain models of wet friction exist (see, for example, [36, 37, 51, 66–69]) but so far it is not possible to prove or disprove the suggestion that snow displacement by the skis can influence the gliding friction.

5.4.3 Friction Dependence on the Loading Pressure

It is commonly accepted that the experimentally measured dependencies of the friction coefficient are due to the changing snow structure and interplay of different friction mechanisms. It is supposed that for the mildly cold conditions gliding happens in presence of thin water film appearing due to the partial ice crystals melting in the area of contact (see, for example, [36, 50, 70, 71]). The exact

Table 5.8 Comparison of experimentally measured friction coefficient values reported in the literature

$\mu_0, \times 10^{-3}$	α, % of μ_0 for T	P (kPa)	T (°C)	v (m/s)	Sliding bodies	Reference
8–38	0.5 @−1 to 1.3 @−15 °C	89–400	−1 to −15	0.5–3	Wood on ice	[39], Table 2
33–80	3 @−5 to 0.4 @−14 °C	1.25–50	−5 to −14	3.6	Skis on snow	[42], Plot 5, 6, Fig 5.2.5
110–300	0.02 @−14 to 0.04 @−5 °C	25–100	−6	0.01–0.2	Skis on snow	[40], Fig. 20
60–83	7.6 @−7 °C	1500–3000	−7	1	PTFE on ice	[43], Fig. 1–10B
9.5–38	0.56 @−1 to 1.36 @−15 °C	870–3900	−1 to −15	0.5	Ice on ice	[43], Fig. 1–10A
40–155	2.8 @−20 °C, 1.7 @−10 °C, 0 @−5 °C,	50–200	−5 to −20	5	Ski on snow	[41], Fig. 2
6.3–120	0.6 @−5–10 to 1.8 @−15 °C	50–300	−5 to −15	5	Skis on snow	[41], Fig. 3
5–9	0.045 @4 to 0.02 @8 °C	?	4–8	1	Ice, hockey blade	[72], Fig. 7
30–100	−71 %*(!)	Up to 1940	0	4	Ice on ice	[38], Fig.2
6–36	4.7 @−27 to 28 @−3.3 °C	Up to 1940	−3.3, −27	4	Ice on ice	[38], Fig. 2

*In this case friction coefficient was growing with increasing loading force
? indicates that no data were available

mechanisms of such melting are still debated, but they are most probably related to the compression and braking of the snow and ice crystals happening in the interface area. With much colder conditions this mechanism becomes less effective, because it is supposed that the heat generated in the interface is no longer able to support the effective melting. With the temperatures increasing and nearing the snow melting point the interface water film becomes much thicker, and the hydrodynamic friction mechanisms start to play a major role. This model is also supported by the experimentally observed dependence of the friction coefficient upon the loading [38–40, 42–44, 56, 72] [also see Eq. (5.10)]. This equation can be also re-written as

$$F_{\mathrm{Fr}} = \mu_0(F_N - \alpha F_N^2) = F_N\mu_0(1 - \alpha F_N) = F_N\mu_x \tag{5.18}$$

Table 5.8 presents the comparison of the corresponding experimentally measured values [in the notation of Eq. (5.18)] reported in the literature.

It should be noted here that for the temperatures close to melting point the pressure dependence dramatically changes, and the friction coefficient starts to grow with increasing pressure (coefficient α in Eq. (5.18) becomes negative). This is in

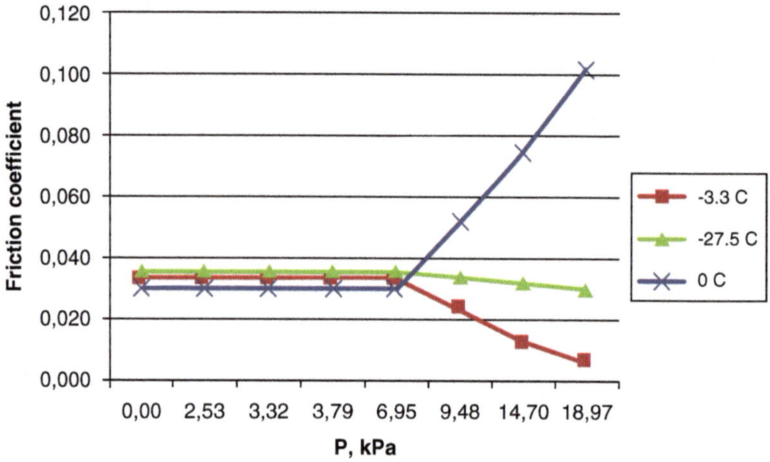

Fig. 5.19 Friction coefficient dependence on the pressure and temperature (adapted from [38])

certain agreement with the generalized model, as the increasing amount of water in the case of hydrodynamic friction would only make the braking forces even larger. See also Fig. 5.19 (adapted from [38]).

5.4.4 Gliding, Friction and Vibration

Another significant influence upon the gliding efficiency is due to the vibrations of the sliding bodies. It is known that average friction forces are often becoming smaller in presence of vibrations [57, 73–80]. Thus it can have significant effect upon the strategy of designing and manufacturing of skis and snowboards. Simple modeling allows one to clarify certain critical issues related to the influence of vibrations upon gliding friction.

If we assume that vibrations can significantly change the dynamic loading force, then basing upon Eq. (5.10) and general thoughts about the nonlinearity of this relation the expressions for the instantaneous values of the normal loading and friction force can be written as:

$$F_N = F_{N0}(1 + k \cdot \sin(\omega t)) \tag{5.19}$$

where the nonlinearity contribution coefficient k reflects the level of vibrations relative to the static loading force. So the equation for the friction force can be re-written as:

$$F_{\mathrm{Fr}} = \mu_0 \cdot F_N \pm \alpha \cdot \mu_0 \cdot F_N^x \tag{5.20}$$

Average friction force then will be given by the time integral over the vibration period:

$$\bar{F_{\mathrm{Fr}}} = \int_0^{2\pi} (\mu_0 \cdot F_N \pm \alpha \cdot \mu_0 \cdot F_N^x) d(\omega t) \tag{5.21}$$

Equations (5.20) and (5.21) are deliberately written in the generalized form, reflecting the loading dependencies for the cold conditions (negative nonlinear contribution) and near-melting temperatures (positive nonlinear contribution). To make the description more general we have also introduced the power parameter ("nonlinearity index") x. With the experimental data acquired so far x nears 2 for the cold temperatures, and varies from -0.5 to $+0.5$ for the near-melting conditions.

Instantaneous and average friction for the specific case of friction coefficient dependence measured in experiments for the cold conditions will then be correspondingly:

$$\bar{F_{\mathrm{Fr}}} = \mu_0 \cdot F_N - \alpha \cdot \mu_0 \cdot F_N^2 = \mu_0 \cdot F_N(1 - \alpha \cdot F_N) \tag{5.22}$$

and

$$\bar{F_{\mathrm{Fr}}} = \int_0^{2\pi} (\mu_0 \cdot F_N - \alpha \cdot \mu_0 \cdot F_N^2) d(\omega t) \tag{5.23}$$

To get an idea of the vibration impact upon the solid body gliding over snow and ice one can use the coefficient values taken from the literature. As indicated in Table 5.8, the spread of the values is rather significant (measurements were done for different environmental conditions, sliding body material and loading): μ_0 [$(10\text{–}100)^{-3}$], and α [0.1–50 % of μ_0]. Following figures are the results of computer modeling, done using the software from the LabVIEW platform by National Instruments [81].

Figure 5.20 presents the time dependence of the friction force [instantaneous friction, Eq. (5.19)] for different vibration levels (coefficient k, Eq. (5.16), in %), in comparison with the cases when there would be no friction coefficient dependence on the loading force. Due to significant nonlinearity of the system the instantaneous friction graphs for the significant vibration levels are quite distorted.

Figure 5.21 presents the dependence of the average friction upon the static loading force for different levels of vibrations in the nonlinear case [calculated from Eq. (5.20)] in comparison with the linear case [Eq. (5.9), or $\alpha = 0$ in Eq. (5.20)]. Note that in linear case the vibrations should not change the average friction force at all.

Interestingly, there is a clear difference in the vibration influence trends in the nonlinear case for low and high static loading forces. This is due to the effect of the nonlinearity on the instantaneous loading force and the fact that this force cannot become less than 0 (sliding surface simply loses the contact with snow). Note also

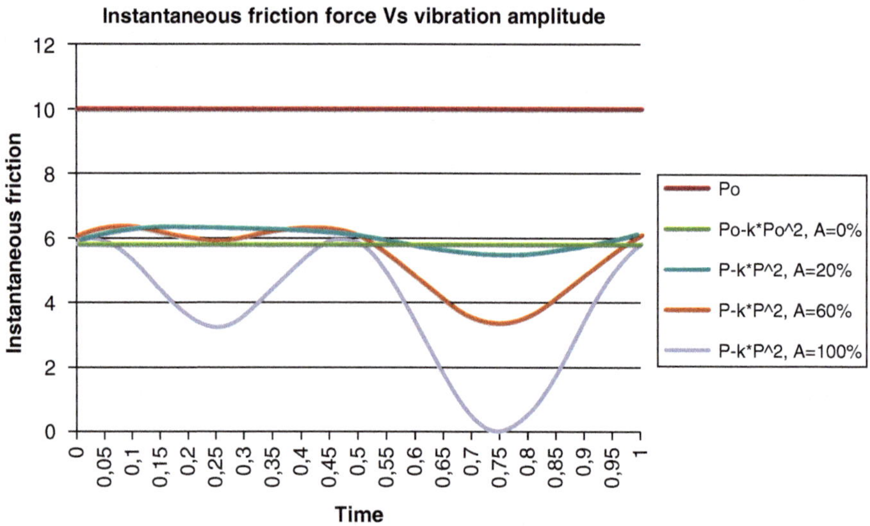

Fig. 5.20 Time dependence of the instantaneous friction force for different vibration levels

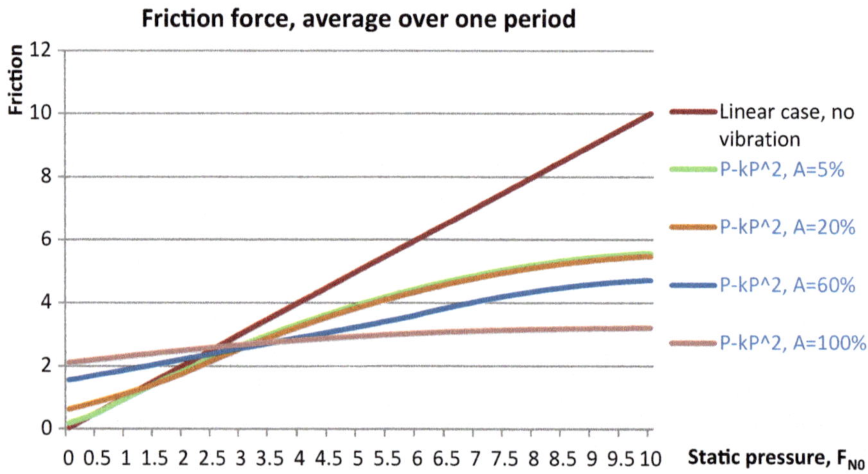

Fig. 5.21 Dependence of the average friction force upon the static loading force for different levels of vibrations

that the vibration amplitude in the case represented by the figure is taken as the fixed percentage of the static pressure value, which to certain extent exaggerates the influence of the vibrations.

Figure 5.22 presents the dependence of the friction coefficients upon the vibration intensity (in % of the static loading pressure) for the linear case (no static load dependence), nonlinear case without vibration and with vibrations [calculated using Eq. (5.20)] for $\alpha = 0.01$.

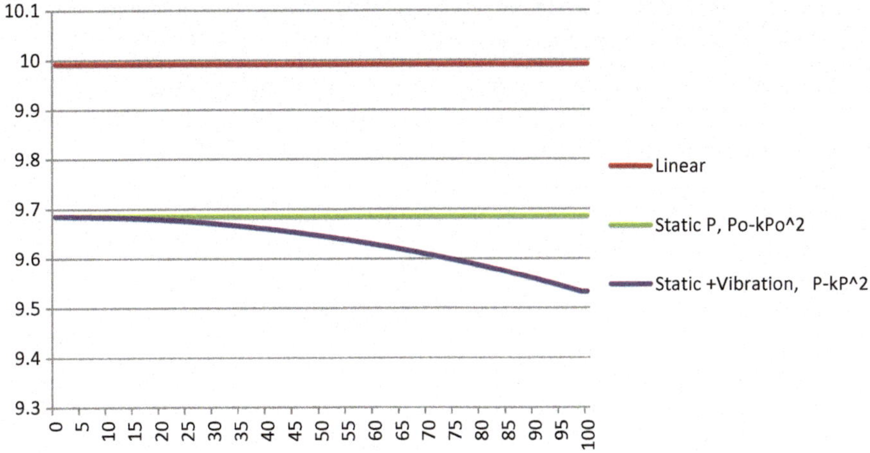

Fig. 5.22 Dependence of the friction force upon the vibration intensity

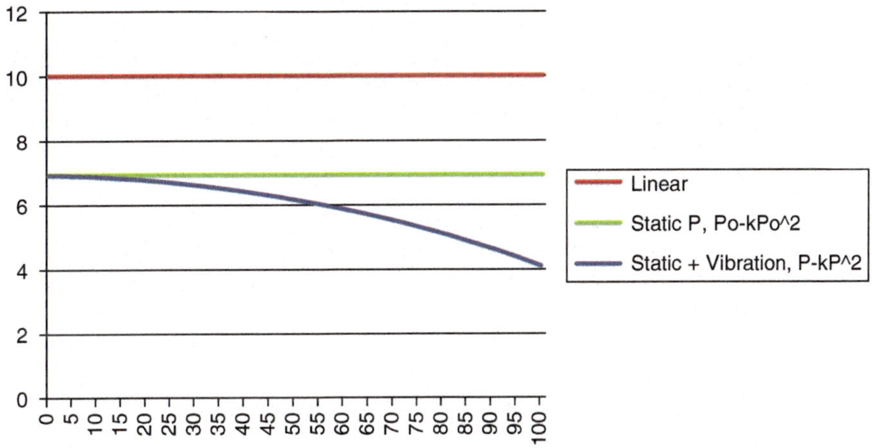

Fig. 5.23 Dependence of the friction force upon the vibration intensity for the higher nonlinearity index

Figure 5.23 presents the dependence of the friction coefficients upon the vibration intensity (in % of the static loading pressure) for the linear case (no static load dependence), nonlinear case without vibration and with vibrations [calculated using Eq. (5.20)] for $\alpha = 0.01$, but for the higher nonlinearity index of 2.5 instead of 2 as in Fig. 5.20. Both the influence of static loading and vibrations is much more pronounced in this case.

Our modeling clearly indicates that higher order terms ($i > 2$) in the polynomial expansion (5.12) could be quite significant for properly accounting for the effect of vibrations upon gliding even when their contribution coefficients (k_i) in the expansion are quite small.

Our conclusions from the simple modeling of the average friction in presence of vibrations performed as described above can be summarized as follows.

- In linear case, when friction force dependence upon the loading force is described by Eq. (5.9) vibrations should not have any impact upon average gliding friction up to the point when the gliding surface starts to lift from the snow (in which case average friction will start to decrease).
- In nonlinear case [described by Eq. (5.20)], mild and intense vibrations can significantly decrease average friction, if the nonlinearity contribution is negative (e.g., if the friction coefficient decreases with increasing loading force). This situation is characteristic for mildly cold conditions, somewhere between −5 and −20 °C.
- In nonlinear case [described by Eq. (5.20)], but in case of the positive nonlinearity contribution (e.g., if the friction coefficient increases with increasing loading force) vibrations are mainly playing a negative role, increasing the average friction force. This situation is characteristic for the near-melting conditions (in some cases around −5 °C and warmer).

Above conclusions are not contradicting the "gut feelings" expressed by the athletes describing the differences between the hard and soft skis. In view of these conclusions it is possible to theorize that the skis for the cold weather should not be very hard and acoustically absorbing and their general design should favor the generation of vibrations in free gliding. On the contrary, the skis for the near-melting conditions should be stiffer and measures for vibration dumping could improve gliding at these conditions.

Another important difference in designing skis for different temperature conditions related to the above discussion should lay in the gliding surface design. As the friction mechanisms are assumed to be significantly different for the cold and near-melting temperatures it is quite feasible that the material and micro-stricture of the gliding surfaces optimized for such conditions should also be different.

For the near-melting temperature friction forces are supposedly dominated by the hydrodynamic influence and capillary and surface tension forces. Thus quite significant research efforts are devoted to the studies of the influence of the wettability of the gliding surfaces upon the friction (see, for example, [37, 82–88]). It is also known that friction coefficient in these cases often increases with increasing loading force (supposedly due to the generation of additional melt water in the interface area). It means that in order to reduce the unwanted additional melting effects one should have the surface without sharp micro-roughness, as it is known that any deformation effects are increasing with increasing localized pressure. Also the hydrodynamic and capillary effects will significantly depend on the surface wetting properties and its micro-roughness [82–88]. The same time, it is feasible that there is a difference in the roughness value (and, possibly, the gliding surface microstructure) impact upon the capillary, hydrodynamic and localized pressure-induced mechanisms of gliding friction. Here one can seek for the analogies from other technical disciplines. For example, vehicle tires optimized for the wet weather often have deeper and wider "grooves" for channeling access water. So optimizing the gliding surface microstructure can improve the gliding properties of skis and snowboards. But though a number of suggestions have been put forward (see, for example, [89]) and patented (see, for example [90–93]) and interesting, rather simple products for modifying ski and snowboard running surfaces are manufactured [94, 95], so far there is no consensus on what microstructure and composition of the running surface is optimal for the ski and snowboard gliding depending on the chosen environmental conditions.

On the other hand, one can try to exploit the friction coefficient nonlinearity at cold temperatures. Again, here one should think of the friction forces decreasing with increasing localized pressure. Thus one can theorize that surfaces with sharper micro-roughness, favoring the additional production of melt water film will be beneficial. In both cases it is rather clear that presence of the roughness features at some different scales simultaneously may be needed to optimize the gliding friction for wider temperature window. Unfortunately the snow structure is also changing depending on the temperatures and the history of temperature changes in previous days and even weeks. Even at the same temperatures interaction of the gliding surfaces with the new fallen and re-crystallized snow will be different. So this question still leaves a wide field for the research.

5.4.5 Loading Pressure

It should be specifically noted that with all effects related to deformations (both elastic and non-elastic) one of the most critical parameters is the loading pressure rather than the loading force. Thus in Table 5.7 and the figures above average pressure values (apparent pressure) calculated from the data given in corresponding references are presented instead of the loading forces. Water-ice system phase equilibrium has quite dramatic dependence upon the local pressure at different temperatures (Fig. 5.24, adapted from [36], Fig. 14.40). Thus the studies of the pressure distribution between the ski running surface and the snow [14, 40, 44, 56, 72, 96, 97] can also contribute to the understanding of the gliding friction mechanisms.

Fig. 5.24 Water-ice system phase equilibrium

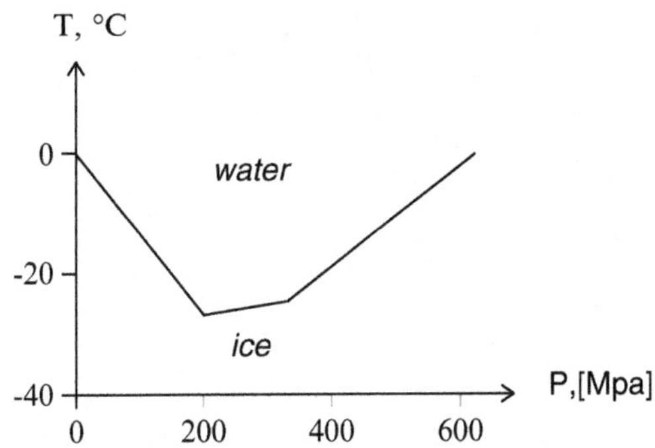

With a pair of cross-country skis in free gliding apparent area of contact corresponds to about $5 \times 20 \times 4 = 400\,\text{cm}^2$, and with the skier weight of 60 kg calculated average contact pressure is only about $1.5\,\text{N/cm}^2$ or 0.015 kPa ($100\,\text{N/cm}^2 = 1\,\text{kPa}$). This is far below the levels needed for the ice-water phase transition for the temperatures below $-5\,°\text{C}$ as predicted by the phase diagram. And according to the existing theories the area of real contact should be used in such calculations instead [36]. This area is strongly dependent on the particular details of the microstructure of the ski or snowboard running surface and of the snow conditions. And the only general conclusions one can draw is that the parameter pairs apparent pressure- microscopic pressure, and apparent area-real area of contact are coherent, e.g. when one increases the other increases also and vice versa.

An obvious drawback of using the rough contact pressure estimates via the loading force and apparent area of contact is related to the uneven pressure distribution under the ski and snowboard gliding surfaces. A number of studies into the contact pressure distribution of the skis were reported [72, 96–98], but so far the main aim of such studies is in assisting for matching ski pair selection. The same time such studies can yield interesting information contributing to the understanding of the ski and snowboard gliding friction.

Figure 5.25 presents the dependence of the normalized average and maximum pressures in the front contact area of the modern composite ski on the loading force. Experiments were carried out using F-Scan VersaTek pressure mapping system by Tekscan Ltd. [99] with two thin film pressure mapping sensors model 3000*E*

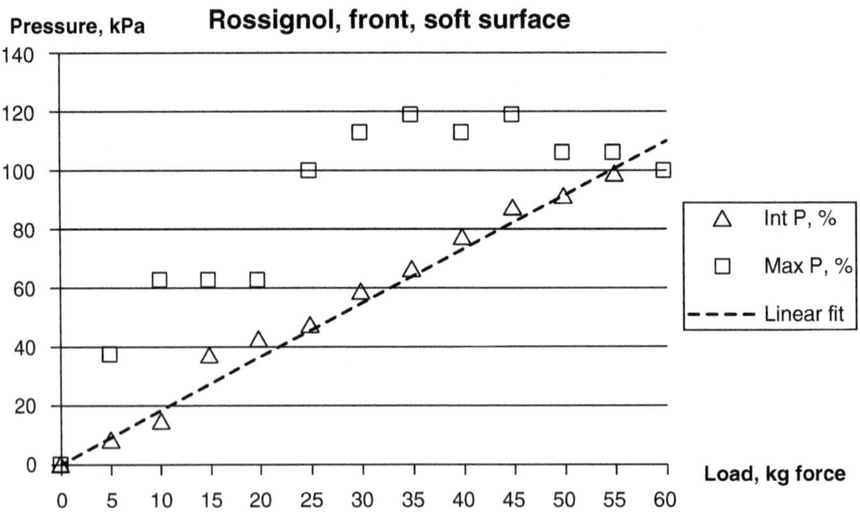

Fig. 5.25 Position of maximum and 50 % pressure outline of the ski contact area depending on the normal load. Pressure is measured under the front contact area of the ski using Tekscan F-scan systems with the sensors model 3000*E*, laying on the soft absorbing mat covering the hard solid base surface. Above 60 kgF loading all ski running surface is touching the rig. Ski: Rossignol SILVER 44 carbon, 195 cm. Soft underlying surface-floor carpeting material

arranged in a single strip and placed under the front of the ski in the contact area. Pressure sensing elements in the sensor are arranged in a rectangular grid with the spacing of 0.2 in. in both coordinates and sensing element size of $0.1 \times 0.1\,\text{in.}^2$ (resolution of 3.9 sensors/cm^2 as quoted by the manufacturer). Sensitivity of the system is about 0.005 Pa.

Integral of the pressure over the full contact area should give the loading force:

$$F = \int P \cdot dA \tag{5.24}$$

And indeed, corresponding integral values are proportional to the loading force up to almost the highest load values. For the loading force above 55 kg the ski is almost laying flat on the surface and contact area is becoming larger than the size of the pressure sensor generating errors. Maximum pressure (absolute peak value for the whole contact area) first grows with the loading force, but then nearly stabilizes starting from about half of the maximum load used. Interestingly, the average pressure calculated via the load force and the apparent area of the contact is almost unchanged above the 5–10 kgF loads due to the gradually increasing apparent area of the contact.

Figure 5.26 illustrates the changes in the position of maximum pressure point and the boundaries corresponding to 50 % of this value in the ski contact area depending on the normal loading force for the soft modern composite ski (position is measured

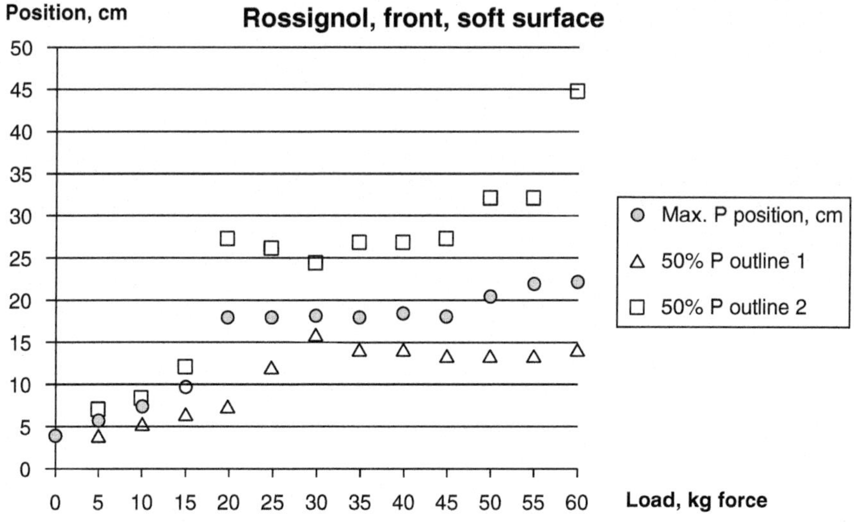

Fig. 5.26 Offset of the maximum pressure point in the ski contact area with changing normal load. Pressure is measured under the front contact area of the ski using Tekscan F-scan systems with the sensors model 3000*E*, laying on the soft absorbing mat covering the hard solid base surface. Above 60 kgF loading all ski running surface is touching the rig. Ski: Rossignol SILVER 44 carbon, 195 cm. Soft underlying surface-floor carpeting material

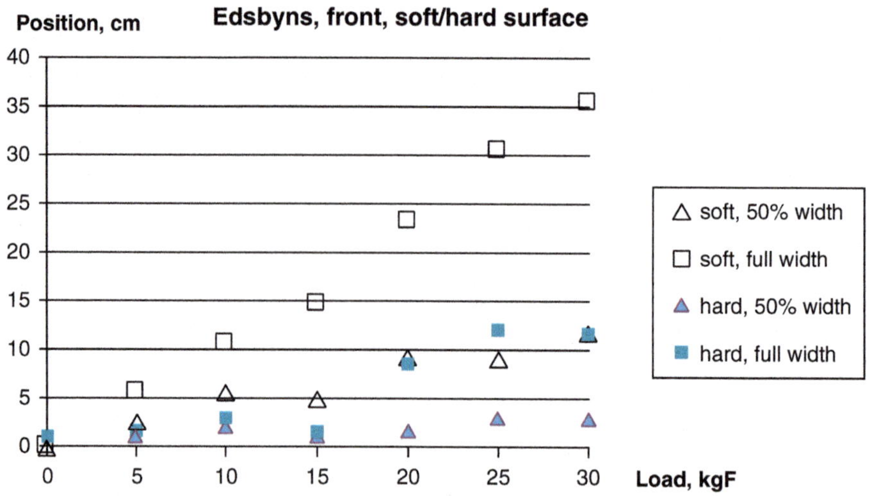

Fig. 5.27 Position of 0.05 Pa and 50 % of maximum pressure outlines in the ski contact area depending on the normal load. Pressure is measured under the front contact area of the ski using Tekscan F-scan systems with the sensors model 3000*E*, laying on the solid base surface (aluminum profile) with and without soft liner. 50 % width is measured as the position of the pressure points corresponding to 50 % of the peak pressure values; full width is measured as the positions where the pressure is above the threshold limit of the measuring system (0.005 Pa). Ski: Edsbyns Racing No748, 201 cm. Above 30 kgF loading all ski running surface is touching the rig. Soft underlying surface-floor carpeting material; hard underlying surface-aluminum base of the rig

from the front tip). Interestingly, the apparent contact width as measured via the contact pressure distribution and the position of the maximum pressure are almost steady for the moderately changing loads. Note that the maximum pressure levels measured in this case are about 10 kPa.

It should be specifically mentioned here that the nature of the second contact surface (which in reality is snow) and its compressibility significantly influences the pressure distribution under the ski running surface. Figure 5.27 presents the dependence of the apparent area of the contact and its position for the old style all-wooden ski on the loading force. Full width corresponds to the pressure falling below the sensitivity limit of the measuring system (0.005 Pa), 50 % width corresponds to the boundaries where the pressure falls to the half of peak value in the contact area.

As expected, the contact pressure area for the hard surface is smaller than for the soft one. But interestingly, the contact area with the hard surface is positioned closer to the front of the ski for all loading forces. Maximum pressures in the contact areas are also quite different, being up to 0.6 kPa for the soft and 2 kPa for the hard surfaces.

It should be specifically noted here that contact area under the ski and snowboard running surfaces is not only dependent on their design and the loading forces, but also on the properties of the snow. Traditionally for the purpose of the ski and snowboard related research snow is characterized by its temperature, humidity,

Table 5.9 Approximate correspondence of the experimentally measured snow hardness and Young's modulus at different temperatures

Snow type	Density (g/cm^3)	Hardness (kg/cm^2)	Young's modulus (MPa)	Temperature (°C)
New snow	0.33		0.15–0.4	0 to −35
Fine-grained snow	0.35		0.15–0.6	0 to −40
	0.28	1–4		0 to −40
Coarse-grained snow	0.38	3–9		0 to −40
	0.47		0.25–1.2	0 to −35
Artificially compacted snow	0.43	3–30		0 to −40
	0.51		0.25–1.7	0 to −35

grain structure, and hardness (latter is mainly measured by the penetrometers of different construction [56, 60, 100–102]). And for the realistic assessment of the pressure distribution and pressure contact area in the laboratory the hardness of the material underlying the pressure mapping sensors should at least roughly correspond to the values characteristic for the snow. Synthetic materials are easier characterized by their elastic properties, and Table 5.9 presents approximate correspondence of the experimentally measured snow hardness and Young's modulus at different temperatures (adopted from [42, 60]).

Soft floor carpeting material used in the layer underlying the pressure mapping sensors in our experiments has the Young's modulus of 7–33 kPa (for the pressures corresponding the loading by the athlete's weight, which is rather low even with the polymer sensor laying over it). On the other hand, 20 mm thick aluminum profile constituting the base of the rig which is used as the "hard" underlying material has the Young's modulus of 69 GPa. And even somewhat reduced by the thin-film pressure mapping sensors it is far too high to be comparable with snow and ice properties. In our experiments such extreme hardness parameters of the underlying layer were used to assess possible spread of the values measured in the ski pressure experiments, which in fact can be very significant. To get more realistic contact pressure measurement conditions one should either use the underlying layers of foam materials or harder carpets, or rubbers (Young's modulus 0.1–50 MPa [103]). Unfortunately for many reported studies of the ski contact pressure elastic properties of the material under the thin-film pressure mapping sensors are not commented upon, making all comparisons with the real case rather far fetching.

References

1. M. Ainegren et al. The influence of grip on oxygen consumption and leg forces when using classical style roller skis. Scand. J. Med. Sci. Sports **24**(2), 301–310 (2014)
2. M. Ainegren, P. Carlsson, M. Tinnsten, Roller ski rolling resistance and its effects on elite athletes' performance. Sports Eng. **11**, 143–157 (2009)

3. D. Buhl, M. Fauve, H. Rhyner, The kinetic friction of polyethylen on snow: the influence of the snow temperature and the load. Cold Reg. Sci. Technol. **33**, 133–140 (2001)
4. S.C. Colbeck, A review of the processes that control snow friction, in *Cold Regions Research and Engineering Laboratory, CRREL Monograph 92–2* (U.S. Army Corps of Engineers, Hanover, 1993)
5. F. Saibene et al. The energy cost of level cross-country skiing and the effect of the friction of the ski. Eur. J. Appl. Physiol. **58**(7), 791–795 (1989)
6. M.D. Hoffman et al. Influence of body-mass on energy-cost of roller skiing. Int. J. Sport Biomech. **6**, 374–385 (1990)
7. M.D. Hoffman et al. Physiological comparison of uphill roller skiing-diagonal stride versus double pole. Med. Sci. Sports Exerc. **26**(10), 1284–9 (1994)
8. G.Y. Millet et al. Effect of rolling resistance on poling forces and metabolic demands of roller skiing. Med. Sci. Sports Exerc. **30**(5), 755–62 (1998)
9. R.W. de Boer et al. Physiological and biomechanical comparison of roller skating and speed skating on ice. Eur. J. Appl. Physiol. Occup. Physiol. **56**(5), 562–9 (1987)
10. M. Ainegren, P. Carlsson, M. Tinnsten, Rolling resistance for treadmill roller skiing. Sports Eng. **11**, 23–29 (2008)
11. M.D. Hoffman et al. Delta efficiency of uphill roller skiing with the double pole and diagonal stride techniques. Can. J. Appl. Physiol. **20**(4), 465–79 (1995)
12. M. Ainegren, P. Carlsson, M. Tinnsten, A portable rolling resistance measurement system, in *The Impact of Technology on Sport V*, ed. by P. Fuss, P. Clifton, K.M. Chan (Subic A., Hong Kong, 2013)
13. M. D. Hoffman et al. Physiological effects of technique and rolling resistance in uphill roller skiing. Med. Sci. Sports Exerc. **30**(2), 311–7 (1998)
14. H. Ekström, Force interplay in cross-country skiing. Scand. J. Sports Sci. **3**, 69–76 (1981)
15. P.V. Komi, Force measurements during cross-country skiing. Int. J. Sport Biomech. **3**, 370–381 (1987)
16. P.V. Komi, Ground reaction forces in cross-country skiing, in *Biomechanics IX-B*, ed. by R. Norman, D. Winter et al. (Human Kinetics, Champaign, 1985)
17. P.V. Komi, R.W. Norman, Preloading of the thrust phase in cross-country skiing. Int. J. Sports **8**, 48–54 (1987)
18. P. Vahasoyrinki et al. Effect of skiing speed on ski and pole forces in cross-country skiing. Med. Sci. Sports Exerc. **40**(6), 1111–6 (2008)
19. G.A. Smith, Cross-country skiing: technique, equipment and environmental factors affecting performance, in *Biomechanics in Sport*, ed. by V. Zatsiorsky (Blackwell Science Ltd, London, England, 2000)
20. M. Ainegren, P. Carlsson, M. Tinnsten, An experimental study to compare the grip of classical style roller skis with on-snow skiing. Sports Eng. **16**, 115–122 (2013)
21. M. Ainegren et al. The multifunctional roller ski, in *Second International Congress on Science and Skiing, Vuokatti*, ed. by S. Lindinger, V. Linnamo, A. Hakkarainen (2012)
22. *FIS Cross-Country Homologation Manual* (International Ski Federation, 2012)
23. S.C. Colbeck, A review of the friction of snow skis. J. Sports Sci. **12**, 285–295 (1994)
24. E. Spring et al. Drag area of a cross country skier. Int. J. Sports Biomech. **4**, 103–113 (1988)
25. J.F. Moxnes, K. Hausken, Cross-country skiing motion equations, locomotive forces and mass scaling laws. Math. Comput. Model. Dyn. Syst. **14**(6), 535–569 (2008)
26. P. Carlsson, M. Tinnsten, M. Ainegren, Numerical simulation of cross-country skiing. Comput. Methods Biomech. Biomed. Eng. **14**(8), 741–746 (2011)
27. J.F. Moxnes, O. Sandbakk, K. Hausken, Using the power balance model to simulate cross-country skiing on varying terrain. Open Access J. Sports Med. **5**, 89–98 (2014)
28. D. Sundström et al. Numerical optimization of pacing strategy in cross-country skiing. Struct. Multidiscip. Optim. **47**(6), 943–950 (2013)
29. R.H. Morton, Modelling human power and endurance. J. Math. Biol. **28**(1), 49–64 (1990)
30. D. Sundström, P. Carlsson, M. Tinnsten, Comparing bioenergetic models for the optimisation of pacing strategy in road cycling. Sports Eng. **17**, 1–9 (2014)

31. *OASIS-ALADDIN User's Manual* (Alfgam Optimering AB, Stockholm, 1990)
32. B. Esping, D. Holm, O. Romell, The OASIS-ALADDIN structural optimization system. Int. Ser. Numer. Math. **110**, 159–186 (1993)
33. D. Zenkert, *An Introduction to Sandwich Construction* (Royal Institute of Technology, Stockholm, 1993)
34. K. Åström, O. Norberg, Förhistoriska och medeltida skidor, in *Västerbotten, Västerbottens läns Hembygdsförbund* (1984)
35. P. Carlsson, B. Esping, *Optimization of a Racing Ski* (Mitthögskolan, 1994)
36. B.N.J. Persson, *Sliding Friction. Physical Principles and Applications*, 2nd edn. (Springer, Berlin, 2000)
37. L. Bocquet, Friction: an introduction with emphasis on some implications in winter sports, in *Sports Physics*, ed. by C. Clanet (2013), pp. 407–427
38. F.P. Bowden, T.P. Hughes, The mechanism of sliding on ice and snow. Proc. R. Soc. Lond. A **172**(949), 280–298 (1939)
39. P. Oksanen, J. Keinonen, The mechanism of friction of ice. Wear **78**, 315–324 (1982)
40. H. Ekström, Force interplay in cross-country skiing. Lic thesis, Linköping University, 1980
41. D. Buhl, M. Fauve, H. Rhyner, The kinetic friction of polyethylen on snow: the influence of the snow temperature and the load. Cold Reg. Sci. Technol. **33**, 133–140 (2001)
42. P. Sturesson, Friction characteristics between ski base and ice-fundamental lab scale tests and practical implications. M.Sc thesis, Uppsala University and Primatera AB, 2008
43. A.-M. Keitzig, Microscopic ice friction. Ph.D. thesis, Technishe Universität Berlin, 2006
44. L. Bäurle, Sliding friction of polyethylene on snow and ice. MA thesis, Swiss Federal Institute of Technology, Zurich, 2006
45. P. Federolf, B. Nigg, Skating performance in ice hockey when using a flared skate blade design. Cold Reg. Sci. Technol. **70**, 12–18 (2012)
46. F.P. Bowden, Friction on snow and ice. Proc. R. Soc. Lond. A **217**, 462–478 (1953)
47. O.M. Braun, A.G. Naumovets, Nanotribology: microscopic mechanisms of friction. Surf. Sci. Rep. **60**(6–7), 79–158 (2006)
48. D. Kuroiwa, The kinetic friction on snow and ice. J. Glaciol. **19**(81), 225–235 (1977)
49. S. Sukhorukov, Ice-ice and ice-steel friction in field and laboratory. Ph.D. thesis, Norwegian University of Science and Technology, 2013
50. S.C. Colbeck, Kinetic friction of snow. J. Glaciol. **34**(116), 78–86 (1988)
51. A.-M. Kietzig, S.G. Hatzikiriakos, P. Englezos, Ice friction: the effects of thermal conductivity. J. Glaciol. **56**(197), 473–479 (2010)
52. P. Kapps, W. Nachbauer, M. Mossner, Determination of kinetic friction and drag area in alpine skiing. Ski Trauma Skiing Saf. **10**, 1–17 (1996)
53. C. Stamboulides, Microscopic ice friction of polymeric substrates. Ph.D. thesis, The University of British Columbia, Vancouver, 2010
54. C. Stamboulides, P. Englezos, S.G. Hatzikiriakos, The ice friction of polymeric substrates. Tribol. Int. **55**, 59–67 (2012)
55. N. Maeno, M. Arakawa, Adhesion shear theory of ice friction at low sliding velocities combined with ice sintering. J. Appl. Phys. **95**(1), 134–139 (2004)
56. D.A. Moldestad, Some aspects of ski base sliding friction and ski base structure. Engineering thesis, Department of Structural Engineering, Norwegian University of Science and Technology, Trondheim, 1999
57. A. Koptyug, L. Kuzmin, Experimental field studies of the cross-country ski running surface interaction with snow. Procedia Eng. **13**, 23–29 (2011)
58. L. Kuzmin, Investigation of the most essential factors influencing ski glide. MA thesis, Licentiate Thesis, Luleå University of Technology, 2006
59. Global speed skydiving database (2014), http://www.gssdb.speedskydiving.eu/hall_of_fame.php
60. S. Colbeck et al. *The International Classification for Seasonal Snow on the Ground* (International Commission for Snow and Ice, World Data Center for Glaciology, University of Colorado, Boulder, Colorado, 1990)

61. T. Katutosi, The temperature dependence of hardness of snow, in *Snow Mechanics. Proceedings of the Grindelwald Symposium*, vol. 114, 1975, pp. 103–109
62. T. Theile et al. Mechanics of the ski-snow contact. Tribol. Lett. **36**, 223–231 (2009)
63. T. Iwase, T. Sakuma, K. Yoshihisa, Measurements of sound propagation characteristics in snow layer, in *Proceedings of 17th International Congress on Acoustics*, Rome, vol. 1, 2001, pp. 274–275
64. R.A. Sommerfeld, A review of snow acoustics. Rev. Geophys. Space Phys. **20**(1), 62–66 (1982)
65. M. Mössner et al. Modeling the ski-snow contact in skiing turns using a hypoplastic vs an elastic force-penetration relation. Scand. J. Med. Sci. Sports **24**, 577–585 (2014)
66. W. Nachbauer, P. Schröcksnadel, B. Lackinger, Effects of snow and air conditions on ski friction. Skiing Trauma Saf. **10**(1266), 178–186 (1996)
67. S.C. Colbeck, A review of the friction of snow skis. J. Sports Sci. **12**, 285–295 (1994)
68. S.C. Colbeck, A review of the friction of snow, in *Physics of Sliding Friction*, ed. by B.N.J Persson, E. Tosatti. NATO ASI Series, vol. 311 (Springer, The Netherlands, 1996)
69. B. Glenne, Sliding friction and boundary lubrication of snow. J. Tribol. **109**(4), 614–617 (1987)
70. B. Bhusam, *Introduction to Tribology* (Wiley, New York, 2002)
71. S.C. Colbeck, Pressure melting and ice skating. Am. J. Phys. **63**, 888–890 (1995)
72. P. Federolf et al. Does a skier's position on the ski affect the results of gliding tests used to assess ski-snow friction?, in *Proceedings of the ISBS-Conference 24 International Symposium on Biomechanics in Sports (2006)*, 2007, pp. 554–557
73. A. Koptyug, M. Bäckström, M. Tinnsten, Cross country ski vibrations: loaded vs. unloaded skis, in *Proceedings of 2nd International Congress on Science and Nordic Skiing, ICSNS 2011*, Vuokatti, Finland, 28–31 May 2012, p. 83
74. A. Koptyug et al. Cross-country ski vibrations and possible mechanisms of their influence on the free gliding. Procedia Eng. **34**, 473–478 (2012)
75. A. Koptyug, M. Bäckström, M. Tinnsten, Studies into the mechanisms of the cross-country ski vibrations and possible models of the phenomenon. Procedia Eng. **60**, 40–45 (2013)
76. A. Koptyug, M. Bäckström, M. Tinnsten, Gliding-induced ski vibrations: approaching proper modeling. Procedia Eng. **72**, 539–544 (2014)
77. A. Shionoya, K. Sato, Development of a simulator generating ski board vibrations in actual skiing. Procedia Eng. **60**, 269–274 (2013)
78. K.W. Buffinton et al. Laboratory, computational and field studies of snowboard dynamics, in *TMS Series. Materials and Science in Sports*, 2001, pp. 171–183
79. M.A. Chowdhury, M. Helali, The effect of relative humidity and roughness on the friction coefficient under horizontal vibration. Open Mech. Eng. J. **2**(1), 128–135 (2008)
80. M.A. Chowdhury, D.V. Nuruzzaman, M.L. Rahman, Influence of frequency and direction of vibration on the coefficient of friction of aluminum, in *Proceedings of the 4th BSME-ASME International Conference on Thermal Engineering, (BSME-ASME 2008)*, 2008, pp. 826–831
81. Lab view Software package LabVIEW-2014, http://www.ni.com/labview/
82. J.L. Giesbrecht, Polymers on snow: towards skiing faster. Ph.D. thesis, ETZ Zurich, 2010
83. A.-M. Kietzig, S.G. Hatzikiriakos, P. Englezos, Ice friction: the effects of surface roughness, structure and hydrophobicity. J. Appl. Phys. **106**, 024303 (2009)
84. A.-M. Kietzig, A biomimetic approach to ice friction, in *Green Tribology*, ed. by M. Nosonovsky, B. Bhushan. Green Energy and Technology (Springer, Berlin/Heidelberg, 2012), pp. 223–264
85. L. Kuzmin, P. Carlsson, M. Tinnsten, The relationship between the type of machining of the ski running surface and its wettability and capillary drag. Sports Technol. **3**(2), 121–130 (2010)
86. L. Bäurle et al. Sliding friction of polyethylene on ice: tribometer measurements. Tribol. Lett. **24**(1), 77–84 (2006)

87. D. Dressler, S.I. Green, An examination of the wettability of the running surface of skis as a means of minimizing snow friction, in *4th Asia Pacific Conference on Sports Technology*, Honolulu, 2009
88. L. Kuzmin, Interfacial kinetic ski friction. Ph.D. thesis, Mid Sweden University, Sweden, 2010
89. F. Breitschädel, Ø. Lund, S. Løset, Cross country ski base tuning with structure imprint tools. Procedia Eng. **2**(2), 2907–2911 (2010)
90. S. Lyng, Plastic non-wax ski base and methods for its manufacture. U.S. Patent 4272577 A, 1981
91. W.D. Danner, H. Woitschatzke, Ski with three-dimensional running surface. Patent EP 0015447 B1, 1984
92. V. Tonel, Ski or snowboard with improved gliding. Patent WO 1995015794 A1, 1995
93. W.D. Danner, I.L. Ver, H. Woitschatzke, Three-dimensional ski surface. U.S. Patent 3858894 A, 1975
94. Kuzmin ski technology AB (2014), http://www.kuzmin.se/pgs/scrapers_engl.html
95. Primateria AB (2014), http://www.primateriasport.se/Scrapers.php
96. K. Pärssinen, Finite element method for the calculation of nominal pressure distribution between ski and track. Finite Elem. Anal. Des. **10**, 1–8 (1991)
97. R.F. Orellana, Experimental methods to measure mechanical properties of cross country skis. M.Sc thesis, Luleå University of Technology, 2012
98. J. Nilsson, L. Karlöf, V. Jakobsen, A new device for measuring ski running surface force and pressure profiles. Sports Eng. **16**, 55–59 (2013)
99. Tekscan Inc (2014), http://www.tekscan.com/medical/system-fscan1.html
100. J. Huang, Speed of a skydiver (terminal velocity), in *The Physics Factbook*, ed. by G. Elert. (Midwood High School, Brooklyn College, Brooklyn, 1999)
101. C. Pielmeier, M. Schneebel, Snow stratigraphy measured by snow hardness and compared to surface section images, in *Proceedings of the International Snow Science Workshop*, vol. 29, 2002
102. M. Schneebeli, C. Pielmeier, J.B. Johnson, Measuring snow microstructure and hardness using a high resolution penetrometer. Cold Regions Science and Technology **30**, 101–114 (1999)
103. S.K. De, J.R. White, *Rubber Technologist's Handbook*, vol. 1 (Rapra Technology Limited, Exeter, 2001). ISBN:9781859572627

Chapter 6
Aerodynamics of Ski Jumping

Mikko Virmavirta

6.1 Introduction

Ski jumping is an exciting and primarily competitive sport involving both ballistic and aerodynamic factors. The ballistic factors are release velocity and release position of a ski jumper from the take-off table, whereas aerodynamic factors include the gliding properties of the jumper/ski system (velocity, suit design, surface area, posture of the jumper/ski system, resisting and lifting forces). Both ballistic and aerodynamic factors place special demands on the jumper so that he/she can, for example, optimally maximize the vertical lift and minimize the drag forces. Ski jumping has attracted spectators 200 years and it has undergone several notable steps in the development of performance technique and equipment. This chapter does not try to provide any complete examination of prior work in this area, but some well-known studies and surveys/reviews will be given.

6.1.1 Historical Background

Ski jumping has been in the Olympic Winter Games since the first Games in Chamonix 1924 where Norwegian Jacob Tullin Thams, probably the world's best ski jumper in the mid-1920s, developed a new jumping style known as the Kongsberger Technique. This technique replaced the former upright style involving jumping with the upper body bent at the hips, a wide forward lean, and with arms extended at the front with the skis parallel to each other. The jumping distance of

M. Virmavirta (✉)
Department of Biology of Physical Activity, University of Jyväskylä,
P.O. Box 35, Jyväskylä 40014, Finland
e-mail: mikko.virmavirta@jyu.fi

F. Braghin et al. (eds.), *The Engineering Approach to Winter Sports*,
DOI 10.1007/978-1-4939-3020-3_6

Fig. 6.1 Flight style used by Josef Bradl in his record jump over 100 m

100 m was first reached with the same style in 1936 (Fig. 6.1). The first ski jumping research in wind tunnel (Fig. 6.2) was carried out already as early as in 1926 by R. Straumann who utilized the equations of motion and developed the first analytical model of ski jumping mechanics in his pioneering work [1]. In the mid-1950s, Swiss jumper Andreas Däscher became the first jumper to hold the arms backwards close to the body with a more extreme forward lean. This was most likely resulted from Straumann's further wind tunnel (see [2]) study for better aerodynamic flight style in 1954 when Däscher himself was a subject in the measurements. As a conclusion from his measurements [2] Straumann introduced a best aerodynamic flight position (Fig. 6.3) which resembles strikingly the best positions obtained many years later. At about the same time another comprehensive theoretical analysis of the entire ski jumping performance was published by König in 1952 [3].

The next important step towards the better understanding of ski jumping aerodynamics was the flight-mechanical investigation of ski jumping by Tani and Iuchi [4]. They performed extensive wind tunnel tests to measure aerodynamic forces and pitching moments acting on a full scale ski jumper model as a function of jumper and ski orientation, and the relative wind direction. They proposed the best flight posture where arms should be set close to the side of the body and they also pointed out that the position with maximum lift-to-drag ratio (L/D) doesn't give the longest distance. Application of this data most likely helped the Japanese to the Olympic triple victory of ski jumping in Sapporo 1972. In the same year with Tani and Iuchi, Grozin [5] provided more wind tunnel data including the relationship between aerodynamic lift (C_L) and drag (C_D) coefficients with angle of

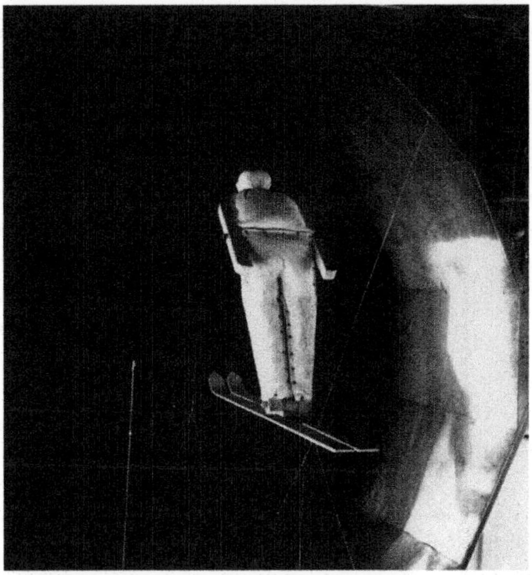

Fig. 6.2 Wind tunnel experiment on a ski jumper model in 1926 ([1] by permission from Dr. Thomas Straumann)

Fig. 6.3 Ski jumper Andreas Däscher in the wind tunnel experiments 1954 ([2] by permission from Dr. Thomas Straumann)

attack (α) as a parameter (Fig. 6.4). Above-mentioned data sets covered nearly the entire spectrum of jumper and ski position found in a real jump and this information could be later used in further calculations and computer simulations of the ski jumping performance. One of these calculations was performed by Remizov [6] whose findings are, at least partly, valid even today. Remizov applied Pontryagin's maximum principle [7] and utilized Grozin's data while solving the problem of optimal flight, i.e. how a jumper must change the angle of attack to obtain the longest jump. It was concluded that "the maximum flight distance is achieved when the angle of attack is gradually increased according to a convex function the form of which depends on the individual aerodynamic parameters" (Fig. 6.4). Similar results

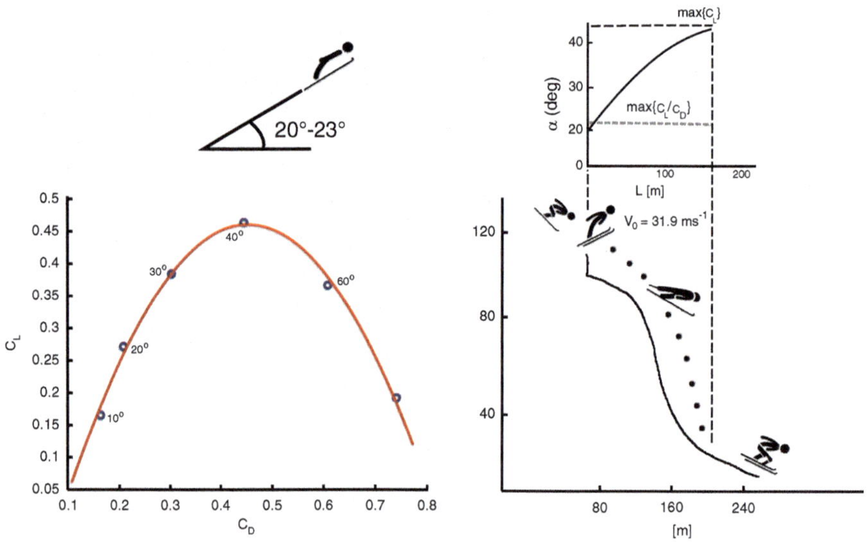

Fig. 6.4 *Left*: C_L/C_D graph according to the angle of attack (10–50°). *Right*: Optimal flight path and corresponding optimal angle of attack (modified from [6])

were found recently by Jung et al. [8] using Pontryagin's method in their work of flight style optimization on normal, large, and ski flying hills in ski jumping. According to Remizov's calculations, maintaining the flight position with maximum C_L/C_D ratio constant during the entire flight would result, for an initial speed of 30 m/s, in a loss of 13 m and the position with maximum C_L in a loss of 10 m as compared to optimal trajectory. The theoretical analysis of Remizov was compared with experimental field studies by Denoth et al. [9] who found the optimization of the flight phase problematic especially during the early stage and they pointed out that optimization must include all parts of a jump. Also the Polish research group (e.g., Maryniak and Krasnowski [10]) has provided a good contribution to the study of the effects of ski jumper configuration on aerodynamics with their strong mathematical approach which was based on a model test in a wind tunnel.

There is not much documented data available about the development of ski jumping equipment (skis and suits) during those early years. However, a lot of innovations have occurred especially in jumper suit design. All the changes in equipment have increased the effects of aerodynamics and increased jump distance. Hubbard et al. [11] mentioned that given a constant set of initial conditions, jump distance should be determined by the overall ratio of aerodynamic to gravity forces. According to [11] this overall ratio appears to have changed gradually over the 17 years since the experimental wind tunnel tests of Tani and Iuchi were completed, perhaps due to several rule changes that have since gone into effect. Thus they adjusted the lift of Tani and Iuchi by a factor of 1.2 and the drag by a factor of 0.85 in all their simulations. One remarkable change in ski jumping suit was introduced in mid-1970s as the material in the back part of the suit was totally air-impermeable giving a huge increase in aerodynamic forces. The idea came probably from Baldur Preiml (head coach of the Austrian ski jumping team 1974–1980), who developed the new suits together with the manufacturer Reinalter. Thereafter the ski jumping suits of laminated foam has become one of the most tightly regulated sport equipments. The Austrian ski jumping team used the new suits first time in the Four Hills Tournament 1974/1975 and took a triple victory in the overall standing which has happened only three times in the 60 years history of the tournament. According to Innauer (2014, personal communication), who was among the first ones to use these new suits, the advantage of the suit was at least 5 m in the large hill. Innauer has said [12] that with the material the national team became at one stroke a class better than what they would have been according to the level of training and development ("Durch das Material war unsere Mannschaft mit einem Schlag eine Klasse besser als sie trainings-und entwicklungsmässig hätte sein können"). Some evidence of the aerodynamic effects of the jumping suits in mid-1970s has been reported by Meier [13]. In the film analysis of [13] the world record jump 176 m of Toni Innauer (AUT) in Oberstdorf 1976 showed much slower landing velocity (25.4 m/s) as compared to the take-off velocity (31.6 m/s), the average of 12 jumpers being 32.0 and 30.5 m/s, respectively. However, Innauer himself was not very happy with these new suits as his flight technique was quite sensitive to the changes in pitching moment.

6.1.2 V-Style Replaced the Classic Flight Style

"I was too weak in my groin, which is why what happened, happened. It was not planned. It's totally by chance." This is what Swedish Boklöv (2014, personal communication) said about the initiation of his revolutionary flight style in ski jumping. Although there have been other jumpers using V-style as well (Jiří Malec, Olympic bronze medalist 1988 and Polish Miroslaw Graf, who has been said to demonstrate the V-style first [14]), Boklöv's success attracted much more attention towards the more detailed research on the advantage of this flight style.

First and probably the most important study of V-style was published by Mahnke and Hochmuth [15]. They performed series of wind-tunnel experiments while investigating the benefits of the V-style compared with the traditional style. The results of this study are summarized in Fig. 6.5. It can be seen that all positions which deviate from the generally accepted traditional one (giving the best style marks as well) give better aerodynamic quality for the jumper/skis system. In the chronological order, the next document on the aerodynamics of the V-style was probably the study of Cutter [16] in which the comprehensive tests were conducted in the subsonic wind tunnel at the United States Air Force Academy Aeronautics Laboratory using a scale model (1:5.5). In this study V-angle of 22.5° with ankle angle of 20° (which was considered the maximum feasible for a ski jumper to turn skis more flat against the direction of motion) provided the best *L*/*D* ratio of 1.55.

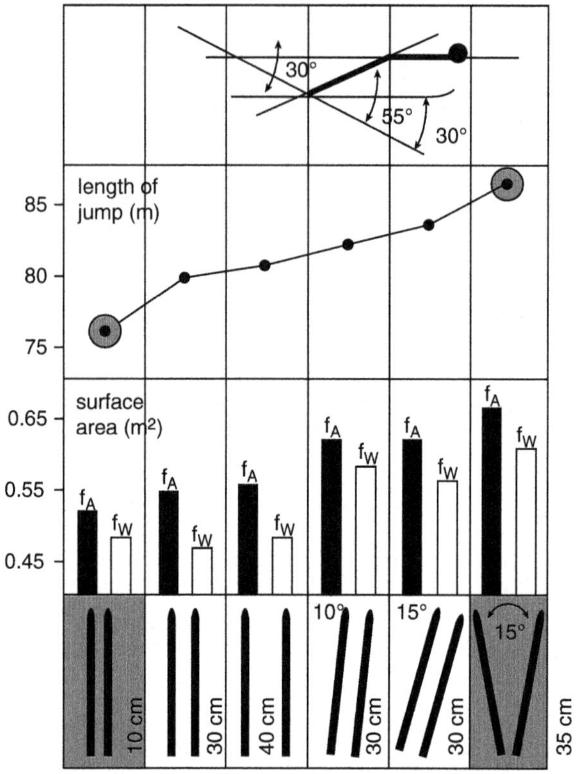

Fig. 6.5 Effect of ski position on the aerodynamic forces and jumping distance (modified from [15])

As there was an urgent need among coaches for more information about the merit of this new flight style, several studies were conducted during these early years of V-style jumping (e.g., [17–22]). In comparison with the old parallel flight style most of the studies clearly showed the better aerodynamic quality for the flight positions with V-style. V-style allowed jumpers to lean more forward which was partly a result from the changed pitching moment (longitudinal rotation of jumper/skis system, i.e. nose up/down). Jin et al. [21] mentioned that V-style is quite sensitive to initial angular velocity, but they assumed that ski jumpers cannot rotate with initial angular velocities over $-17°/s$. Soon after changing to V-style athletes found that they could lean even more forward if mounting the ski bindings further back on their skis. This led to a very high tumbling risk, and many severe accidents occurred due to an unbalanced pitching moment. Thereafter, based on some analyses (Fig. 6.6, [22]), the maximum percentage of front ski (L_1) to total ski (L) was limited to 57 % by the FIS from the season 1994/1995. Several other studies concerning the merit of V-style have been published [23, 24].

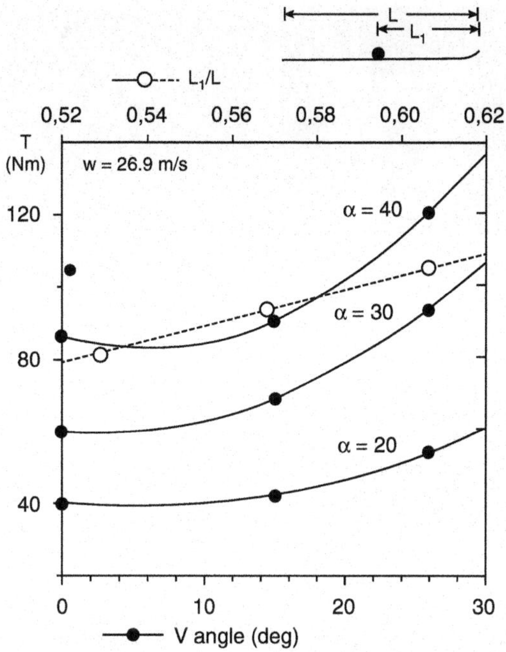

Fig. 6.6 Torque on skis. *Solid lines*: $L_1/L = 0.57$ for different angles of attack (α). *Broken line*: $L_1/L = 0.52$–0.62 ($\alpha = 30°$, $V = 26°$) (modified from [22])

6.2 Aerodynamics in Different Phases of Ski Jumping Performance

The study of ski jumping aerodynamics can be considered to involve wind tunnel measurements, computer simulation with mathematical equations, and field measurements to provide an appropriate input data for the computer simulation. Although *computational fluid dynamics (CFD)* is generally considered as a future tool in ski jumping aerodynamics, only few studies have been published so far [25–27]. Based on their numerical simulations [25] concluded that "current CFD tools do not seem to be capable of simulating the aerodynamic forces adequately, at least within the tested framework." They found that the lift and drag areas differed from the experimental results and therefore the flight paths could not be predicted with the sufficient accuracy. Nørstrud and Øye [26] presented a numerical analysis for the flow over a ski jumper in various positions and provided several results from non-viscous (Euler) and viscous (Navier–Stokes) computations. It was concluded that the V-style represents improved aerodynamics and that the skis play a major part in producing favorable lift (Fig. 6.7). The angle of attack and the body angle relative to the ski were selected as the design variables for the optimization work of Lee et al. [27] where the posture of a ski jumper was numerically optimized based on the Kriging optimization method coupled with flow analysis using 3-D RANS equations. The *L/D* ratio was used as the objective function, and the optimized posture showed a 28.8 % improvement in the *L/D* ratio compared to the reference

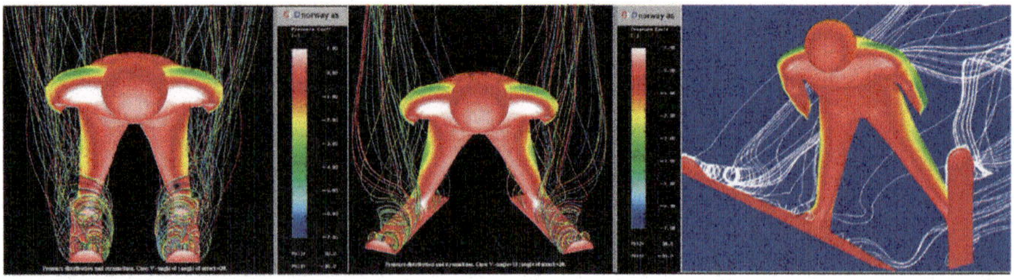

Fig. 6.7 Pressure distribution and flow visualization around a ski jumper. Important vortex generation can be seen on the upper side of the skis [26]

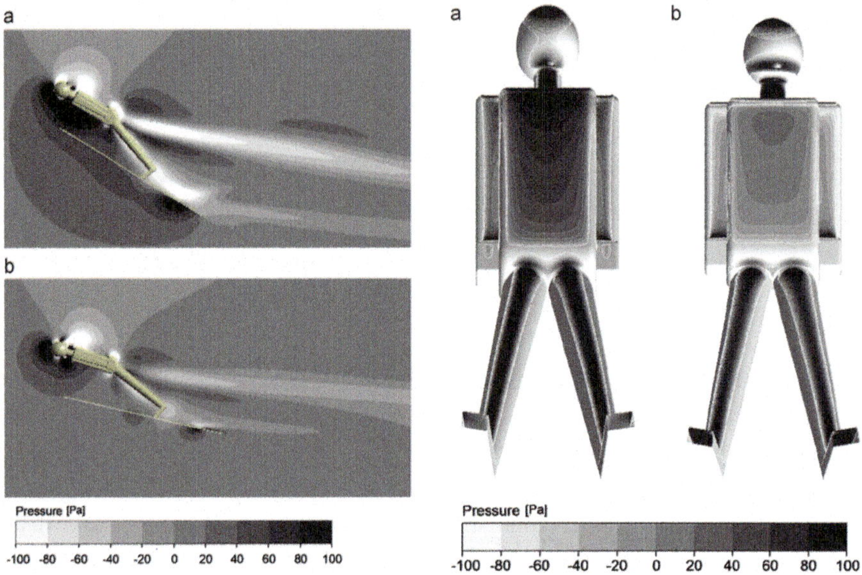

Fig. 6.8 Pressure distribution on *x*–*z* plane (*left*) and on the front surface of body (*right*). (**a**) Reference and (**b**) optimized [27]

pose (Fig. 6.8). CFD was also used in several phases of the project in which the Holmenkollen ski jump underwent a total rebuild with permanent wind shielding integrated in the architectural design [28].

Ski jumping involves several phases—inrun, take-off, flight and landing—each of which has importance to the length of the jump. Take-off is probably the most crucial phase for the entire ski jumping performance as it also determines the subsequent flight phase and consequently the length of the jump (Fig. 6.9). Aerodynamics plays important role during each of these phases although the flight phase has been studied most.

A good way to assess the role of aerodynamics in different sport disciplines is to relate the involved aerodynamic forces to the weight of athlete/equipment. The term "aerodynamicity" (β) was used by Hubbard et al. [11] and Hubbard [29] to give an

Fig. 6.9 Transition phase from the take-off to the flight phase

Table 6.1 Sensitivity analysis of the effect of selected parameters on jumping distance

Parameter	Change	134 m	Difference (m)
Official inrun speed	+1 km/h	138.4	4.5
Jumper's mass	−1 kg	135.7	1.7 (1.4[a])
C_L flight phase	+1 %	135.7	1.7
C_D flight phase	−1 %	134.8	0.8
Steady head wind	+1 m/s	140.1	6.1

[a]The effect of mass on V_z and V_{inrun} ignored

idea of the overall importance of aerodynamic forces relative to gravitational ones (F_{aero}/F_{grav}). For instance, a shot put is not affected much by the air it moves through and therefore, effects like air drag or aerodynamic lift can be ignored ($\beta_{max} = 0.0086$). On the other hand, the very light badminton shuttlecock experiences aerodynamic forces 100 times its weight ($\beta_{max} = 100$). A Formula 1 car uses aerodynamics to generate its own weight in downforce—so they could theoretically drive upside down on the roof of a tunnel at only 160 km/h (e.g., [30]). In ski jumping, and especially in ski flying, it is sometimes seen that the aerodynamic forces compensate the gravitational ones and jumpers fly parallel to the landing slope.

Table 6.1 shows the effect of selected aerodynamic factors on jumping distance in ski jumping. This sensitivity analysis was done for the reference jump of 134 m by using the Aquila ski jumping simulator [31] which is a combination of first- and second-order integration for ski jump performance with a discrete time step of 0.02 s. The computer simulation was performed in four stages (Fig. 6.10) including (1) inrun linear section and curve; (2) linear take-off table section at the end of the inrun where the actual take-off takes place; (3) transient flight (typically 0.25–0.50 s long period after take-off); (4) fully developed flight with angle of attack dynamically adjustable. The reference jump in Table 6.1 is used several times in the following text.

The aerodynamic forces can be described by the following equations:

$$\begin{aligned} F_D &= \frac{1}{2}\rho v^2 C_D A \\ F_L &= \frac{1}{2}\rho v^2 C_L A \end{aligned} \tag{6.1}$$

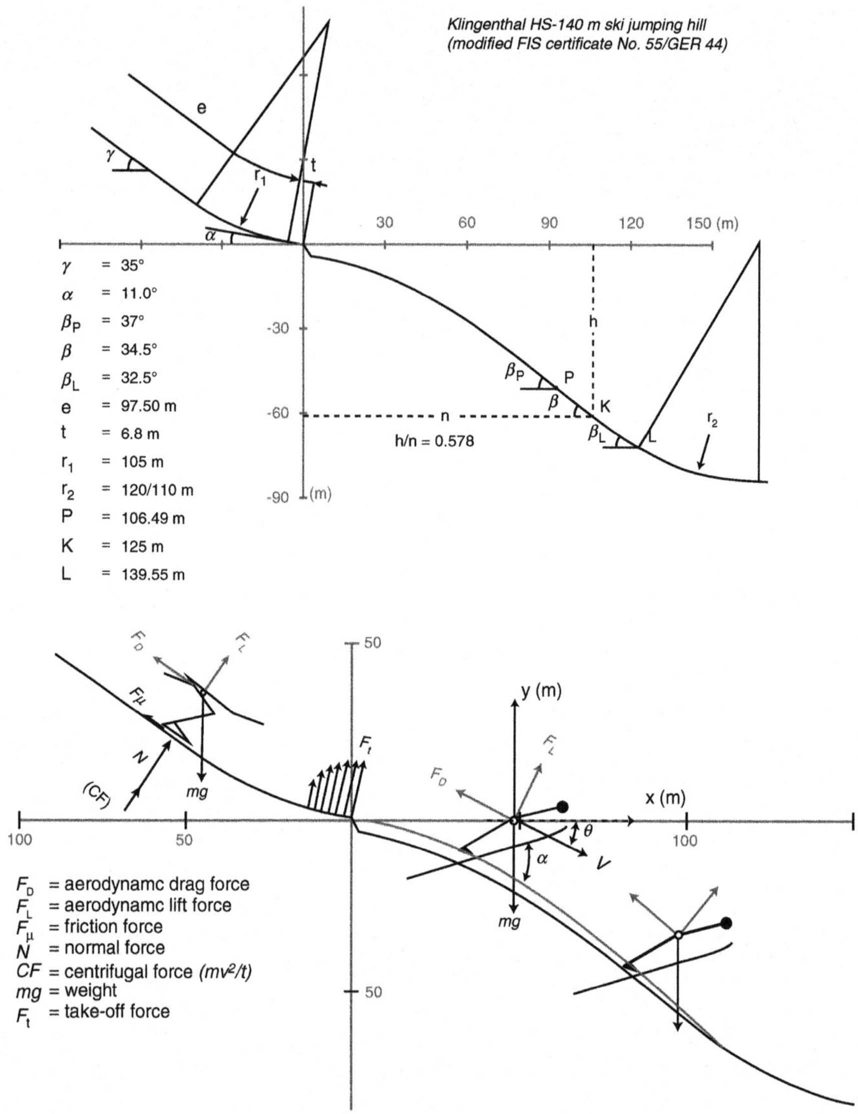

Fig. 6.10 Official jumping hill profile (*left*) used in the computer simulation of the entire jumping performance (*right*)

where: F_D aerodynamic drag force, F_L aerodynamic lift force, ρ air density, v velocity, A projected area, C_D drag coefficient, C_L lift coefficient.

6.2.1 Inrun

Table 6.1 shows clearly that inrun speed is the most important factor affecting the jumping distance. However, the inrun phase has not been the subject of extensive investigation [32]. During the inrun jumpers try to maximize acceleration by minimizing the aerodynamic drag (F_L) which is primarily dependent on jumper's posture and drag area (usually expressed as a product C_DA, where A is a representative drag area of the object, and C_D is the drag coefficient, which represents its shape and how streamlined it is). If the aerodynamic forces during the inrun of the reference jump in Table 6.1 are ignored, the official inrun speed increases by 3.6 % (3.3 km/h) and the jumping distance by 18.2 m. Figure 6.11 shows an example of the measurement of ski jumper's aerodynamic forces in a wind tunnel. The results of these forces during the inrun position of three jumpers can be seen in Fig. 6.12. These force values give an estimation of CDA values of 0.084 and 0.150 for the jumpers A and B, respectively (see also [33]). It can be seen that jumper's bigger size does

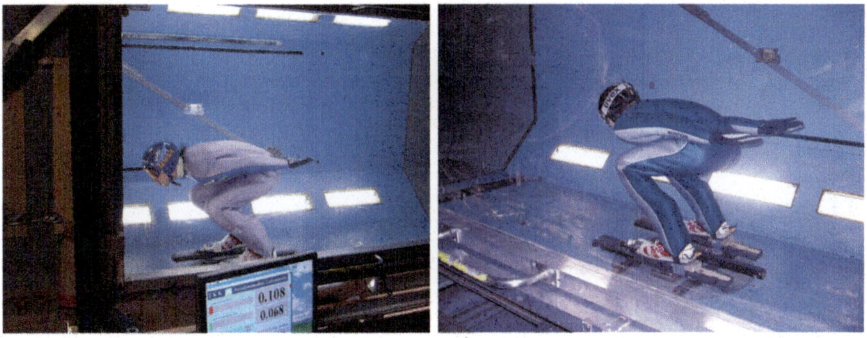

Fig. 6.11 Example of the measurement of the aerodynamic forces in the inrun position of a ski jumper in wind tunnel

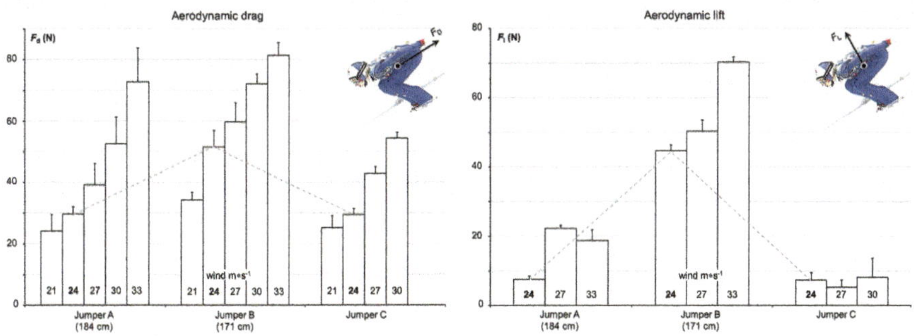

Fig. 6.12 Aerodynamic forces during the inrun position of three jumpers measured in wind tunnel

not necessarily mean bigger air resistance. With estimated drag area of 0.20 m^2 the above-mentioned C_DA values give the drag coefficients of 0.42 and 0.75 indicating a big difference in the aerodynamic quality of the inrun position between these two jumpers. Even bigger differences can be seen in the aerodynamic lift forces (F_L). In the inrun of ski jumping the aerodynamic lift should most likely be as close to zero as possible whereas, for instance, in speed-skiing (speed around 250 km/h) the aerodynamic lift may be important as it reduces ski friction [34]. Thompson et al. [35] obtained a drag coefficient of 0.160 for speed skiers in a wind tunnel. As jumpers' inrun position in ski jumping must also provide optimal initial conditions for a take-off, aerodynamics is not the only factor to be considered in this phase. These initial conditions under the effect of aerodynamic drag were discussed by Ettema et al. [36].

The average inrun speed of the jumpers participating in the large hill competition of the Olympic Games 2014 was 93.1 $\pm$ 0.4 km/h (92.1–94.2 km/h, $n = 48$) from the same starting gate and the correlation between the inrun speed and jumping distance was 0.434 ($p < 0.01$). The relationship between jumpers' body mass and inrun speed in the same Olympic competition can be seen in Fig. 6.13 which gives some insight of the effect of aerodynamics on different masses.

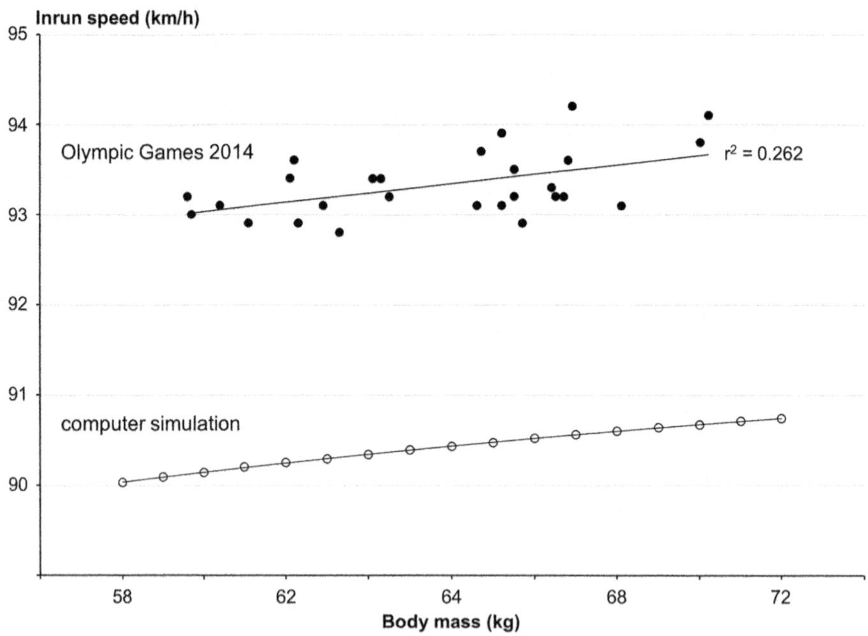

Fig. 6.13 The effect of body mass on jumpers' inrun speed (Olympic Games and computer simulation)

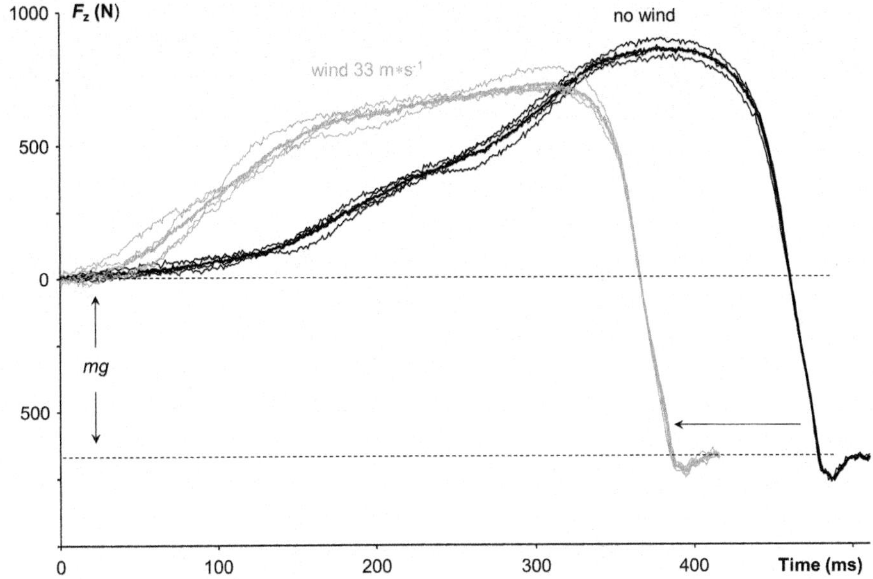

Fig. 6.14 Vertical take-off force curves measured in consecutive trials of one jumper in wind tunnel

6.2.2 *Take-Off Aerodynamics*

Due to the ballistic features of the ski jumping take-off, the role of aerodynamics during take-off has not been studied extensively (e.g., [33, 37, 38]), although some speculations had been presented earlier (e.g., [39]). From the aerodynamic point of view the ski jumping take-off is very crucial and always a compromise between maximizing vertical acceleration and minimizing air resistance to which jumpers are rapidly exposed during the take-off. Based on a series of wind tunnel experiments, Virmavirta et al. [33] concluded that aerodynamic lift (F_L) assists jumpers during the take-off by reducing the load (i.e., bodyweight mg—F_L) against which the jumpers are working in real ski jumping take-off movement. The reduced load results in decreased take-off time and higher rate of force development (Figs. 6.14 and 6.15). The measurements of Virmavirta et al. [33] were based on the ground reaction forces which include both take-off forces and aerodynamic forces and therefore, the instantaneous behavior of aerodynamic lift during take-off remains masked. However, by using equations for average acceleration and thereafter for average force it was possible to solve the average lift force during take-off from the difference in average force between the non-wind and wind conditions. The calculated average aerodynamic lift forces were 72 and 100 N for two different jumpers. Virmavirta et al. [37] confirmed their earlier findings by repeating the measurements of ski jumping take-off in a wind tunnel with skis (Figs. 6.16 and 6.17). Similar wind tunnel measurements were done by Müller [38]. In these measurements the low friction conditions between the inrun tracks and skis were simulated by using rollers under the skis.

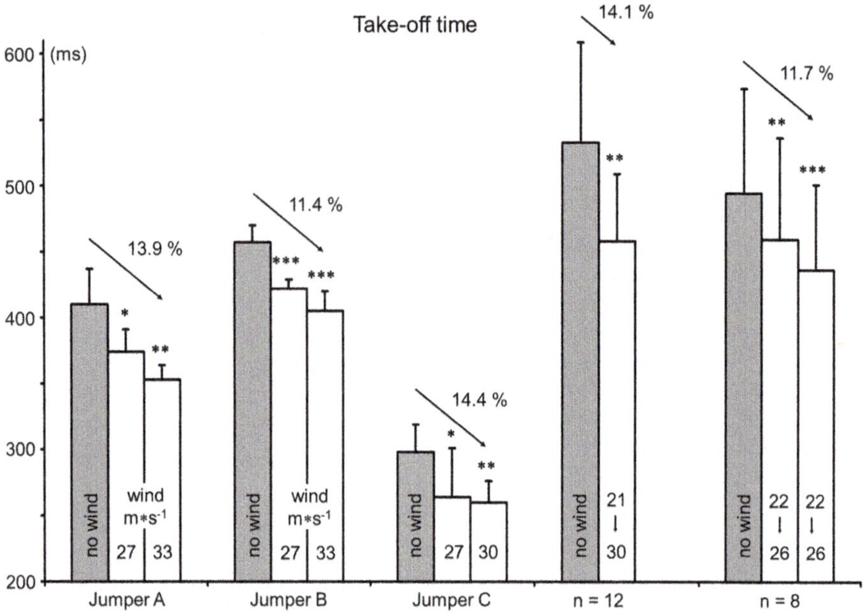

Fig. 6.15 Take-off time of several jumpers in different wind conditions in wind tunnel

Fig. 6.16 Example of the ski jumping take-off performed with skis in wind tunnel at wind speed of 105 km/h

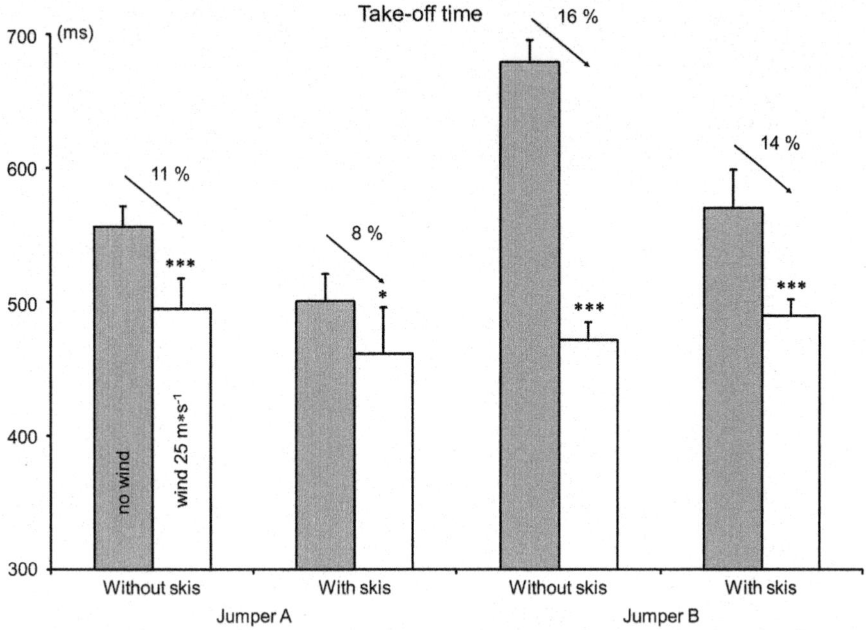

Fig. 6.17 Take-off times of two jumpers in different wind conditions with and without skis in wind tunnel

As the air resistance tends to produce backward rotation on a jumper during the latter part of the take-off and early flight phase, ski jumper must balance this rotation by generating a forward-rotating angular momentum during the take-off. This is well demonstrated in Fig. 6.18 by Schwameder [32].

6.2.3 Flight Phase

It is said that jumpers should achieve a stable flight position (Fig. 6.19) as quickly as possible after the take-off and small drag during the initial flight phase is a key for long flight distance (e.g., [40]). According to Schwameder [32] this is done when the backward- and forward-rotating angular momentums (Fig. 6.18) have been completely balanced during the early flight phase. Jumpers' steady flight position is completed within 0.5 s (Fig. 6.20, [41]) and during this short time period ($\sim$15 m) jumpers are, most likely, not able to do any intentional adjustments and therefore, the take-off phase determines the subsequent flight position. This means also that mistakes during take-off cannot be corrected during the flight phase, but the benefits of a successful take-off action can be lost by mistakes during the flight. In general the flight phase can be considered as the most "aerodynamic" part of the entire ski jumping performance.

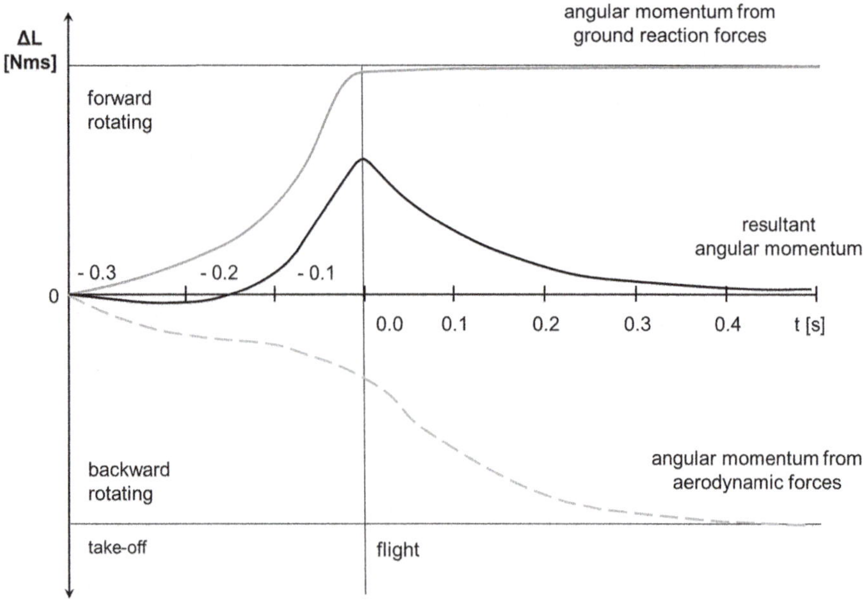

Fig. 6.18 Schematic illustration of angular momentums of a ski jumper during the take-off and flight (modified from [32])

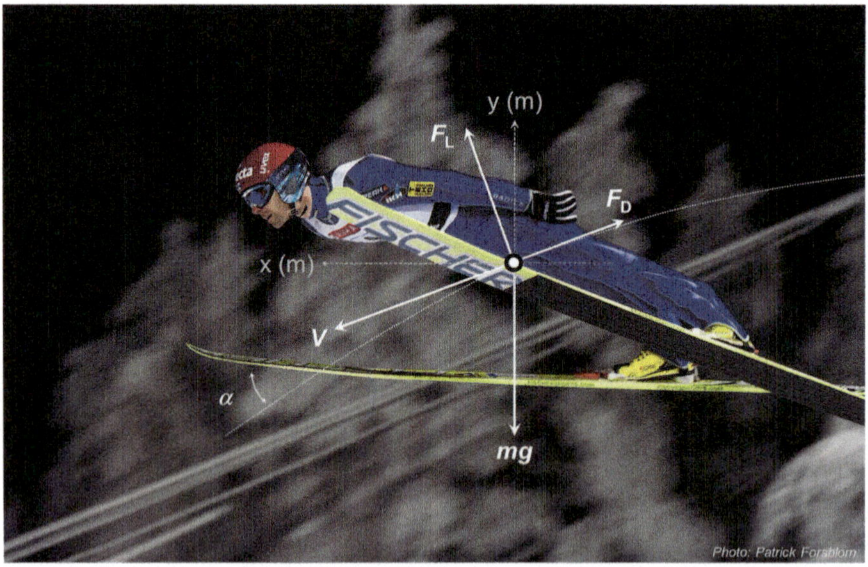

Fig. 6.19 Ski jumper at a stable flight phase (photo by Patrick Forsblom)

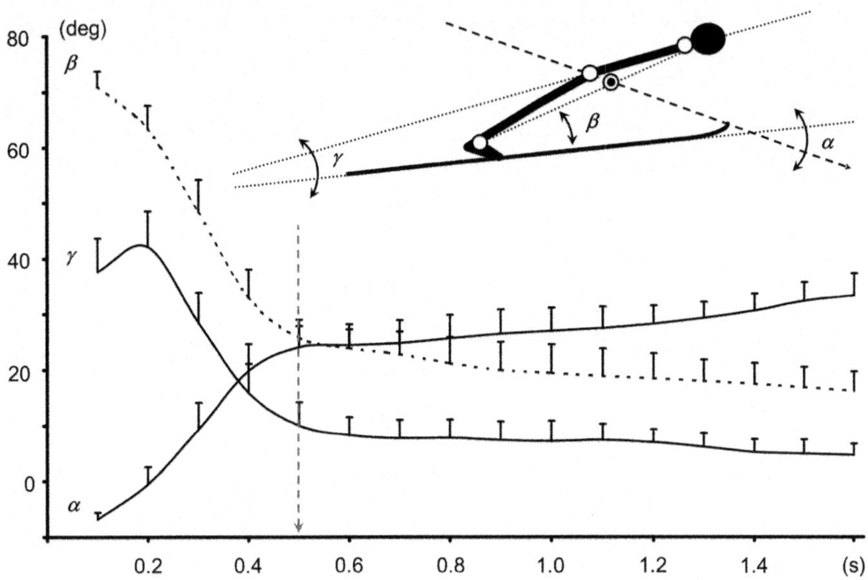

Fig. 6.20 Ski jumpers' average angular parameters during the early flight phase in the Olympic large hill competition 2002 [41]

Aerodynamics of the flight phase of the reference jump presented in the sensitivity analysis in Table 6.1 was based on the "aerodynamic polar" (Fig. 6.21) which shows the relationship between the lift (C_L) and drag (C_D) coefficients used in the computer simulation. The polar curve in Fig. 6.21 represents a jump with a good aerodynamic quality as it follows the upper part of the practical limits obtained from the series of the wind tunnel measurements. The C_D and C_L were swept in a wind tunnel with a model covering an extended range of α (angle of attack, 18–46°). The effect of aerodynamics on jumping distance of the reference jump (134 m) in Table 6.1 can be estimated by fully ignoring the aerodynamic forces during the flight phase. The jumping distance of this purely "ballistic" performance is only 45.2 m indicating a huge effect of aerodynamics in the flight phase of ski jumping. Figure 6.21 shows also a 10 % reduction in the C_L during the flight phase which shortened the jump distance by 20 m in the computer simulation. The aerodynamic lift and drag forces of the reference jump (Table 6.1) during the entire flight phase (4.7 s) are presented in Fig. 6.22 where the aerodynamic forces from the other studies [42, 43] are also presented. The forces of different studies are about same order of magnitude but their behavior during the flight differs due to the given $C_{L,D}$ values in different phases of the jump. Uhlář and Janura used Pontryagin's maximum principle and they said that the drag force should be minimized during the first 0.5 s of the flight and then after clear increase within 2–3 s, there is a strong requirement to minimize the drag force towards the end of the flight phase. They also mentioned that the best option is to minimize the lift force up to 0.5 s and then

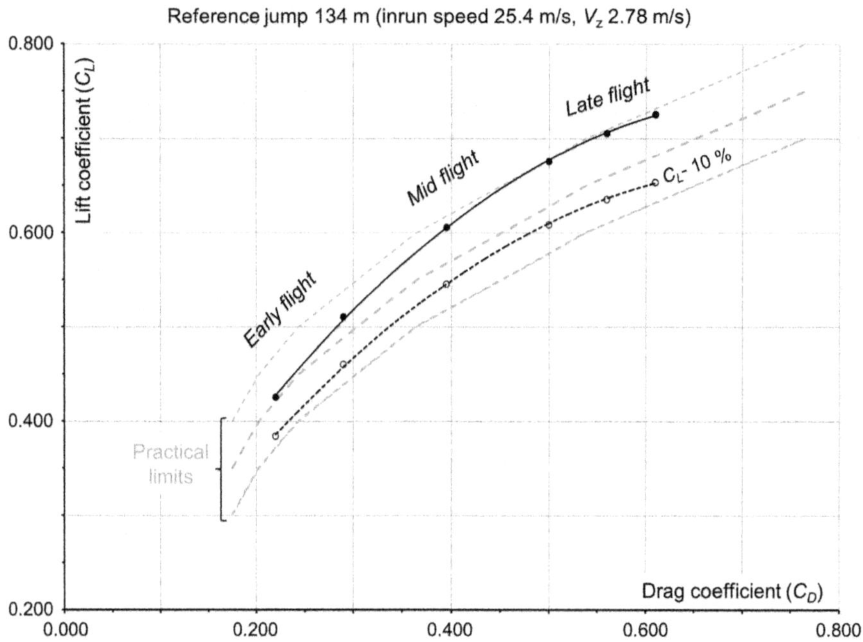

Fig. 6.21 Aerodynamic polar representing lift and drag coefficients used in the different phases of the flight in the computer simulation of Table 6.1

subsequently maximize it. According to this and other studies it is quite evident that the optimal time course of jumpers' flight position is a compromise between maximizing and minimizing different factors and it strongly depends on individual athletes.

A good contribution to the scientific literature of the flight phase in ski jumping has been provided by Müller and his co-authors [22, 26, 38, 42, 44, 45]. This research group has used a good mixture of wind tunnel measurements, computer simulations and field studies in their experiments. Typical example of the field study results can be seen in Fig. 6.23 which shows that the position angles of the world class ski jumpers change very little after the early flight phase [22, 45]. Chardonnens et al. [46] have introduced a new 3D technology based on inertial sensors to evaluate, e.g. the total aerodynamic force of stable flight. According to the authors, all dynamic parameters provided by this system were within the range of prior studies based on stationary measurement systems, except for the center of mass mean force which was slightly lower. The critical transition from the early phase to the stable flight phase is well demonstrated in Fig. 6.20. The mean values of the velocity components of jumpers during the early flight phase in the Olympic large hill competition are presented in Fig. 6.24. The importance of maintaining speed during the flight phase can be seen in a detailed comparison of the early flight phase between two best jumpers in the Olympic large hill competition 2002 (Fig. 6.25). It can be seen that

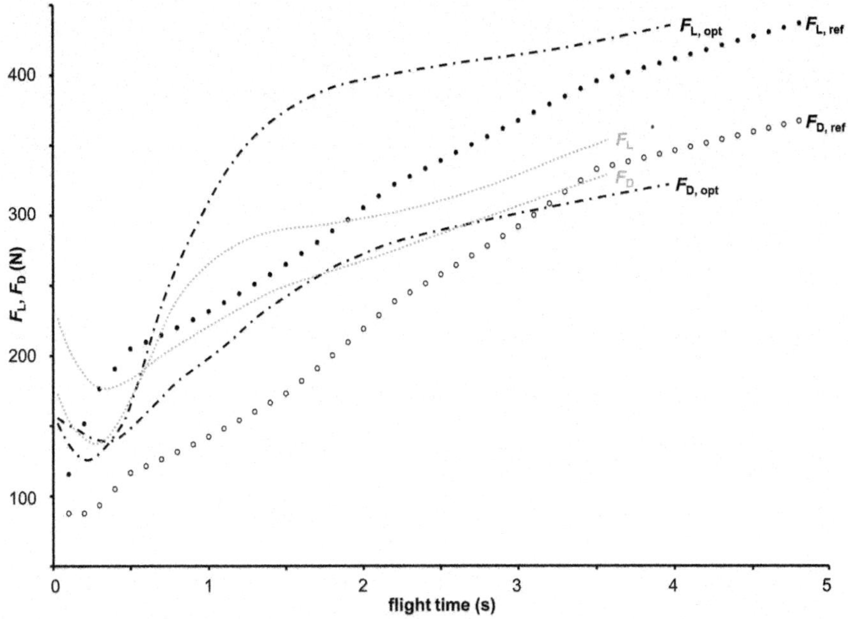

Fig. 6.22 Aerodynamic lift and drag forces during the entire flight phase from the computer simulation ($F_{L,\mathrm{ref}}$, $F_{D,\mathrm{ref}}$) and optimization study of Uhlář and Janura [43] ($F_{L,\mathrm{opt}}$, $F_{D,\mathrm{opt}}$) compared with the data of Schmölzer and Müller [42] (F_L, F_D)

jumper M was almost 80 cm higher at the end point of the analysis area (about 1.5 s after the take-off), but jumper A finished 5 m further with this jump. Although it is not seen in this analysis, the jumpers' velocities most likely explain the final result. With small air resistance right after the take-off jumper A certainly lost some lift but by maintaining the resultant velocity he probably had a necessary lift for the rest of the jump. The velocity difference of these jumpers at approx. 38 m at the x-axis gives 3.9 % higher aerodynamic forces for jumper A.

As the importance of being light is well known in sport events in which gravitational factors restrict performance, in ski jumping low body weight development resulted in some serious underweight problems in the early 2000s (e.g., [42]). Therefore, in 2004 International Ski Federation (FIS) decided to solve the problem by relating maximum ski length to body mass index (BMI). It was believed that it would no longer be attractive to be underweight. At this moment the maximum ski length is limited to 145 % of jumper's body height, provided that jumper's BMI is at least 21. For athletes with less than this minimum, a grading table of 0.125 BMI per 0.5 % of ski length will be applied. Also the size of the jumping suit has become less aerodynamic as the tolerance of measured circumference of the suit relative to the respective body measurement has been limited. The *effect of body mass* on jumping distance is presented in Fig. 6.26 where the simulated jump of 134 m in Table 6.1 was used as a reference. Figure shows that the advantage of low body weight has

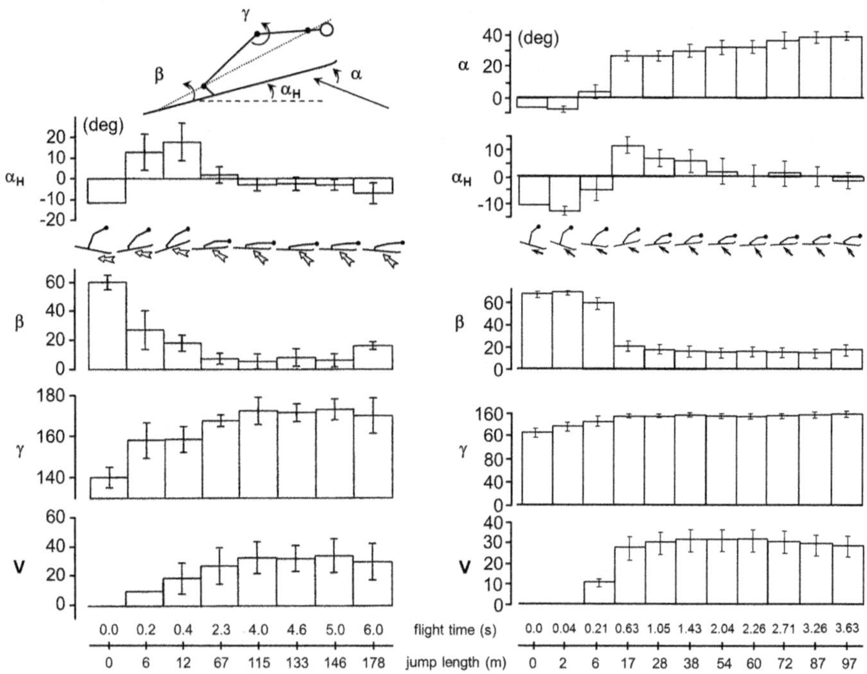

Fig. 6.23 Position angles of the world class ski jumpers in two different field measurements (modified from [22, 45])

not been defeated by the above-mentioned ski length reduction. The advantage in jumping distance is still about 1 m per every 0.5 BMI units below 21. BMI regulation compensates ~35–38 % of the advantage of low body weight. This is because skis cover only part of the aerodynamic area of the jumper/skis system. The low body weight problem in ski jumping has been used as a heuristic example for solution strategies for other sports as well [47].

Due to jumpers' continuous search for more aerodynamic equipment, regulations for ski jumping skis and suits have undergone several modifications during past 20 years. Much attention has been paid recently on the ski position during the flight phase in order to turn skis more flat relative to air flow. In V-style skis turn naturally on their medial edge around the longitudinal axis (Fig. 6.19). For avoiding this turn and thus improving aerodynamics of their ski position, jumpers have started to use the curved sticks as a link between the rear part of the ski boots and bindings. The setup for measuring the effect of ski position in a wind tunnel can be seen in Fig. 6.27. According to those measurements the lift force increases with increasing V-angle (ski opening) up to 30° and a harmful effect of "edge angle" seems to increase with increasing V-angle. The overall aerodynamic stability in

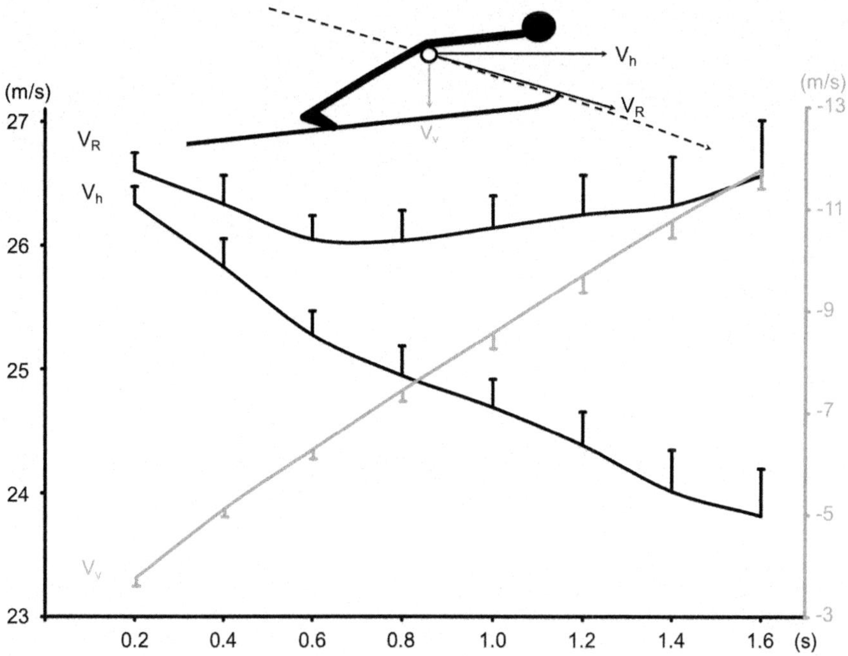

Fig. 6.24 Jumpers' average velocity components during the early flight phase in the Olympic large hill competition 2002 [41]

pitch, roll, and yaw during the flight phase is examined in detail by Marqués-Bruna and Grimshaw [48, 49]. The pitching moment on the jumper/skis system was also studied by Seo et al. [23].

As mentioned in the introduction, a lot of modifications have occurred in jumper *suit design*. All the changes in the past have increased the effects of aerodynamics and increased jump distance, but thereafter, when the dramatic effect of size and thickness of the jumping suit was fully understood (Fig. 6.28), the changes in suit regulations have tried to limit its aerodynamic quality because of too long jumps. Since the reproducibility of ski jumpers' position in consecutive trials in wind tunnel is fairly low, doll models should be used in suit research. Jumpers can be used only for rough estimation of the fit and behavior of suit in the air flow (Fig. 6.29). Within and sometimes outside the regulation limits, jumpers have always tried to modify their suits to improve their aerodynamics. Today the tolerance of measured circumference of the suit relative to the respective body measurement has been limited to maximum of 3 cm (men) and 4 cm (women) and the maximum thickness between 4 and 6 mm. Usually the changes between the modified suits are so small that the expected difference will be masked due to the difference between the consecutive dressing of the models in a wind tunnel. One example of the "ineffective" suit modification (neckline) used by ski jumpers can be seen in Fig. 6.30. Reducing aerodynamic effects is often compensated by

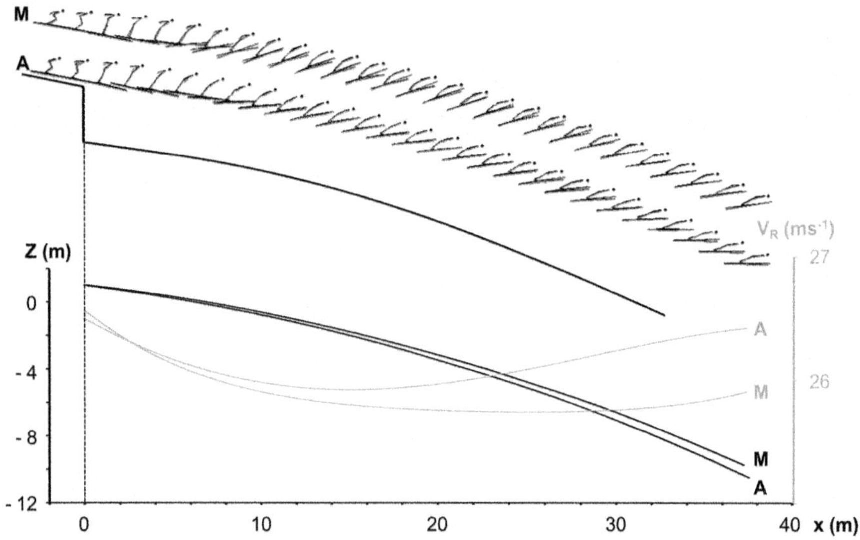

Fig. 6.25 Comparison of the early flight phase between two best jumpers in the Olympic large hill competitions 2002 [41]

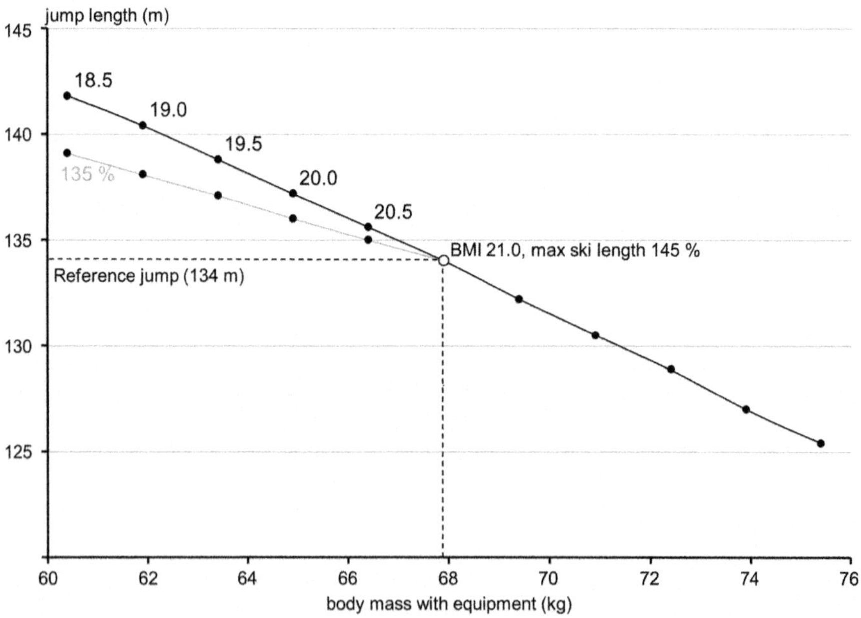

Fig. 6.26 The effect of body mass on jumping distance with the effect of ski length reduction for BMI values under 21 (18.5–20.5)

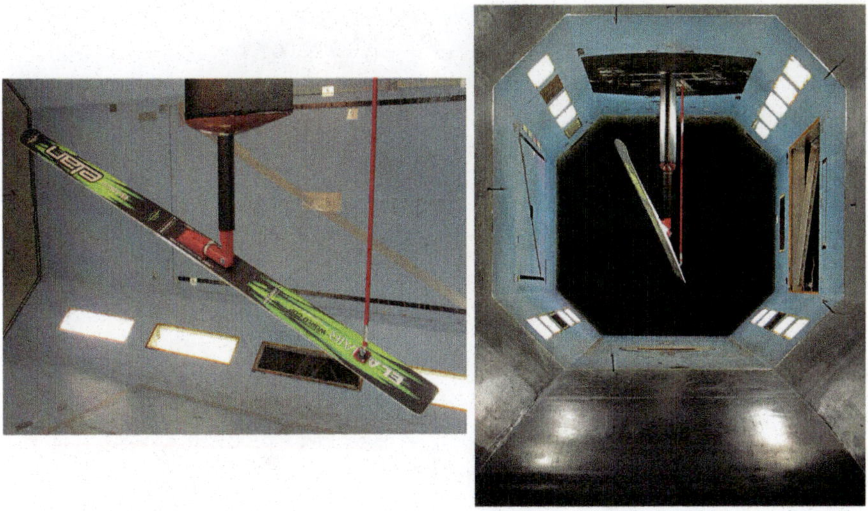

Fig. 6.27 Example of the aerodynamic study on ski position (edge and V angle)

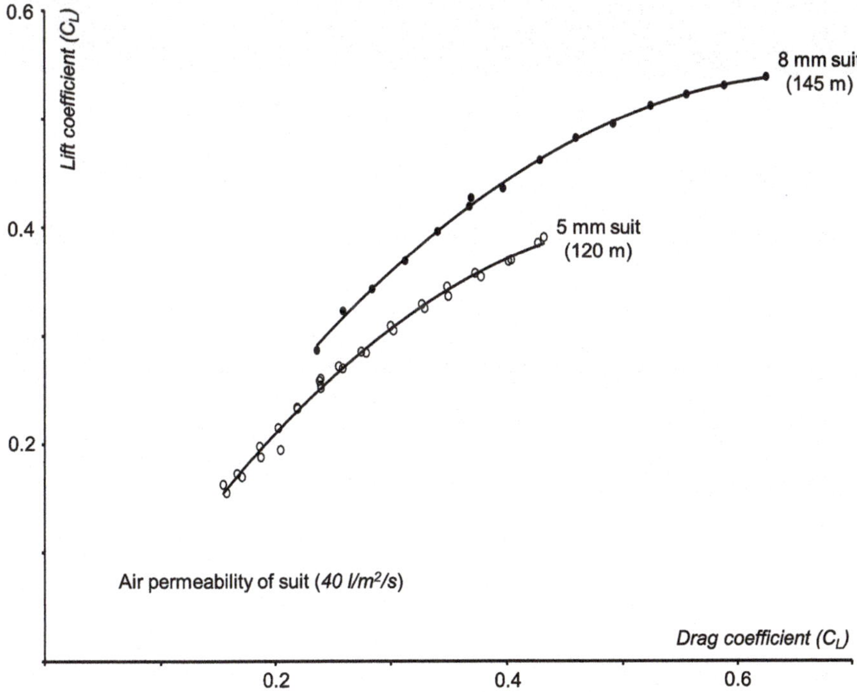

Fig. 6.28 Aerodynamic polars for two different suits of different size with corresponding difference in jumping distance [51]

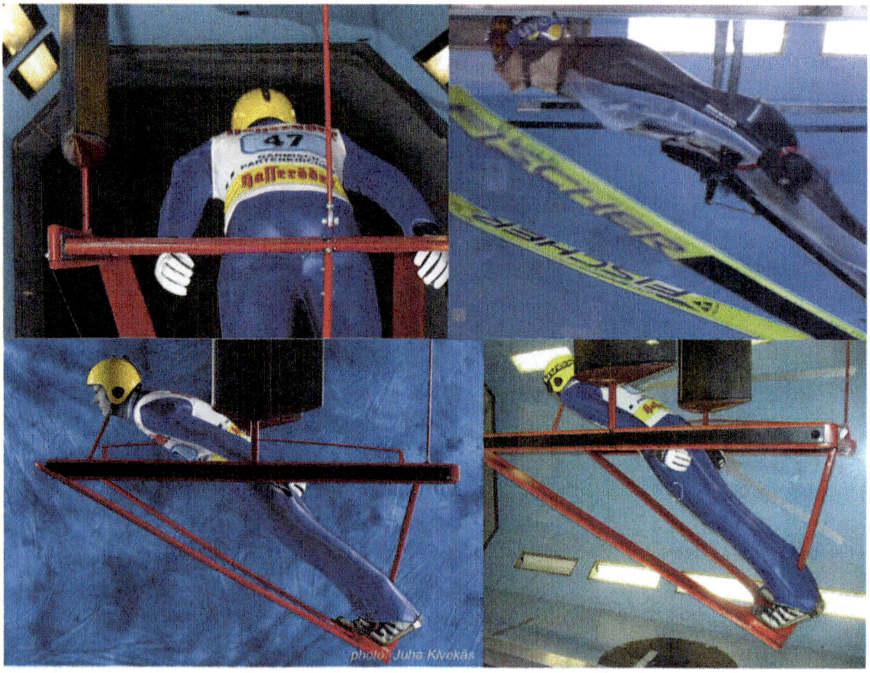

Fig. 6.29 Wind tunnel measurements of a ski jumper and a full-sized model

increased inrun speed which inevitably results in some problems experienced in ski jumping in previous years when best jumpers accidentally fly too far. The present situation requires a good understanding of the combined effects of different factors on jumping distance which is only possible if wind tunnel measurements and the proper computer simulation can be applied. At this moment only few published articles concerning the suit research are available (e.g., [25, 50]). Meile et al. [25] focused on three different jumping suits: (1) a skin-tight suit, (2) one suit in compliance with current regulations, and (3) one suit with moderate oversize. They concluded that "the influence of the suit might be of competitive relevance, because the observed small differences may have decisive influence on the length of the flight path."

Few years ago, the Fédération Internationale de Ski (FIS; International Ski Federation) adopted new solutions for smooth running of the events carried out in changing wind conditions which have a huge effect on aerodynamics in ski jumping. Müller et al. [22] calculated that a constant wind speed of 3 m/s from an advantageous direction increases jump length of 185 m by 16 m and reduces jump length by 23.7 m when blowing from a disadvantageous direction. The effect of wind on jumping distance was also mentioned by Seo et al. [24] who reported a relative gain or loss of 4 m in flight distance of about 130 m for a wind speed of 1 m/s. In 2010 FIS started to use the wind and gate (inrun length) compensation

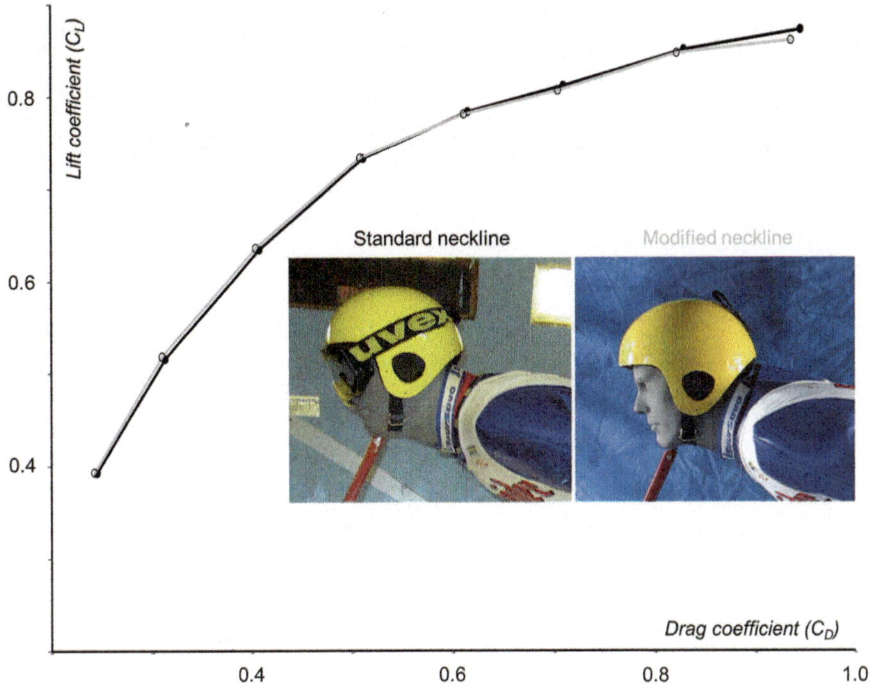

Fig. 6.30 Example of the jumping suit modification (neck "scoop") [51]

systems in ski jumping based on wind tunnel data and computer simulations. In order to create an easy-to-use table for each hill, wind compensation factors were calculated according to the following formula ([52]; Gasser 2009 unpublished FIS correspondence): $\Delta\omega$ =TWS(HS−36/20), where $\Delta\omega$ is the wind effect on jumping distance (m), TWS is the averaged tangential wind speed (m/s), and HS is the hill size (m). It is good to notice that the above-mentioned equation is linear with regard to wind speed (average measure of several anemometers). The compensation is calculated for one generic jump and therefore, it does not take into account the point of landing on the hill profile (Fig. 6.31). The effect of wind is naturally greatest in jumps landing on the area where the landing slope profile is steepest. The effect of wind and compensation system in ski jumping was examined in detail by Virmavirta and Kivekäs [31].

6.2.4 Landing

Landing in ski jumping is characterized by "Telemark technique" in which the knees are bent with one leg in front of the other. This is important for style marks because the jumper, who does not use Telemark, may lose points corresponding to

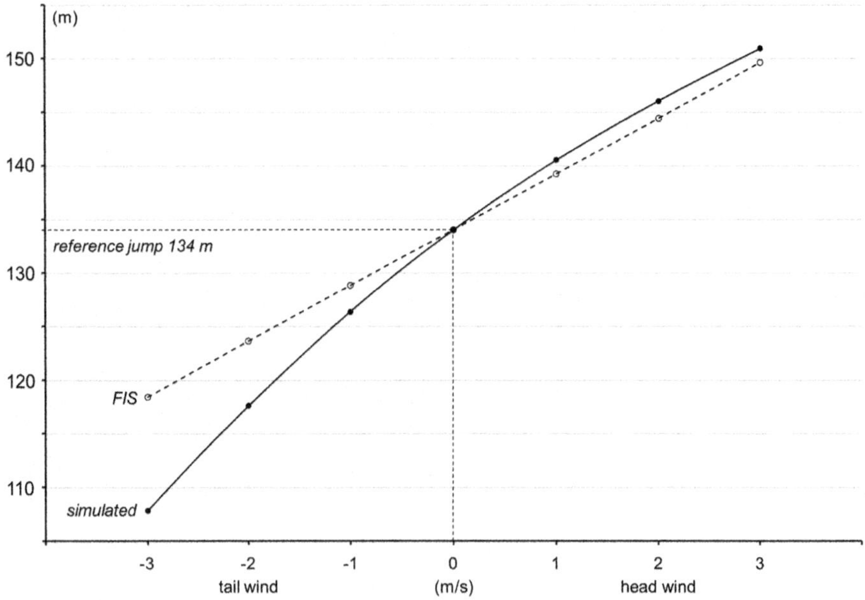

Fig. 6.31 The effect of wind on jumping distance, and the linear compensation used by the FIS

several meters in jumping distance. From the aerodynamic point of view, when the jumper approaches the ground in landing, he/she experiences a boundary condition which may modify the air flow past the body. This ground effect is important in the performance of aircraft over runways and aquatic birds over a water surface and therefore, Ward-Smith and Clements [53] thought that it could be important in landing phase of ski jumping as well. To simulate the effect of the ground on the ski jumper, they used a rolling ground and a scaled model in a wind tunnel with the adjustable ground clearance height. As a conclusion Ward-Smith and Clements [53, 54] mentioned that the ski jumper does not experience any in-creased lift from ground effect. According to Seo et al. [24, 55] the Reynolds number used by Ward-Smith and Clements was smaller than the practical one and they stated that their own aerodynamic data of the ground effect for the V style was in the range of the practical Reynolds number. In the ground effect setup with a full-size model a ground plate was set over the whole test section of the wind tunnel at 0.66 m and the clearance of the ski tails above the ground plate was several centimeters at $\alpha = 35°$. However, the authors mentioned that a rolling ground should be used in order to simulate the flight near the ground. Seo et al. [23] found that the lift area with the ground plate was always larger than that without ground plate. These results are presented in Fig. 6.32 where the other aerodynamic data of Seo et al. [24] is also presented. The difference of the lift area between the situations with and without ground plate increased with the increase of the ski-ski angle. Similar ground effect was not seen in the old flight style with skis parallel.

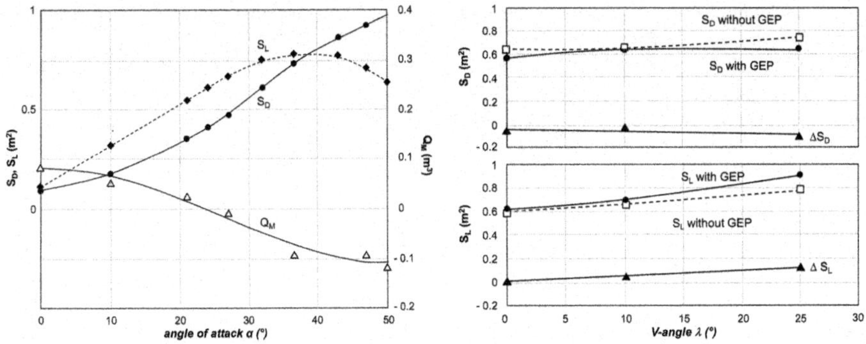

Fig. 6.32 *Left*: lift (S_L) and drag (S_D) areas and moment volume (Q_M) as a function of the angle of attack. *Right*: lift (S_L) and drag (S_D) areas as a function of ski-opening (V-angle) with and without ground effect plate (GEP), $\alpha = 37$. The differences are denoted by ΔS_D and ΔS_L (modified from [24])

Acknowledgements The author acknowledges Mr. Juha Kivekäs (Arteform Ltd.) for his expertise in aerodynamics and the computer simulations.

References

1. R. Straumann, Vom Skiweitsprung und seiner Mechanik, in *Jahrbuch des Schweize-rischen Ski Verbandes* (Selbstverlag des SSV, Bern, 1927), pp. 34–64
2. R. Straumann, Vom Skisprung zum Skiflieg. Sport, Zürich **63**, 7–8 (1955)
3. H. König, Theorie des Skispringens angewandt auf die Flugschanze in Oberstdorf, in *Uhrentechnische Forschung* (Steinkopf Verlag, Stuttgart, 1952), pp. 235–253
4. I. Tani, M. Iuchi, Flight mechanical investigation of ski jumping, in *Scientific Study of Skiing in Japan, Tokyo, Hitachi*, ed. by K. Kinoshita, Hitachi, Tokyo (1971), pp. 35–52
5. E.A. Grozin, Ski-jumping (in Russian), (Phyzcultura i sport, Moscow, 1971)
6. L.P. Remizov, Biomechanics of optimal flight in ski-jumping. J. Biomech. **17**(3), 167–171 (1984)
7. L.S. Pontryagin et al., *Mathematical Theory of Optimal Process* (Wiley Inter-science, New York/London, 1962)
8. A. Jung, M. Staat, W. Müller, Flight style optimization in ski jumping on normal, large, and ski flying hills. J. Biomech. **47**(3), 716–722 (2014)
9. J. Denoth, S.M. Luethi, H. Gasser, Methodological problems in optimisation of the flight phase in ski jumping. Int. J. Sport Biomech. **3**, 404–418 (1987)
10. J. Maryniak, B. Krasnowski, Balance and longitudinal stability of a ski jumper in flight [in Polish]. Mechanika Teoretyczna i Stosowana **2**(3), 351–373 (1974)
11. M. Hubbard et al., A multisegment dynamic model of ski jumping. Int. J. Sport Biomech. **5**, 258–274 (1989)
12. E. Pfaff, Interview mit Toni Innauer, Trainerforum. Leistunsport **6**, 34–38 (2008)
13. R. Meier, *Skifliegen -Schanzenbau* (Diplomarbeit in Biomechanik, ETH, Zürich, 1977)
14. J. Maryniak, E. Ladyzyńska-Kozdraś, S. Tomczak, Configurations of the Graf-Boklev (v-style) ski jumper model and aerodynamic parameters in a wind tunnel. Hum. Mov. **10**(2), 130–136 (2009)
15. R. Mahnke, G. Hochmuth, Neue Erkenntnisse zur Luftkraftwirkung beim Ski-springen. Research Report for Forschungsinstitut für Körperkultur und Sport, Leipzig (1990)

16. D.A. Cutter. *Nordic Ski Jumping Aerodynamics* (AIAA, Aeronautical Engineering, United States Air Force Academy, Colorado Springs, CO, 1993)
17. K. Watanabe, I. Watanabe, Aerodynamics of ski-jumping: effect of 'Vstyle' to distance, in *XIVth Congress of the International Society of Biomechanics* (1993), pp. 1452–1453
18. M. Tavernier, P. Cosserat, Flight simulation of ski-jumping comparison of two styles of flight, in *Proceedings of the 14th Congress of ISB Congress* (Paris, 1993), pp. 1328–1329
19. B. Jošt, Differences in some kinematic flight parameters between the classical and the new so called "V" technique in ski jumping. Kinesiology **26**(1–2), 18–21 (1994)
20. H. Schwameder, E. Müller, Biomechanische Besreibung und Analyse der V-Technik im Skispringen. Spectrum der Sportwissenschaften **7**, 5–36 (1995)
21. H. Jin et al., Desirable gliding styles and techniques in ski jumping. J. Appl. Biomech. **11**, 460–474 (1995)
22. W. Müller, D. Platzer, B. Schmölzer, Dynamics of human flight on skis: improvements in safety and fairness in ski jumping. J. Biomech. **29**(8), 1061–1068 (1996)
23. K. Seo, I. Watanabe, M. Murakami, Aerodynamic force data for a V-style ski jumping flight. Sports Eng. **7**, 31–39 (2004)
24. K. Seo, M. Murakami, K. Yoshida, Optimal flight technique for V-style ski jumping. Sports Eng. **7**, 97–104 (2004)
25. W. Meile et al., Aerodynamics of ski jumping: experiments and CFD simulations. Exp. Fluids **41**, 949–964 (2006)
26. H. Nørstrud, I.J. Øye, On CFD simulation of ski jumping, in *Computational Fluid Dynamics for Sport Simulation*, ed. by M. Peters. Lecture Notes in Computational Science and Engineering, vol. 72 (Springer, Berlin/Heidelberg, 2009), pp. 63–82
27. K.-D. Lee, M.-J. Park, K.-Y. Kim, Optimization of ski jumper's posture considering lift-to-drag ratio and stability. J. Biomech. **45**(12), 2125–2132 (2012)
28. B. Venås, I. Øye, Integrated wind shielding for the new Holmenkollen ski jump, in *The Fifth International Symposium on Computational Wind Engineering (CWE2010)*, Chapel Hill, NC, 23–27 May 2010
29. M. Hubbard, The flight of sports projectiles, in *Biomechanics in Sport*, ed. by V. Zatsiorsky (Blackwell Science Ltd, Oxford, UK, 2008), pp. 381–400
30. P.G. Wright, The influence of aerodynamics on the design of formula one racing cars. Int. J. Veh. Des. **3**(4), 383–397 (1982)
31. M. Virmavirta, J. Kivekäs, The effect of wind on jumping distance in ski jumping – fairness assessed. Sport Biomech. **11**(3), 358–369 (2012)
32. H. Schwameder, Biomechanics research in ski jumping, 1991–2006. Sport Biomech. **7**(1), 114–136 (2008)
33. M. Virmavirta, J. Kivekäs, P.V. Komi, Take-off aerodynamics in ski jumping. J. Biomech. **34**, 465–470 (2001)
34. H. Nørstrud, Alpine downhill and speed-skiing. Sport Aerodyn. **506**, 131–138 (2008)
35. B.E. Thompson, W.A. Friess, K.N. Knapp II, Aerodynamics of speed skiers. Sports Eng. **4**, 103–112 (2001)
36. G.J.C. Ettema, S. Bräten, M.F. Bobbert, Dynamics of the in-run in ski jumping: a simulation study. J. Appl. Biomech. **21**, 247–259 (2005)
37. M. Virmavirta, J. Kivekäs, P.V. Komi, Ski jumping take-off in a wind tunnel with skis. J. Appl. Biomech. **27**(4) 375–379 (2011)
38. W. Müller, Performance factors in ski jumping, in *Sport Aerodynamics*, ed. by H. Nørstrud. CISM International Centre for Mechanical Sciences, vol. 506 (Springer, Vienna, 2008), pp. 139–160
39. M. Virmavirta, P.V. Komi, Measurements of the take-off forces in ski-jumping. Part I and II. Scand. J. Med. Sci. Sports **3**, 229–243 (1993)
40. M. Murakami et al., Ski jumping flight skill analysis based on high-speed video image. Procedia Eng. **2**, 2381–2386 (2010)
41. M. Virmavirta et al., Characteristics of the early flight phase in the Olympic ski jumping competition. J. Biomech. **38**, 2157–2163 (2005)

42. B. Schmölzer, W. Müller, The importance of being light: aerodynamic forces and weight in ski jumping. J. Biomech. **35**, 1059–1069 (2002)
43. R. Uhlář, M. Janura, Pontryagin's maximum principle and optimization of the flight phase in ski jumping. Acta Univ. Palacki. Olomuc. Gymnica **39**, 61–68 (2009)
44. W. Müller, D. Platzer, B. Schmölzer, Scientific approach to ski safety. Nature **375**, 455 (1995)
45. B. Schmölzer, W. Müller, Individual flight styles in ski jumping: results obtained during Olympic games competitions. J. Biomech. **38**, 1055–1065 (2005)
46. J. Chardonnens et al., Measurement of the dynamics in ski jumping using a wearable inertial sensor-based system. J. Sports Sci. **32**(6), 591–600 (2014)
47. W. Müller, Towards research-based approaches for solving body composition problems in sports: ski jumping as a heuristic example. Br. J. Sports Med. **43**, 1013–1019 (2009)
48. P. Marqués-Bruna, P. Grimshaw, Mechanics of flight in ski jumping: aero-dynamic stability in pitch. Sports Technol. **2**(1–2), 24–31 (2009)
49. P. Marqués-Bruna, P. Grimshaw, Mechanics of flight in ski jumping: aero-dynamic stability in roll and yaw. Sports Technol. **2**(3–4), 111–120 (2009)
50. H. Chowdhury, F. Alama, D. Mainwaring, Aerodynamic study of ski jumping suits. Procedia Eng. **13**, 376–381 (2011)
51. M. Virmavirta, J. Kivekäs, Effective use of a wind tunnel for ski jumping suit research, in *Proceedings of the XXIIth ISB Congress*, Cape Town, 2009
52. FIS Fact sheet, Important new rules in ski jumping and nordic combined summer grand prix (2009), http://www.fis-ski.com/mm/Document/document/General/04/23/02/July292009NewrulesatSkiJumpingandNordicCombinedSummerGrandPrix2009_Neutral.pdf. 28 July 2009
53. A.J. Ward-Smith, D. Clements, Experimental determination of the aerodynamic characteristics of ski-jumpers. Aeronaut. J. **86**, 384–391 (1982)
54. A.J. Ward-Smith, D. Clements, Numerical evaluation of the flight mechanics and trajectory of a ski-jumper. Acta Appl. Math. **1**(3), 301–314 (1983)
55. K. Seo et al., Aerodynamic study for the ground effect of ski jumping, in *Proceedings of 19th ISBS Conference*, San Francisco, ed. by J.R. Blackwell, R.H. Sanders (2001), pp. 128–130

Chapter 7
Bobsleigh and Skeleton

Edoardo Sabbioni, Stefano Melzi, Federico Cheli, and Francesco Braghin

The aim of this chapter is to give a complete overview about the bobsleigh (or bobsled) and the related discipline of skeleton. Although not the focus of the chapter, some analyses will also be extended to luge. Before starting to analyze specific topics about these sports, it is interesting to give a brief introduction about the history of these disciplines.

7.1 Introduction

For this part we mainly refer to [1, 2].

We can correctly argue that the sled has been for centuries a mode of transportation, but the sport of bobsleigh didn't begin until the late nineteenth century when the Swiss attached a steering mechanism to a toboggan.[1] The bobsleigh has humble beginning, it born thanks to the idea of Caspar Badrutt, an hotelier of St. Moritz (Switzerland). In fact St. Moritz in the nineteenth century was known as a town where rich people went to took mineral cures during the months of May through September, that is for the two thirds of the year there were no guests in the local hotels. This fact stimulates Mr. Badrutt to invent a "winter resorting," that is to

---

[1]A toboggan is a simple sled which is a traditional form of transport used by the Innu and Cree of northern Canada. The big difference with respect to the most part of sleds is that it has no runners nor skis on the underside, but it rides directly on the snow. The traditional toboggan is made of bound, parallel wood slats, all bent forward at the front to form a sideways "J" shape.

E. Sabbioni (✉) • S. Melzi • F. Cheli • F. Braghin
Department of Mechanical Engineering, Politecnico di Milano, Milano, Italy
e-mail: edoardo.sabbioni@polimi.it; stefano.melzi@polimi.it; federico.cheli@polimi.it; francesco.braghin@polimi.it

F. Braghin et al. (eds.), *The Engineering Approach to Winter Sports*,
DOI 10.1007/978-1-4939-3020-3_7

Fig. 7.1 The evolution of the bobsled during time. *From left to right*: the Swiss bobsleigh team in 1910; an East German bobsleigh in 1951; the USA bobsleigh team at 2010 Winter Olympics. Adapted from [2]

provide to his guests (in a large part coming from England) some activities during the winter seasons which were the beginning of winter sports. The first sleds were created at the beginning of 1870s, when some Englishmen adapted a type of delivery sled for daring dashes down twisting narrow streets of St. Moritz. Careening around the town's streets became increasing popular, but the incidence and frequency of pedestrian collisions and risk to life grew proportionately. Therefore Mr. Badrutt stepped in and created the first purpose-built half-pipe track: it is still in operation today and has served as a host track during two Winter Olympics.

The first racing sleds were made of wood but were soon replaced by steel sleds that came to be known as bobsleds, so named because of the way crews bobbed back and forth to increase their speed on the straightaways. In 1897, the world's first bobsleigh club was founded in St. Moritz, Switzerland, spurring the growth of the sport in winter resorts throughout Europe. By 1914, bobsled races were taking place on a wide variety of natural ice courses.

The bobsled popularity and diffusion increased during the years (in Fig. 7.1 the big evolution of the discipline is depicted).

The *Fédération internationale de bobsleigh et de tobogganing* (FIBT) was founded in 1923 and the following year a four-man race took place at the first ever Winter Olympics in Chamonix, France. A two-man event was added at the 1932 Olympics in Lake Placid, USA, a format that has remained to the present. For years this discipline remained an activity for rich and adventurous people. Only by the 1950s the sport as we know it today had begun to take shape. Until the advent of World Cup competition in the mid-1980s, bobsleigh success was determined solely by performance at the Olympics, World and European Championships. Another stage in the evolution of the sport came in the early 1990s with the debut of women bobsledders at events in Europe and North America.

Apart from the British influence in the sport's infancy, a strong US presence from 1928 to 1956, and recent advances by other countries, bobsledding has been dominated largely by Europe's alpine nations over the years. By far the most successful bobsledding nations have been Switzerland and Germany. The Swiss have won more medals in Olympic, World, and European championships and World Cup competitions than any other nation. East Germany emerged as the sport's major power in the mid-1970s with its emphasis on sled design and construction. Since reunification, German bobsledders have remained a formidable group, winning numerous Olympic medals and World Championship titles since 1990. Italy also has a long and successful track record in the sport, particularly from the mid-1950s to late 1960s, and Austria has had its shining moments. In World Cup competition the Swiss and Germans have won the most medals, followed closely by Canadian teams.

Also the skeleton origins are related to English people. In 1882 English soldiers constructed a toboggan track leading from Davos to Kloster in Switzerland: it was used to descending on toboggans and in order to make it more challenging, curves were inserted in the Davos track. In 1892, an Englishman, Mr. Child, surprised his sports friends with a new sled made mostly from metal. Some speculate that since the sled looked like a skeleton, it was thusly called. Others claim that the name "Skele" derives from an incorrect Anglicization of the Norwegian word "Kjaelke." In 1905, skeleton was practiced for the first time outside Switzerland. In 1923, the FIBT was formally constituted. According to the minutes of the FIBT Paris Congress in 1926, the International Olympic Committee declared bobsleigh and skeleton Olympic sports. Another important date of skeleton history is 1970, when a new sled was created from the traditional skeleton: it was specifically designed for use on bobsleigh runs. On Oct. 2, 1999, the FIBT achieved its goal to have skeleton included on the Olympic Games program.

7.2 Rules

This section deals with the basic rules of bobsleigh. In particular we want to focus on the guidelines about sled construction and track building, because the most part of engineering innovations to this sport must abide by the restriction imposed by the regulations. Moreover, because of the high risks related to this sport (basically due to high velocities and relatively low protections), some construction criteria must be known to guarantee the safety of the athletes. In the end, it is also useful to remind that this sport implies substantial costs to build and maintain the track both during competitions and "down times": therefore rules are important to always provide to athletes a good experience and to do a suitable maintenance in order to preserve the track from deterioration and damages.

Although an update of the rules was issued by the Federation in October 2014 (the update is freely accessible by FIBT [1]), the rules we refer to in this book are taken from [3]. It was the latest set of rules issued by the Federation at the beginning of book writing. It must however be noticed that updated rules are only slightly different with respect to the ones described in the following.

We remember that the bobsled is both for men and for women. For men there are two kinds of competition: 2-man bobsled and 4-man bobsled. For women there is only the 2-athletes competition. In skeleton too, the competitions are both for men and women, but a skeleton sled is driven by a single person. There are also particular "team competitions" in which the final score is calculated by the sum of a bobsled and a skeleton run.

The track is prepared after agreement among the Jury, the Race Director, and the Chief of Track. In case of snowfall, it shall be made certain that a regularly scheduled race is possible. Particular attention shall be paid to the ice surface at the start-up to the start time measurement (50 m), which shall be swept after every sled. The track should be swept at regular intervals.

Each part of the sled is strictly subjected to rules and limits. We are now listing the most important ones both for bobsled and for skeleton; if one is interested in a specific dimension or parameter can refer to [3]. The common purpose of these rules is the safety of athletes and to guarantee a fair and honest competition to all people involved in. We also give some comments on these rules and limitations, to underline how they affect and restrict final performances.

7.2.1 Bobsleigh Weights (Rule 14.1.3.1)

All the dimensions of the bobsled are subjected to norms, and must respect some minimum and maximum values: in Fig. 7.2 there is an example. All dimensions refer to bobsleds without any load on a flat surface, if not specified to the contrary.

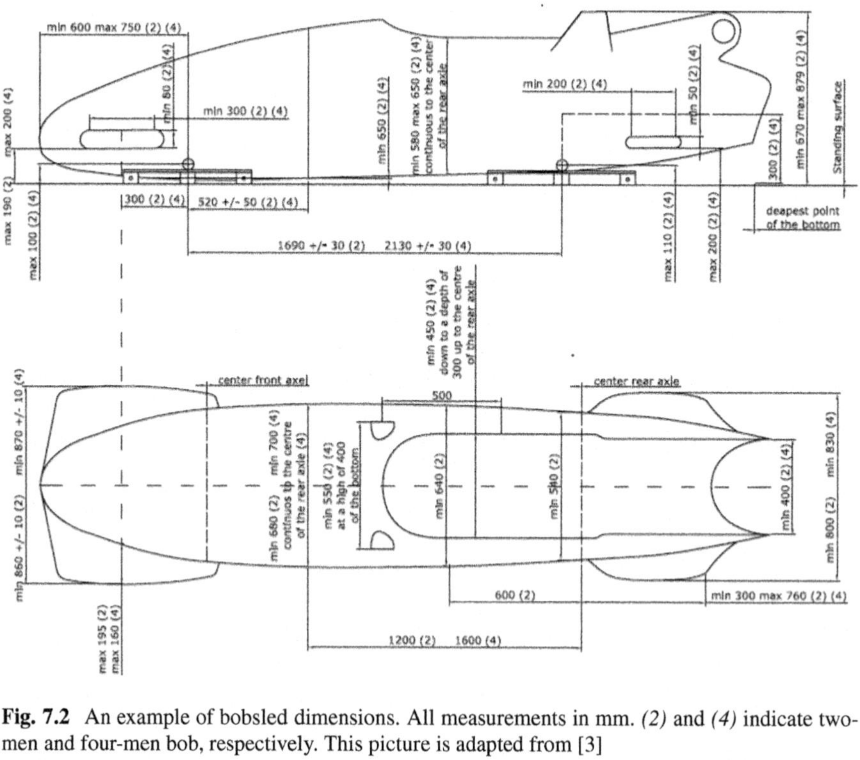

Fig. 7.2 An example of bobsled dimensions. All measurements in mm. *(2)* and *(4)* indicate two-men and four-men bob, respectively. This picture is adapted from [3]

Also the weight of the sled must respect some limits:

1. Minimum weight (bob including runners, excluding the crew):
   - two-men bob and women's bob: 170 kg;
   - four-men bob: 210 kg.
2. Maximum weight (including crew members and equipment required):
   - two-men bob: 390 kg;
   - four-men bob: 630 kg;
   - women's bob: 340 kg.

Bobs have four runners (one pair of front runners and one pair of rear runners), and only the pushing force of the team at the starting point and the gravity during the run are allowed to propel the bob. Due to the fact that a bob is propelled by athletes pushing at the start and by gravity during the race course, both limited weights are crucial on the final performance. Obviously having the bob at minimum weight, while at its maximum with the crew will be the best configuration, because a lighter bob means lower inertia during pushing and a heavier combination of bob and crew means a higher potential energy during the course.

Moreover, athletes usually have muscle mass that means that more weight means more force at the start.

If the athlete's weight doesn't fulfill the difference from minimum and maximum weight, ballast could be used. If for this purpose, ballast is used that is not firmly welded, bolted, or clamped to the frame by steel fasteners, it must be contained in a steel hollow profile; moreover no liquids or liquid-like materials may be used as ballast materials. The optimal ballast is calculated from the best compromise between pushing inertia and potential energy available. Ballast role is also crucial in front/rear weight distribution definition. The best weight distribution is the one that optimizes runners contact pressure during the whole course, considering that loads could vary in turns up to five times due to centripetal acceleration. Different runners usually require different weight distribution to maximize their performance. This means that designing a bob it is recommended to have a minimum weight that is less than the one allowed by Rules, with an average value of weight distribution (around 37–38 % on the front axle) and then use some ballast to adjust the weight distribution to the one required by track conditions and runners choice.

7.2.2 Bobsleigh Functional Dimensions (Rule 14.1.3.2)

7.2.2.1 Gauge (Rule 14.1.3.2.1)

The gauge is the same for the front and rear runners, 670 mm $\pm$ 1 mm (see Fig. 7.3).

To optimize pushing performance it is mandatory to have gauge at 670 mm because grooves in the first part of the track, made to guide bobs in the first pushing

Fig. 7.3 Gauge dimensions

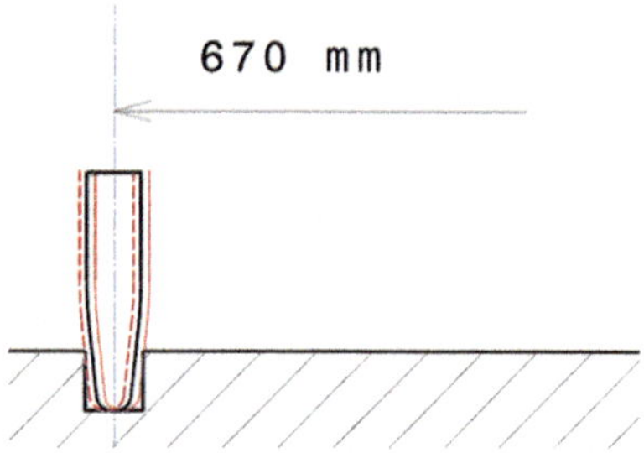

Fig. 7.4 Bobsleigh axle base

part where crew is outside the sled, are made with a standard machine and runners must avoid to contact grooves on the sides causing an increase in friction. Gauge is achieved by means of metallic spacers inserted between runners and U shaped runner brackets. The spacers purpose is not only to obtain optimal gauge, but also to perfectly align runners on the same axle and between axles, because any misalignment between runners increases friction during the whole race course.

7.2.2.2 Axle Base (Rule 14.1.3.2.2)

Distance from the center of the front axle to the center of the rear axle (referred to the plane of symmetry of the bob, see Fig. 7.4): 1690 mm ± 30 mm (two-men bob); 2130 mm ± 30 mm (four-men bob).

A longer axle base means a more stable bob, but less drivable and less reactive in turn changes. A shorter axle base means a more reactive and drivable bob, but less stable in turn changes. Even in this case the optimal axle base is a compromise between track layout and pilot driving technique.

7.2.3 Bobsleigh Construction and Functioning (Rule 14.2)

The purpose of the Rules governing the construction and functioning of the equipment used is to promote that the equipment used in FIBT events is able to withstand the stresses of training and competition and that the equipment used provides adequate protection for those using the equipment. Furthermore the Rules aim to promote a fair and honest competition and moreover are directed at controlling the cost of competing in FIBT events.

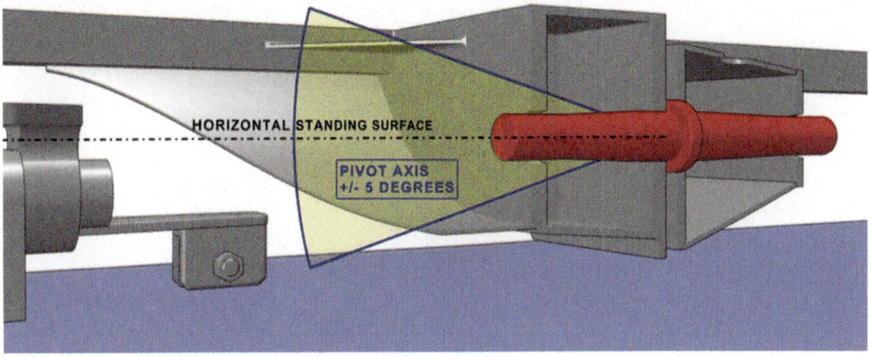

Fig. 7.5 Pivot bolt and pivot axis

7.2.3.1 Articulation (Rule 14.2.2.1)

During transitions between straights and turns, front runners and rear runners could be on different planes. This rotation around longitudinal axis is essentially defined and restricted inside Rules.

Every bob frame must have a transverse division between the front axle and rear axle. The front portion and the rear portion of the bob are pivotally connected, the longitudinal pivot axis being essentially parallel to the horizontal standing surface. The angle between said longitudinal pivot axis and the horizontal standing surface must not exceed 5° (see Fig. 7.5). Having an angle, usually downward, between front and rear portions of the bob generates a self-steering behavior during transitions. The greater is the pivot angle, the greater is the self-steering effect. This angle is chosen as a function of pilot driving technique.

The longitudinal pivot axis is embodied by a longitudinal pivot bolt of steel. This pivot bolt is defined in its minimum dimensions in order to control its strength and stiffness, and may have means of rotational elastic stabilization. These stabilizing elements must not contribute to any vertical movement. No hydraulic or pneumatic damping system is allowed. A pivot bolt that is softer in flexibility could be useful because it could absorb and compensate vertical movements of the bob, but it may be that the front cowling, that is connected with the front portion of the frame, collides with the rear cowling, that is connected with the rear portion of the frame. Even in this case, pivot bolt stiffness may be defined as a compromise between these considerations, in association with its strength calculation.

7.2.4 Frame and Running Gear (Rule 14.2.3)

The purpose of the paragraph is to give some rules to control the sturdiness of the frame in the interest of safety and to control the amount of suspension provided to

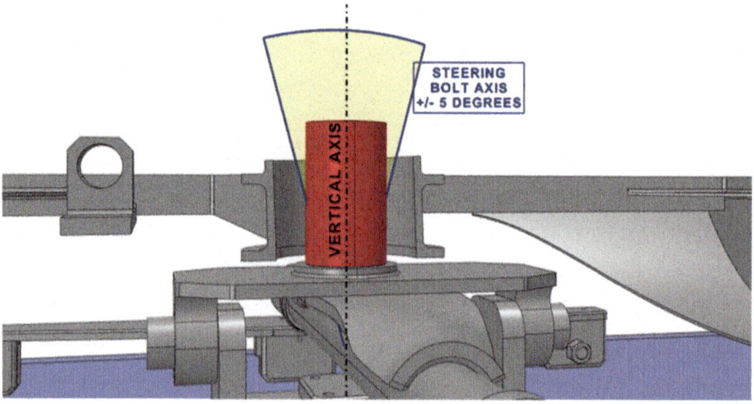

Fig. 7.6 Steering bolt and steering bolt axis

bobsleigh and crew. All joints in the frame must be rigid. The front and rear frame must not be adjustable in stiffness during heats or between heats.

7.2.4.1 Front Portion of the Frame (Rule 14.2.3.3)

The front portion of the bob must be constructed with a continuous welded supporting frame made of steel and extending from the longitudinal bolt at the transversal division line to the bearing housing of the steering bolt. This rotating steering bolt is fitted to the frame and rotates by means of a steering mechanism. The said steering bolt must not exceed 5° from vertical (see Fig. 7.6). A change in steering bolt angle changes runners mechanical trail (that is the distance in a side view between rotation axis and runners contact point), varying the self-alignment forces acting on the steering system.

The lower end of the steering bolt is completely welded perpendicular to a steel steering plate. On this steering plate, at a bolt lying in front of the axle and at a bolt lying behind the axle, steel flanges are pivoted on a pivot axis that is perpendicular to the front axle and perpendicular to the steering bolt. The swivel angle of the front axle around this pivot axis is limited to maximally $\pm 12°$ in a vertical plane (see Fig. 7.7). A continuous, one-piece steel bottom plate is mounted below the axle; this bottom plate is firmly bolted, with no free play, to both steel flanges. Lengths, widths, and thicknesses of these said components are regulated by Rules and their values define the flexural stiffness of the steering plates assembly. Defining these dimensions means finding the optimal compromise between stiffness, strength, and fundamental frequencies of the entire assembly.

For the purpose of returning the runners in neutral position, rubber or rubber-like material may be used in the steering mechanism. Usually two lever arms, connected to the steering bolt by said mechanism, are kept in position by two or more rubber bands. Mechanical properties of this set of rubber bands define stiffness

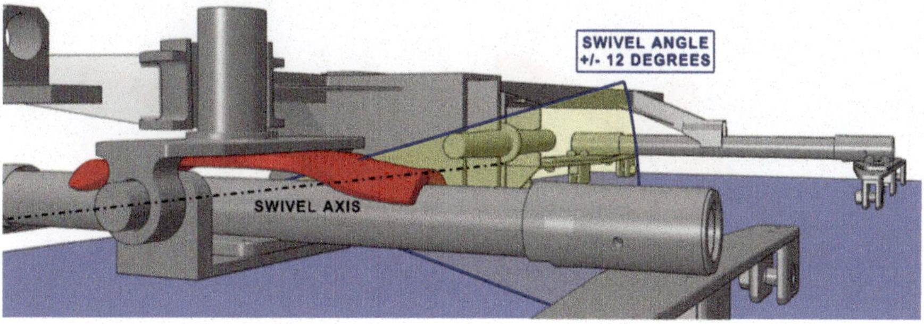

Fig. 7.7 Front leaf spring and swivel axis

and damping properties of the steering system and their definition may satisfy pilot driving feeling and technique. Between the steering plate and the front axle there is a continuous steel leaf spring that defines stiffness characteristic of front swivel angle.

7.2.4.2 Rear Portion of the Frame (Rule 14.2.3.4)

The rear bob portion must be constructed with a continuous (welded) supporting frame, made of steel and extending from the longitudinal pivot bolt at the transversal division line to the fastening means for the rear axle. All load bearing frame members and bearing housing(s) of the longitudinal bolt must be welded to the frame. From the division line to the rear axle, the longitudinal frame members must be made of closed steel profile. The longitudinal frame members are defined in their minimum section. Usually the rear portion of the frame shares with the cowling almost all the crew load: this means that longitudinal frame members section definition is crucial to define the whole bobsleigh stiffness and strength.

The rear axle may be fastened to the rear frame by means of bolts or by means of clamps. In both cases Rules define a set of mandatory dimensions in order to avoid compliance between rear axle and rear frame.

7.2.4.3 Front Runner Carriers (Rule 14.2.3.3.4) and Rear Runner Carriers (Rule 14.2.3.4.4)

Each front-runner carrier consists of a straight leaf spring with a thickness of no less than 7.5 mm throughout. A bearing housing for the front axle must be welded to the upper side of the leaf spring. On the lower part of the leaf spring there will be a guiding bracket welded in the center, while on either end there will be holding brackets welded in position. The middle guiding bracket has the same dimensions as the holding brackets, with the exception that U legs are connected by welded metal straps, front and rear, to hold a layer of rubber or rubber-like material at the bottom of the U.

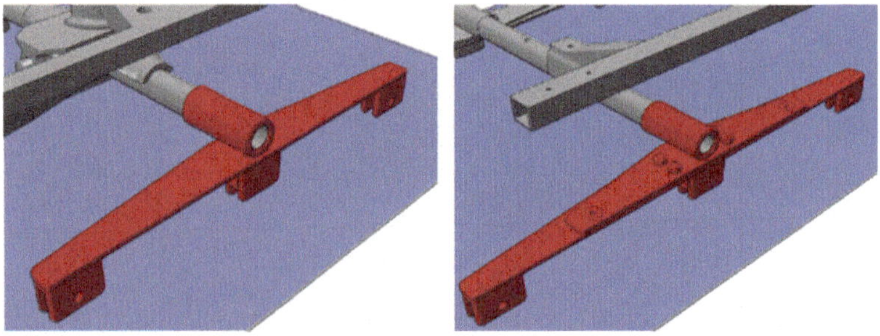

Fig. 7.8 Front runner carrier (**a**) and rear runner carrier (**b**)

This consideration is also applied to rear runner carriers, with the exception that each rear runner carrier consists of an upper and a lower straight steel leaf spring with a thickness of no less than 7.5 mm throughout. The bearing housing is welded to the upper leaf spring, while the holding brackets are welded to the lower leaf spring. The upper and lower leaf springs lie flat on one another and are bolted together by no less than six and no more than eight bolts with hexagonal nuts (see Fig. 7.8).

The runner carriers must have a means of rotational elastic stabilization.

The rubber or rubber-like material layer between runner carriers and runners is crucial for performance, because it is the only defined bob suspension. Rubber changes its properties with temperature, so it is reasonable to have different configuration for different track conditions. The optimal rubber or rubber-like material choice is the one that optimizes runner contact pressure during the whole race course, considering the specific track conditions, runners choice, and pilot driving technique.

7.2.5 *Cowling (Rule 14.2.4)*

The purpose of the Rules is to promote the safety of the athletes competing in FIBT events by attempting to ensure that the cowling of a bobsleigh will provide sufficient protection in case of crashes. Furthermore, the Rules aim to promote competition without unfair aerodynamic advantages.

7.2.5.1 General Cowling Stipulations (Rule 14.2.4.2)

- It is not allowed to use any additional material on the cowling cut-out;
- the brake cut-out must be open;
- no attempt may be made to reduce the area of the cut-out with any material;

- any unusual additions to the shape that are clearly vortex generators are not allowed;
- additional holes that may give an improved aerodynamic effect are also not allowed;
- the cowling provides adequate protection for the athletes.

7.2.5.2 Cowling Shapes (Rule 14.2.4.4)

The cowling shape must be convex with the following exceptions:

- the bumpers (front and back);
- the brake cut-out;
- the impression for the driver's push handle;
- the articulation joint;
- the holes for the front and back axles;
- the transition from the side push bar into the main body of the cowling;
- the area around the brakeman's push handles.

Within the Rules are defined limits on main dimensions for two-men and four-men bobsleigh cowlings. The said convex rule is quite limiting defining the cowling shape. Furthermore, deigning a bobsleigh cowling, it is fundamental to consider driver's ergonomic position during pushing and so the shape of driver push handle when open defines the lateral cowling shape once closed, because in this case it should essentially flush with the cowling surface (see Fig. 7.9).

Even in this case the cowling shape arises from the best compromise between aerodynamic optimization and driver's pushing ergonomic optimization.

7.2.6 Connection Between Frame and Cowling (Rule 14.2.5)

The purpose of the Rule is to control the amount of suspension that can be achieved in the connection between cowling and frame. All elements that mechanically attach the cowling to the frame are considered to be joints between cowling and frame and are controlled.

Fig. 7.9 Cowling

The cowling must be attached to the frame by means of at least four mounting brackets that restrict the vertical movement of the cowling relative to the frame. The movement of the cowling relative to the frame, allowed by the mounts of the cowling must be limited to 5 mm in the downward direction by use of hard travel stops of sufficient rigidity to guarantee effective limitation of cowling movement.

In the mounts between the cowling and the frame, homogeneous rubber or rubber-like material may be used. Different connection between cowling and the rear portion of the frame could control the stiffness and strength properties of bobsleigh, changing the runners' load transfer distribution during transition and/or during vertical movement.

7.2.7 Bumpers (Rule 14.2.6)

The purpose of the Rules is to attempt to ensure that the elements of the bobsleigh that come into contact with track walls are suitable for that purpose. Furthermore, Rules aims to promote that no unfair aerodynamic advantages can be achieved. Finally, Rules are directed to preserve track integrity.

Rules define a set of limited dimensions for front and rear bumpers. Within these dimensions it is possible to minimize aerodynamic resistance changing bumpers size, wetted surface, angle of attack, transitions to cowling and alignment between front bumpers and rear bumpers. Moreover it is important to try to minimize friction between bumpers and track walls in case of contact.

7.2.8 Runners (Rule 14.2.7)

Purpose of the Rule is to promote a fair competition while controlling the costs of participating and administering in FIBT events by banning the use of exotic material and (surface) treatment. All runners must be produced in a solid piece of standard material according to FIBT specifications (see Fig. 7.10 for runners' dimensions). These material specifications, selected by FIBT, will guarantee that all pre-machined pieces of steel are the same. All type of treatment are forbidden, including those that even cause only a local variation of the physical characteristics and/or the composition and/or the material structure. No plating and/or coating is allowed. Moreover, Rules define a set of limited dimensions for front and rear runners, both for two-men and four-men bobsleighs.

With a standard material, and with no chance to do any treatment, it is possible to play only with these limited dimensions in order to optimize runners' friction characteristics during the whole course, that means optimize runners contact area from gravity load to the maximum centrifugal load and during transitions. This means that it is fundamental to find the optimal longitudinal contact curve and the profile radius along the said longitudinal curve in relation with track and ice characteristics and bob suspensions.

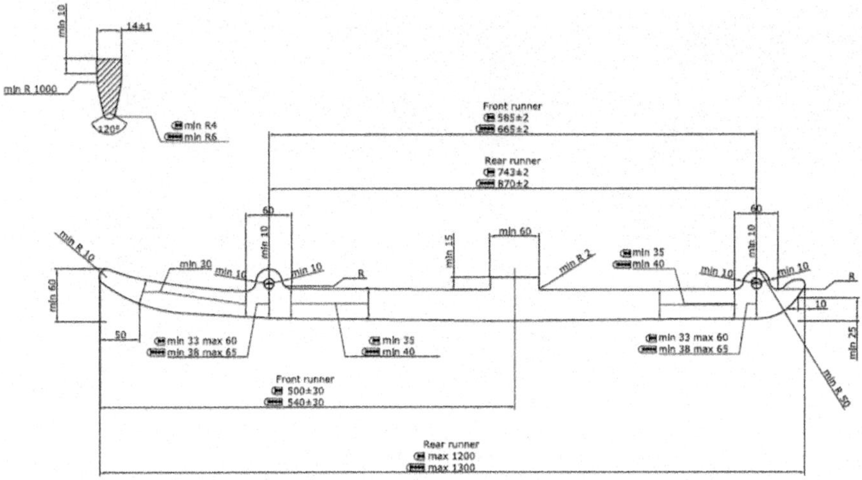

Fig. 7.10 Runner dimensions. All measurements in mm. Adapted from [1]

Each set of runners is inspected and validated before its use in official FIBT events. During inspection, hardness and material composition (through spectrograph analysis) may be verified.

In order to avoid surface treatment, FIBT cleans for all the competitors all the runners with a standard procedure prior to any race, using acetone and standardized sandpaper (usually 400 μm grade) that is passed back and forward five times with constant pressure by means of a custom machine. In case that residual left on sandpaper is not enough, Jury could increase sandpaper grain to remove more material or, in other cases, disqualify the competitor and ask for more specialized controls. This cleaning operation is made 45 min prior to race start, and from cleaning to the end of the race it is not possible to touch bobsleighs anymore. Even between two consecutive races it is possible to clean runners only with acetone and sandpapers officially furnished by FIBT. Even if final surface is ruined by FIBT sandpaper procedure, teams spend a lot of time polishing their runners with very fine sandpaper and diamond paste.

7.2.8.1 Reference Runner and Runners Temperature

The purpose of the Rules is to avoid any kind of advantage due to runners heating. A reference runner must be hung in open air one hour before race start, in the same climatic weather conditions as the sled included in the race. When the runner is positioned in the air, its temperature must be between +18 °C and +22 °C. At standardized intervals before and during race reference runner temperature must be updated on a clearly visible board in the start area. Runners being used on the race must be mounted on the sled at least 45 min before race start. After this

Fig. 7.11 (**a**) The athlete giving a boost to the sled at the starting point. (**b**) The athlete during the jump on the skeleton sled. Adapted from [1]

time point, all teams (athletes and support staff) must leave *parc fermé*, and no work may be performed on either the sled or the runners; only Jury and FIBT Materials Committee members may enter the *parc fermé*. Prior to any competitor start, temperature of runners that are mounted on the sled being used in the race may exceed the reference runner temperature by a maximum of 4 °C. If the temperature of the reference runner is less than $-14\,°C$, the runners on the sled may have a maximum temperature of $-10\,°C$. Runners temperature greater than the limit leads to disqualification.

7.2.9 Skeleton

Skeleton is an individual sport: on this kind of sled only one person can stay during the competition. The position of the driver is totally different with respect to the bobsled previously described (one can see an example in Fig. 7.11). Also for the skeleton sled, all its parts are subjected to norms and regulations: dimensions indicated in the drawings (Fig. 7.12) are compulsory. Protruding items that could cause injury are prohibited as well as fairings, spoilers, and aerodynamics elements. The sled must be on "steel," meaning an alloy of iron and carbon with an iron (Fe) content of more than 50 % or an alloy that reacts positively to the FIBT chemical test. The *construction frame* consists of longitudinal and transverse bars and runner supports, which must be made of steel. Each of the two longitudinal bars must be constructed of a continuous piece of steel. Any additional elements on the construction frame must be bolted or welded. A skeleton sled has only two runners, which must be bolted or welded on the construction frame. The runner supports must be rigidly attached to the construction frame. The runner supports are defined as the parts with slots and do not include any elements that have been welded on.

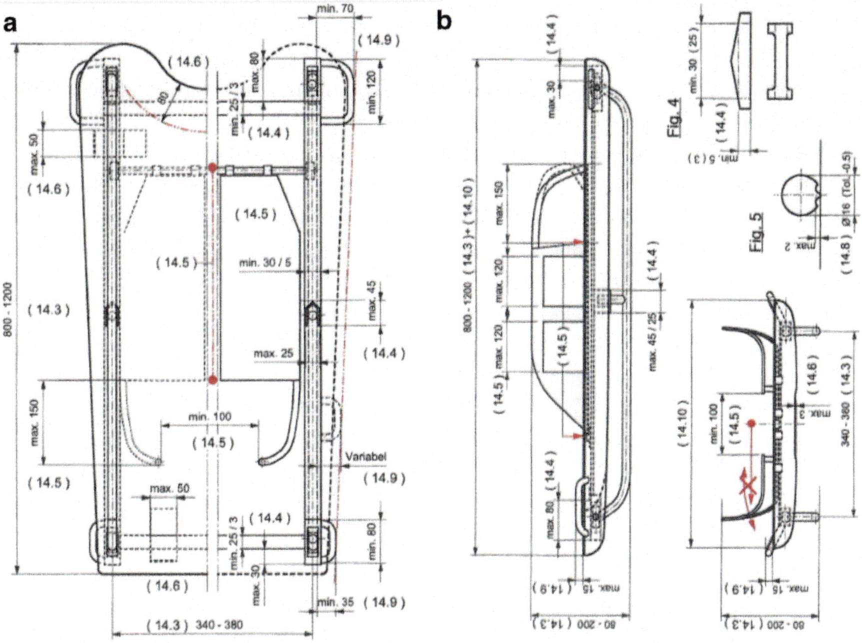

Fig. 7.12 Skeleton sled dimensions (expressed in mm): (**a**) upper view; (**b**) side views. Adapted from [4]

The *saddle* with handles must be made of steel, mounted on the construction frame and divided into a maximum of two pieces. No additional handles, howsoever they are constructed, are permitted on the saddle: secure handles must be firmly attached to it at both ends. The athlete's arms must be outside the saddle during each run. The *base plate* is constructed in one single piece and is convex throughout. It is mounted to the construction frame by welding, gluing, bolting, or riveting. Its is allowed to affix securely to the frame some ballasts, in order to reach the maximum permitted weight limit. Also in this case, there are really restrictive criteria for the *runners*. To make them a standard material is chosen by FIBT and is supplied as polished round bars with a diameter of 16 mm (h9 tolerance class) and a length of 1000 mm, –0/+50 mm. Markings are present on these materials, and they must be present on the runners, without any alteration. All types of treatment on the material are forbidden. The grooves and milled areas on the runners must be formed in such a way as to avoid any excessive damage to the ice surface of the track. The depths of the millings and grooves, measured from the surface of the ice, must not be greater than 2 mm. It is compulsory to include security bumpers in the front pate of the sled (on both sides), and between the saddle and the rear end of the sled.

7.2.10 Track

The rules about the track are the same both for bobsled and for skeleton (and for luge too).

The directives, which track constructors must observe, shall be designed in such a way as to keep the costs of the tracks as low as possible and to guarantee satisfactory use of the installations after the Olympic Winter Games. For sure the athlete safety has a leading importance for the track project and construction.[2] Moreover a great attention should be paid to the environment and its protections. A dedicated commission, nominated by FIBT, has the assignment to supervise the preparatory phase of the design, the design itself, the construction and the commissioning of each track.

First of all a Technical Commission shall be called upon to effect an inspection and to provide consultancy before a decision is made regarding the terrain on which the track will be constructed. Then they have to follow all the phases of track design. Generally the tracks are located close to a city or a densely populated area to guarantee its optimal use also after Olympic Winter Games. If possible a track is built on the north side of the slope and follows, as far as possible, the natural development of the terrain. The new artificial tracks shall be between 1200 and 1650 m long, with a maximum gradient which cannot exceed 12 %.

The track shall include elements of varying technical difficulty. At design level, it is to be foreseen to be able to reach a speed of between 80 and 100 km/h after the first 250 m. Bends, combinations of bends and straight stretches of a suitable length shall be inserted into the track. The bends must be constructed in such a way as to allow sleds to move not only along a single trajectory, but to provide a band of possible trajectories from which to choose. In the central part of the bend, the trajectory should extend along the upper half of the bend. Entries and exits from bends must be rounded so that the sleds can take them smoothly with no risk of capsizing, if no mistake is made in driving. Guardrails must be constructed in such a way that they return the sleds on to the track. Also the centrifugal force an athlete can receive is subjected to rules: the maximum time for which a centrifugal force of "4g" may occur is 3 s. The maximum centrifugal force allowed is "5g" and must not last for more than 2 s. In the iced construction, maximum width is 140 cm. The external part of the sidewalls of the straight stretches should not be higher than 80–100 cm, including any raised section. In the iced state, the inside part of the sidewalls must be at least 50 cm high. The transition between the sidewall and the base of the track must be provided with a channel: in the iced state its radius must be 10 cm.

[2]Due to velocity and relatively few protection devices, the safety of the track is really critical. As an example, one can remember the fatal accident happened to the 21 years old Georgian luge athlete Nodar Kumaritashvili during the Vancouver Winter Olympic Games in 2010: the high velocity of the luge, in this case around 140 km/h, and a wrong trajectory in a bend brought off road the sled. The luge, running off the track, crashed against a not-protected pole on the side of the track, causing the athlete death few hours later.

The deceleration stretch must be constructed in such a way that a sled can stop without applying the brakes: the gradient of the deceleration stretch must not be in excess of 20 %. At the end of the iced stretch, sleds must be able to continue without any obstacle. If considered necessary, at speeds less than 30 km/h, special cushioning devices may be installed, in order to avoid any injury to people and to restrict sled damages to a minimum.

7.3 Experimental Tests

In sports involving vehicles (car, bicycle, motorbike races, etc.), full scale experimental tests with instrumented vehicles and tracks are often considered a privileged way to:

1. evaluate the influence of athletes and equipment on overall race performance;
2. test innovative technical solutions;
3. assess vehicle dynamics and athletes' response;
4. identify unknown parameters/phenomena affecting overall race performance.

Concerned with items 1–3, measurements in fact provide objective information to trainers, coaches, engineers, and technicians. Athletes' feedback about vehicle performance and track difficulties is subjective by its own nature and thus it must be integrated to become independent on athletes' skills and preferences.

Moreover measurements allow the setup and/or the validation of numerical models (items 3 and 4), which can be profitably used at a design stage to develop innovative technical solutions or equipment. However in bobsleigh, luge or skeleton, performing on-board vehicle measurements and comparing results collected in different test sessions represents a very complex task, due to:

- difficulties in placing on-board instrumentation without hampering athletes (especially during boarding);
- environmental factors:
  - temperature can significantly vary from one test/race to another as well as weather conditions. These variations may affect results of race/test much more than the studied technical solution, being, as well known, ice temperature and weather key-factors in determining performances in bobsled competitions;
  - ice degeneration consequent to repeated runs;
- limited number of tests which can be performed by the athletes during a single session before their biomechanical (especially during the push phase) performance decreases.

On the other hand, laboratory tests do not often allow the reproduction of working conditions experienced during races. To address these issues, approaches and experimental setups of different complexity and aim were proposed in the literature and described in the following for skeleton, bobsleigh, and luge.

7.3.1 Statistical Analysis of Luge, Bobsleigh, and Skeleton Competitions

The simplest experimental setup for evaluating athletes' influence on overall performance in skeleton, bobsleigh, and luge competitions is represented by photocells (or timing eyes) and speed traps placed along the track, which, respectively, allow to record split/overall times and speed of the athletes while they are running down the track. This setup can possibly be integrated with the measurement of track ice temperature to assess influence of environment and video recordings to assess influence of athletes' movements and/or position inside/on the sled on overall performance.

The main advantage of this experimental setup is that all the tracks of World Cup Circuit are equipped with facilities allowing these kind of measurements, thus a huge amount of data is available for analysis. Moreover, this setup can be used during official races since it does not hamper athletes during their runs, so available data include top level competitions.

Statistical analysis can then be applied to find relationships between measured quantities (related to athletes behavior and environmental factors) and overall performances.

Using this approach, data collected during four-men bobsleigh competition of Calgary Olympic Games were analyzed in [5], the ones concerned with bobsleigh (both two-men and four-man) and luge competitions during Lillehammer Olympic Games were investigated in [6], while the ones relevant to major skeleton international competitions during 2003–2004 World Cup were studied in [7].

A good push start is referred in [5, 6] to be a prerequisite in order to achieve excellent overall performance in bobsleigh.

A correlation between push start times and final descent times of 77 % is shown in [6], when all the teams and all the heats are included into the statistical analysis. This result is consistent with the analysis reported in [5], when push time, start number, and ice temperature are combined. This means that, under given external conditions, the result of a bobsleigh run is already strongly determined after about 10 % of the total run time. In [6] it is however highlighted that the final positions for the top 15 competitors in bobsleigh at the Lillehammer Winter Olympics 1994 were determined through driving ability in the most difficult sections of the track (especially during four-men bobsleigh competition) and not the push start alone. In [5] it is instead shown that ice temperature and start number are significantly correlated (about 30 % of explained variance) with overall performance. These two variables, during Calgary Olympic Games, influenced each other strongly, since temperature was not constant during the heats. Eliminating influence of ice temperature reduced the correlation between start number and finish time for the analysis of each heat separately and vice-versa.

The statistical correlation for push start and descent times in luge presented in [6] confirms that even in this competition push start is essential for a competitive overall performance. Specifically, a correlation of 55 % between push start times and final descent times was found. As in bobsleigh, within top-level group (top 15 competitors), push start time does not constitute alone a selection criterion for final ranking, but it has to be combined with driving skills and athlete-dependent positioning on the sled.

The statistical analysis of skeleton World Cup races presented in [7] found elite athletes in this competition to have less correlation between push start times and final descent times: 23 % for the men and 40 % for women. This highlights how unforgiving driving errors are on a skeleton sled. Importance of push start phase however does not have to be underestimated.

In [8] the skeleton push start performance is analyzed in detail. Using video footage acceleration over 0–15 m and velocity at the 15 m, 45 m marks were determined. It was found that the 15 m velocity is more important in assessing overall performance of push start, additionally acknowledging that the results at 45 m may be useful when considering each tracks unique characteristics. Similarly, skeleton push start performance of US national team was investigated in [9]. The importance of the 0–15 m split time was confirmed, although only the 0–30 m time was referred as a positive indicator of performance. This is perhaps due to those with a taller physical stature affecting ability to accelerate from a standstill are still able to generate a high loading velocity by the 30 m without sacrificing an irrecoverable time deficit in the earlier stages.

7.3.2 Performance Analysis

To support the previously presented statistical analysis and further investigate the parameters more affecting performance of bobsleigh and skeleton, simple mathematical models can be used. In the following a simple mathematical model of a bobsled (which can be easily adapted for skeleton) is introduced, which includes its most important parameters. One can focus on four parameters: the friction coefficient between runners and ice μ, the aerodynamic drag C_D, the bobsleigh mass m_{BOB}, and the initial velocity v_{START}.[3]

The simple model we are referring to is based on a solid body sliding on a plane inclined by angle α. The reader may refer to Fig. 7.13, where:

- **m** bobsleigh total mass (kg);
- **g** gravitational acceleration (m s^{-1});
- $\boldsymbol{F_Z}$ total normal force (N);
- $\boldsymbol{F_R}$ total resistance force (N);
- $\boldsymbol{\alpha}$ inclined plane angle (α);

Starting from this very simple model, one can easily write the dynamic equation along the x-axis:

$$m\frac{dv}{dt} = m \cdot g \cdot \sin\alpha - F_R \tag{7.1}$$

[3]We neglect the very first part of the race, when the athletes are pushing the sled: this velocity is considered at race time zero, 15 m after start.

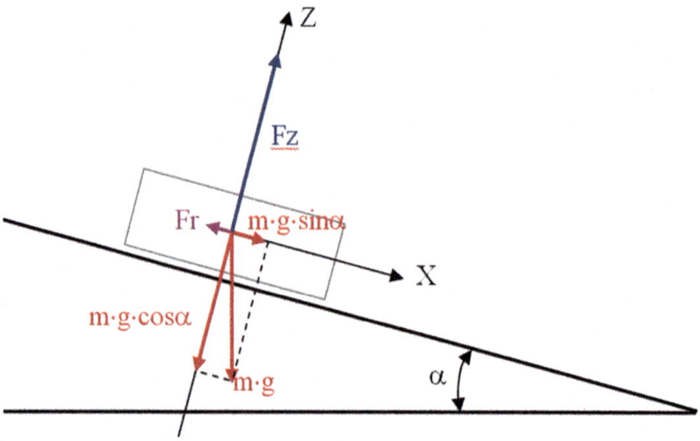

Fig. 7.13 Simplified model scheme

The total resistance force is $F_R = F_{\text{DRAG}} + F_{\text{FRICTION}}$, where:

$$F_{\text{DRAG}} = \frac{1}{2} \cdot \rho \cdot C_D \cdot A \cdot v^2 \qquad F_{\text{FRICTION}} = \mu \cdot F_Z \tag{7.2}$$

Along the y-axis, the total vertical force is

$$F_Z = m \cdot \gamma_Z - F_{\text{LIFT}}$$
$$F_{\text{LIFT}} = \frac{1}{2} \cdot \rho \cdot C_L \cdot A \cdot v^2 \tag{7.3}$$

In the above equations we have used these symbols: ρ air density (kg m^{-3}), A bob frontal area (m^2), C_L aerodynamic lift coefficient, and γ_Z average vertical acceleration, function of gravitational and centrifugal accelerations (m s^{-2}).

Let us suppose some average input parameters (referring to a standard configuration) in order to analyze the model behavior:

- $A = 0.4\,\text{m}^2$;
- $C_D = 0.3$;
- $C_L = -0.07$;
- $\mu = 0.01$;
- $\gamma_Z = 15.2\,\text{m}\,\text{s}^{-2}$;
- $m_{\text{BOB}} = 390\,\text{kg}$ (we consider 170 kg of bobsleigh and a crew of 220 kg, the reader may refer to Sect. 7.2);
- $v_{\text{START}} = 5\,\text{m}\,\text{s}^{-1}$;

Table 7.1 A synthesis of the effects of parameter changing

Parameters	Ref. value	Modified value	Ref. time (s)	Modified time (s)	Time difference
C_D	0.3	0.33 (+10 %)	65.461	65.740	0.279 s (0.43 %)
C_L	−0.07	−0.077 (+10 %)		65.462	0.001 s (0.00 %)
A	0.4 m²	0.44 m² (+10 %)		65.741	0.280 s (0.43 %)
μ	0.01	0.011 (+10 %)		66.143	0.682 s (1.04 %)
$m = m_{\text{bob}} + m_{\text{crew}}$	390 kg	351 kg (−10 %)		65.772	0.311 s (0.48 %)
v_{START}	$5\,\text{m}\,\text{s}^{-1}$	$4.95\,\text{m}\,\text{s}^{-1}$ (−1 %)		65.621	0.160 s (0.24 %)

Referring now to previous equations (7.1) and (7.3), with few mathematical operations it is possible to obtain:

$$\begin{aligned} m\frac{dv}{dt} &= mg\sin\alpha - F_R = mg\sin\alpha - \left(\frac{1}{2}\rho C_D A v^2 + \mu F_z\right) \\ &= mg\sin\alpha - \left(\frac{1}{2}\rho C_D A\right)v^2 - \mu\left(m\gamma_Z - \frac{1}{2}\rho C_L A v^2\right) \end{aligned} \tag{7.4}$$

If now we define the coefficients $K_1 = 1/2\rho A C_D$, $K_2 = 1/2\rho A C_L$, $K_3 = (K_1 - \mu K_2)/m$ and $K_4 = \mu\gamma_Z - g\sin\alpha$, Eq. (7.4) can be expressed in a more compact fashion:

$$\frac{dv}{dt} + K_3 v^2 + K_4 = 0 \tag{7.5}$$

We have obtained a first order differential equation, and it can be integrated numerically step by step using, for instance, the explicit Euler method.

$$\frac{dv}{dt} \approx \frac{\Delta v}{\Delta t} \Rightarrow v(t + \Delta t) = -(K_3 v(t)^2 + K_4)\Delta t + v(t) \tag{7.6}$$

Considering a track length of 1500 m with a 5° of slope, it is possible to use the above described model to evaluate race track final time varying each parameter by 10 %. In Table 7.1 the reader can check the results. Further in Fig. 7.14 it is possible to visually appreciate the four-parameter contribution to the performances.

Let us analyze the parameters in a more specific way.

- **Aerodynamic drag coefficient**: it modifies the resistance force linearly with speed squared. Its influence grows in faster tracks. In the considered case there is a loss of time of 0.279 s, that is to say an increase of +0.43 %. The design of a bobsleigh should try to reduce this coefficient as much as possible, with respect of Rules restrictions (see Sect. 7.2).
- **Aerodynamic lift coefficient**: it modifies the vertical forces acting on bobsleigh by an increase of friction forces due to the friction coefficient μ_{ICE}. The final

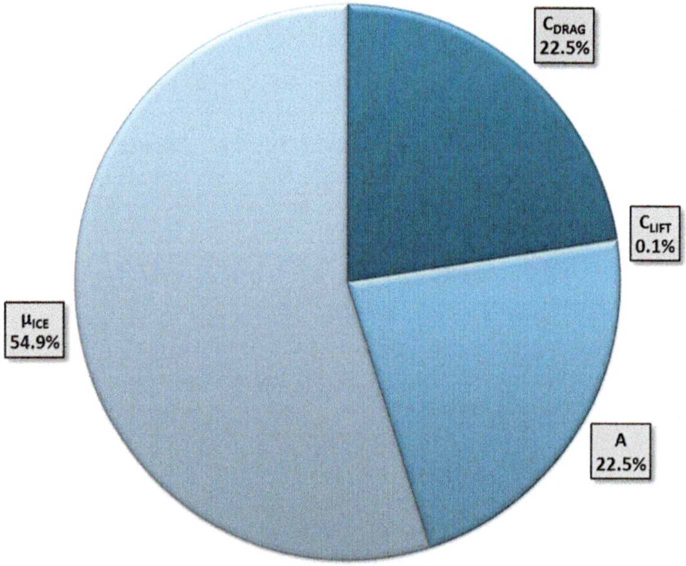

Fig. 7.14 The weight of the four parameters considered for the model

effect of this parameter is negligible (see Fig. 7.14). Anyway, even if its contribution is little, the minimization of this coefficient is beneficial, in fact generally also C_D coefficient is affected by its trend.

- **Bob frontal area**: as seen in Eq. (7.2), the frontal area A works both on drag and lift coefficients. Neglecting the lift coefficient, we can state that it could be compared with C_D: its influence on final time is linear and provokes an increase by 0.43 % on final time. The problem of frontal area minimization tackles not only Rules restrictions, but must guarantee also crew ergonomic and fitting inside the sled.
- **Friction coefficient**: it acts linearly on resistance forces. Apart from the mentioned above lift minimal contribution, in this simple model its effect is not influenced by speed. Runners and bobsleigh suspension should be designed trying to optimize contact areas during the whole race course so that friction between runners and ice is reduced working with optimal contact pressure.
- **Bobsleigh total mass**: we consider both the weight of sled and the one of crew: its influence is not linear inside the model. The aim of example is to demonstrate that there is a loss of performances with a mass reduction, due to the fact that gravity is the only positive force during the race descent. The only reason to reduce total mass could be considered when the crew weight is not at the maximum, and so ballast may be used to increase bobsleigh weight. In this case a lighter bob could achieve a greater start velocity that could compensate the loss in potential energy available, that means:

$$E_{K,\text{start}} - E_{P,\text{start}} \longrightarrow \text{maximized}$$

where $E_{K,\mathrm{start}}$ and $E_{P,\mathrm{start}}$ are the kinetic and potential energy in starting position.
- **Start velocity**: its effect is to rise up the total energy at the start, acting on kinetic energy. Due to the fact that potential energy is the same for all the competitors, initial velocity could provide a fundamental contribution to final result. In our case, a starting velocity decrease of only 1 % causes a loss of time equal to 0.160 s (that is +0.24 %). Bobsleigh should be designed considering athletes ergonomics while pushing and in a manner that entering the sled at the end of pushing phase could be fast and easy.

What we have done in this section is a brief overview of some important parameters affecting bobsleigh descent. This is not a complete list of parameters nor an exhaustive discussion on peculiarities of each of them. Along the chapter they will be discussed with a more complete analysis (for example, a more sophisticated model of sled descent and its aerodynamic optimization will be treated in Sect. 7.4).

It must however be noticed that, although based on a simple model, results offered by the performed analysis are consistent with the ones of the previously presented statistical investigation (Sect. 7.3.1).

7.3.3 Full Scale Tests on Skeleton

In this section, the main full scale experimental activities aimed at investigating push start and skeleton sled dynamics are reported.

7.3.3.1 Analysis of Skeleton Start Phase

A skeleton push start can be viewed as being divided into four phases, which are shown in Fig. 7.15 [10]. During the early stages of the descent, the athlete:

- drives off a fixed starting block with as much force as possible;
- continues to accelerate by running alongside the sled until they can run no faster;
- loads onto the sled with a dive, landing in a head-first, face-down position, flat on the sled;
- settles into the race position, in which subtle movements and flexing of the sled are used to control the remainder of the descent.

Although push start is only a small part of overall run, interaction between sled and athlete during start has a strong influence on final descent time and consequently it is a particular focus of a skeleton athlete's training. A detailed analysis of starting phase is reported in [10]. Four male skeleton athletes took part to the study, three of whom were experienced internationally competitive athletes, while the fourth was a recreational skeleton athlete, who has been regularly involved in the sport, but who has not undergone rigorous push start training. Tests were performed on the push tracks of the Calgary Ice House Track and of the Torino Olympic Park Track.

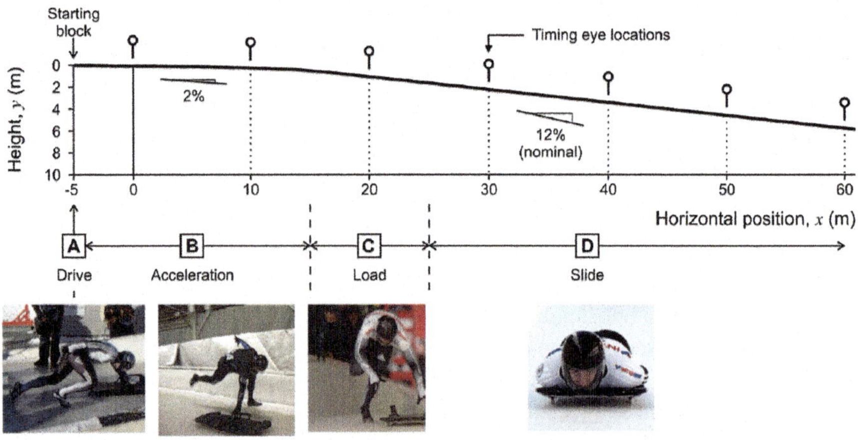

Fig. 7.15 The vertical profile of a push track, showing the four phases of a push start, [10]

Both tracks are equipped with timing eyes at 10 m intervals, as shown in Fig. 7.15. Velocity is measured at the final timing eye. A sled instrumented with a tri-axial capacitive accelerometer attached towards the rear part of the frame was used during the tests. Data were logged at 100 Hz per channel using Race Technology DL1 data logger. Acceleration along the horizontal axis was compensated for the shift in gravitational contribution with track gradient and then numerically integrated to calculate velocity throughout the push start. Recorded split times were used to calibrate the accelerometer and the double integration process. Figure 7.16 shows an example of resulting time histories of acceleration, velocity and position of the sled during start. The four phases of push start are highlighted. Footfall events, recorded through video footage, are also reported. As it can be seen, the athlete pushes the sled off the starting block (phase A) where the peak acceleration is reached, then continues to accelerate the sled during phase B whilst running along side it. The acceleration diminishes until the maximum speed is reached, then the athlete loads on the sled (phase C). A successful loading will increase the velocity of the sled. In phase D, starting once the athlete is settled into race position on the sled, is governed by gravity, aerodynamic resistance, and friction between ice and runners.

Analysis of data collected throughout the tests led to the following considerations [10]:

- a strong initial acceleration off the starting block is key-part in push start, correlating well with overall performance. However acceleration must be sustained in the following phase (phase B) during which the athlete runs along the sled;
- two-handed push start (which is a more traditional technique) is more stable than a single-handed start allowing a higher acceleration off the starting block. However grasping the sled with both hands whilst running to one side makes more difficult to push the sled forward and thus sustain acceleration;

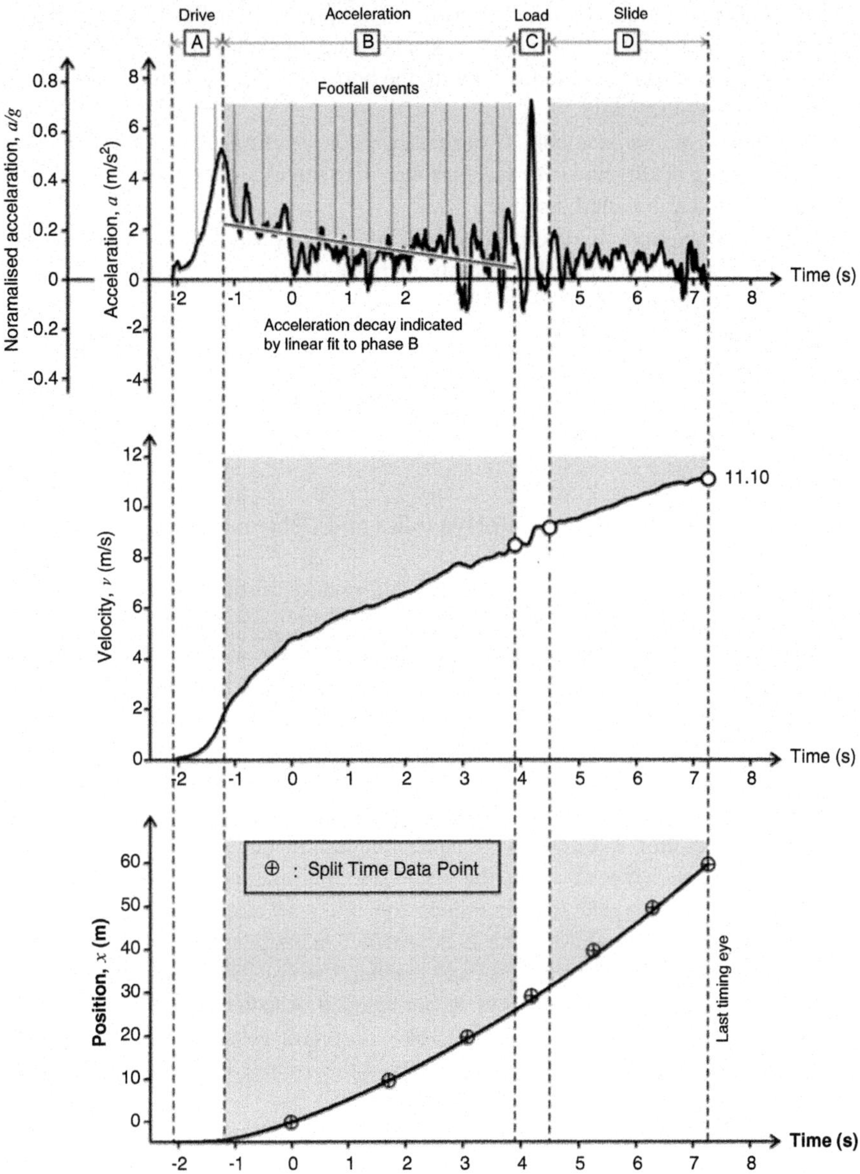

Fig. 7.16 Example of the recorded acceleration data and its double integration to determine the velocity and position of the athlete with time [10]

- velocity at the second timing eye, which is commonly used as an indicator of quality of drive out of the starting block does not correlate so well with overall performance of push start;

- while athlete footfall frequency during 30 m sprints on a flat track correlates well with the overall performance, this is not true for skeleton push start. This is because a track sprint is conducted with the body upright, while the body is bent over during a push start;
- best athletes are the ones able to run faster before loading. Time and position at which loading occurs however varies from run to run as higher velocity generally results in further traveled distance;
- loading is a key phase in determining performance of push start. Best athletes are the ones able to maximize the velocity increase during loading. This parameter is however difficult to be consistently assessed;
- maximum (due to transfer forward momentum) and minimum (resulting from landing) acceleration during the loading phase correlates well with overall performance. These data are however not available during standard training and races although they could be of great benefit to skeleton athletes.

7.3.3.2 Analysis of Skeleton Sled Dynamics and Deformation

An analysis of skeleton sled dynamics and deformation during timed runs on race tracks is proposed in [10].

To assess the sled dynamics a tri-axial accelerometer was rigidly attached to the sled saddle (see Fig. 7.17a) and several runs were performed at the Koenigsee international race track, which is one of the most technical and difficult tracks of the World Cup. Figure 7.17b shows an example of the measured accelerations along the x, y, and z axes during a whole run down the Koenigsee track. As it can be seen, as the sled travels through the banked curves, the contribution to measured acceleration due to gravity shifts one axis to the other. Looking at the longitudinal acceleration (x-axis), the push start phase (previously analyzed) is clearly visible. This signal then becomes too noisy to allow gaining additional information. Acceleration along the z-axis, which is perpendicular to the sled, is representative of the g-force experienced by athletes within the corners during a run: as it can be seen peak accelerations of $4g$ are reached. Acceleration along the y-axis (transverse acceleration) is the most related with driver behavior as it is different from zero only when the sled enters and exits the curves, i.e. when the driver's action takes place.

From a detailed analysis of the most critical sections of the track, the following considerations were drawn in [10], with the aim of finding indications for optimizing the sled trajectory:

- the athlete with the fastest push start time does not consistently achieve the fastest descent time. This confirms the statistical analysis performed in [7] and highlights that in skeleton competitions, driving errors compromise overall performance more than a slow push start, especially in difficult tracks like Koenigsee;

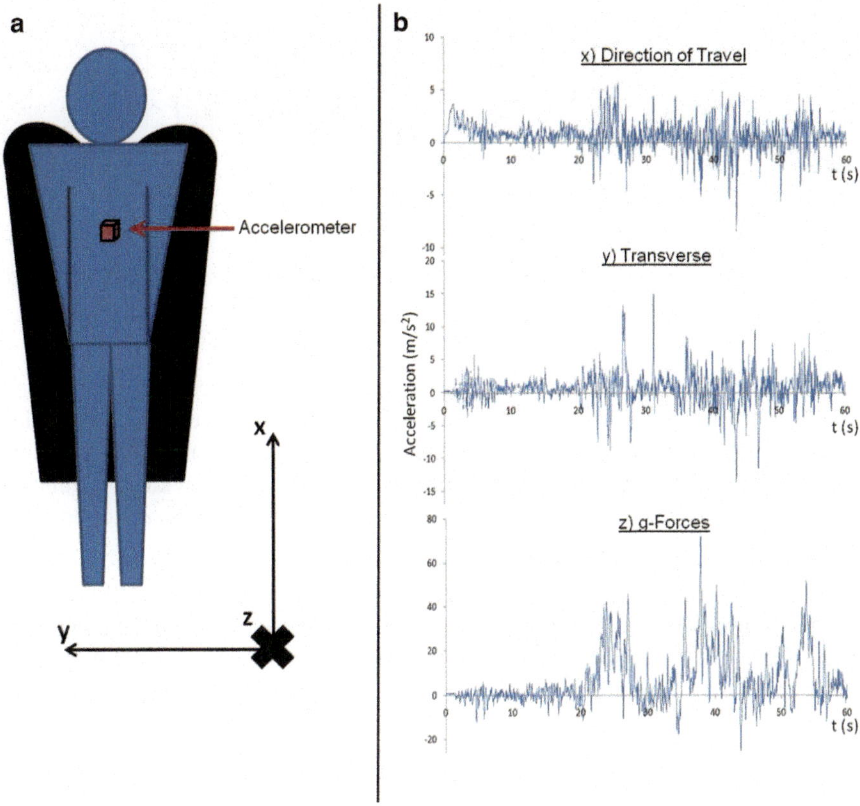

Fig. 7.17 (**a**) Tri-axial accelerometer position and reference system; (**b**) accelerations along axes x, y, z measured during a whole run down the Koenigsee track, [10]

- the measurement providing most useful information for analyzing the sled dynamics and trajectory (and so driving behavior) is the acceleration along the y-axis as it shows significant changes as the sled travels in and out of the corner, where the athlete is more likely to alter the trajectory;
- during S-bends, the y-axis acceleration measurement enables to calculate the ratio between sled rise (t_1 in Fig. 7.18) and fall (t_2 in Fig. 7.18) time. This ratio allows to infer where the highest point of the trajectory is achieved within the curve. By setting a particular entry angle and using a particular entry steer into the curve (trajectory 2 in Fig. 7.18a), the athlete can anticipate the highest point of the trajectory (not necessarily higher than the apex), resulting in a longer, straighter exit and smoother transition into the following curve with less energy dissipation. From the analysis of the performed runs (Fig. 7.18b) resulted that the lower the ratio t_1/t_2 the lower is the time needed to travel the S-bends;
- whenever it is needed to impact the wall of the track to change the orientation of the sled (technical gesture called "threading the needle"), as in the Bendway/Kink

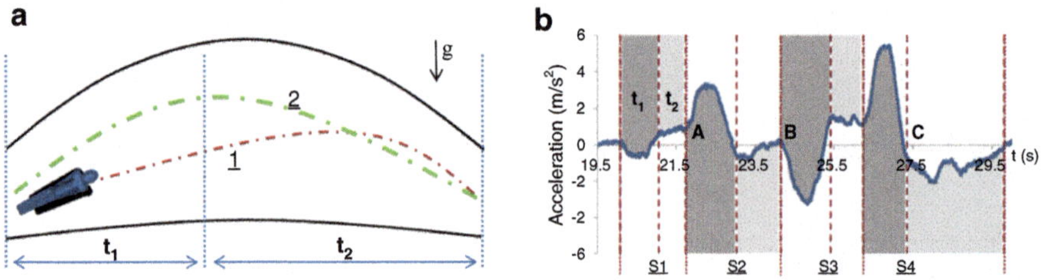

Fig. 7.18 (**a**) Vertical view of the curve: trajectory 1-uncontrolled entry, trajectory 2-controlled entry; (**b**) filtered acceleration along the *y*-axis between S-bends at Koenisee race track, [10]

section of the Koenigssee track (see the positive transverse acceleration peak after about 30 s in Fig. 7.17b), it is essential that the impact point is unique to minimize energy dissipation and path deviations. Transverse acceleration measurement allows to quantify magnitude and timing of the impact, providing significant information whether the athlete had successfully *threaded the needle* or not.

In order to correlate the dynamic analysis of the sled motion with its deformations, a dual stiffness sled was instrumented with strain gauges in [10]. The two stiffness settings, referred as *stiff* and *loose*, were achieved by modifying the bolts fixing the rear of saddle plates. Rosettes of strain gauges were applied to front and rear crossbars (positions 1, 2 and 3 in Fig. 7.19b) and to the longitudinal bars (positions 4 and 5 in Fig. 7.19b) of the sled frame. A configuration of three train gauges orientated at 45° to each other was used. Each gauge was incorporated into a half Wheatstone bridge circuit, using a dummy gauge on an isolated steel plate to compensate for any thermal expansion/contraction.

Laboratory tests were performed to correlate forces applied to the sled with strain gauges deformations. After calibration, the instrumented sled was used to perform several timed runs at the Lake Placid Track. Deformations of crossbars turned out to provide the most useful information as they are related with the twist angle of the frame imposed by the driver's action to control the sled trajectory. Combined with athlete perception, collected data showed the stiff frame to have less response than the loose frame and a harsher ride. However, it was the stiff frame that resulted in the fastest descents indicating the smoothness of the ride and ability to position the sled more accurately was excessive and sacrificed acceleration.

7.3.4 Full Scale Tests on Bobsleigh

This section describes the main full scale experimental activities carried out with the aim of investigating bobsled dynamics, driver behavior, and ice-runners contact forces.

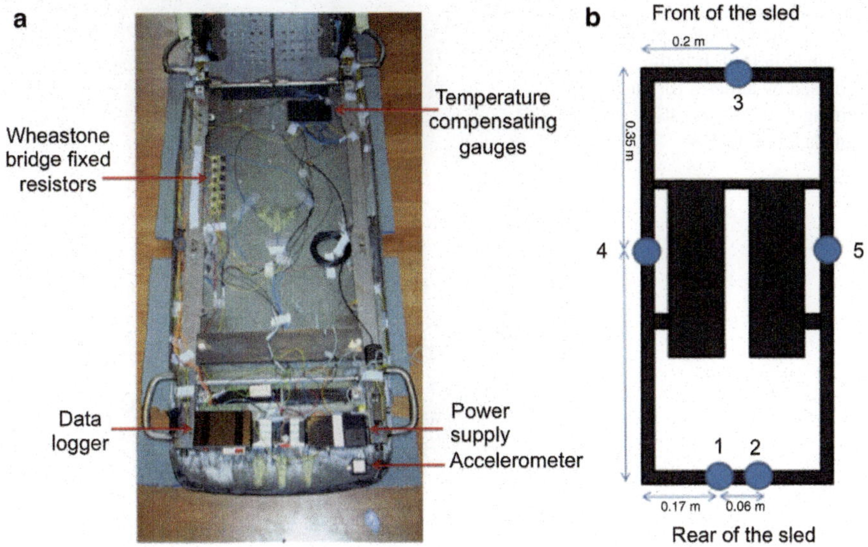

Fig. 7.19 (**a**) Experimental setup used in [10] within the sled. (**b**) Position of the strain gauges of the sled frame

7.3.4.1 Full Scale Tests to Assess Ice-Runners Friction Coefficient

During ice friction laboratory tests (see Chap. 3), specimens are subjected to loads lower than 100 N and sliding speeds not exceeding 10 m/s [11–14], while during bobsleigh competitions load on skates exceeds 4000 N and peak velocities of 36–40 m/s are reached. Thus full scale experimental tests are often preferred to reproduce real working conditions. A simple experimental setup for estimation of runner-ice friction coefficient in conditions similar to real working ones is proposed in [15, 16]. In those experiments, the sled is launched either on a flat and sloped ice surface with different initial speeds. The friction coefficient is then determined from the rate of acceleration/deceleration of the sled.

During the tests described in [16], where a larger range of speeds was investigated, the sled shown in Fig. 7.20 was pushed by professional athletes to an initial speed ranging from 1.5 to 6 m/s, then it was released free to glide for about 30 m either on a flat or on a constant slope ice surface. Tests were repeated adding different weights to the sled. Flat ice surface tests were performed in the hockey rink at Calgary Olympic Oval, while tests on constant slop ice surface were carried out in the push training facility at Canada Olympic Park. During the tests, the time history of sled speed was measured by a radar gun placed behind the sled and parallel to its motion. Moreover, temperature was measured to verify that its variation were contained in a range (specifically temperature was between −2.2° and −4.6° during the tests) not affecting ice hardness and thus obtained results. On the purpose of estimating the friction coefficient μ from performed measurements, the sled can be

Fig. 7.20 Front view of the sled used in the experiments described in [16]. It is the sled used by athletes for push training in Ice House at Canada Olympic Par

approximated as a single degree of freedom particle (see Sect. 7.3.2) on which are acting the weight force, ice friction, and aerodynamic drag:

$$ma = mg\sin\phi - \mu mg\cos\phi - \frac{1}{2}\rho AC_D v^2 \tag{7.7}$$

where m and a are the sled mass and acceleration, ϕ is the ice surface slope (constant and equal to 6.8° at Canada Olympic Park and null at Calgary Olympic Oval), ρ is the air density, A is the cross section of the sled, C_D is the sled aerodynamic drag coefficient, and g is the gravitational acceleration. Assuming that the acceleration of the sled is constant, average speed ($\bar{V}$) and acceleration ($\bar{a}$) can be used to estimate ice-runners friction coefficient and sled aerodynamic drag. Terms of Eq. (7.7) can in fact be re-arranged as:

$$\bar{a}^* = \bar{a} - g\sin\phi = -\mu g\cos\phi - \zeta\frac{\bar{V}^2}{m} \tag{7.8}$$

where the aerodynamic coefficient $\zeta = \frac{1}{2}\rho AC_D$ and the mean net sled acceleration (i.e., the value of the sled acceleration once subtracted the gravitational contribution) $\bar{a}^*$ are introduced.

The assumption of constant acceleration can be assured by limiting the data acquisition time, as shown in Fig. 7.21, where an example of six speed measurements carried out on flat ice surface are shown.

Equation (7.8) points out a linear relation between the mean net sled acceleration $\bar{a}^*$ and the term $\bar{V}^2/m$. Thus experimental data can be fitted with a least squares linear regression. The best fit line and the 95 % confidence limits are reported in Fig. 7.22, together with the experimental data: white allow circles refer to the flat ice surface tests, while grey filled circles are concerned with constant slope ice surface experiments. The slope of best fit curve allowed to determine the aerodynamic

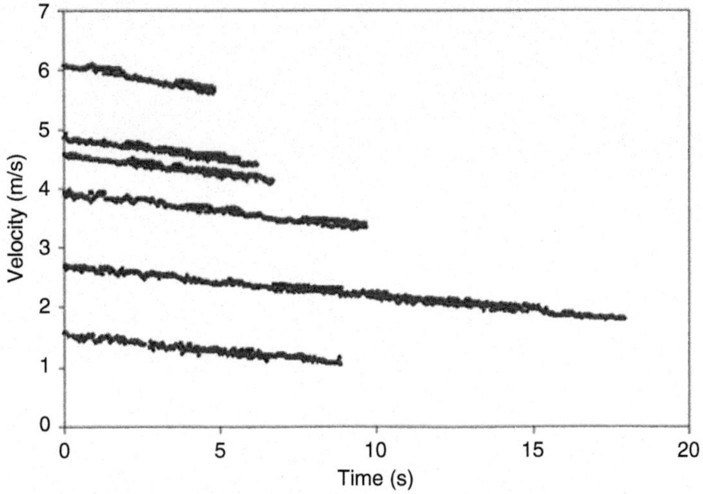

Fig. 7.21 Sample of the radar speed data collected in [16]. Six runs with different initial speed on flat ice surface

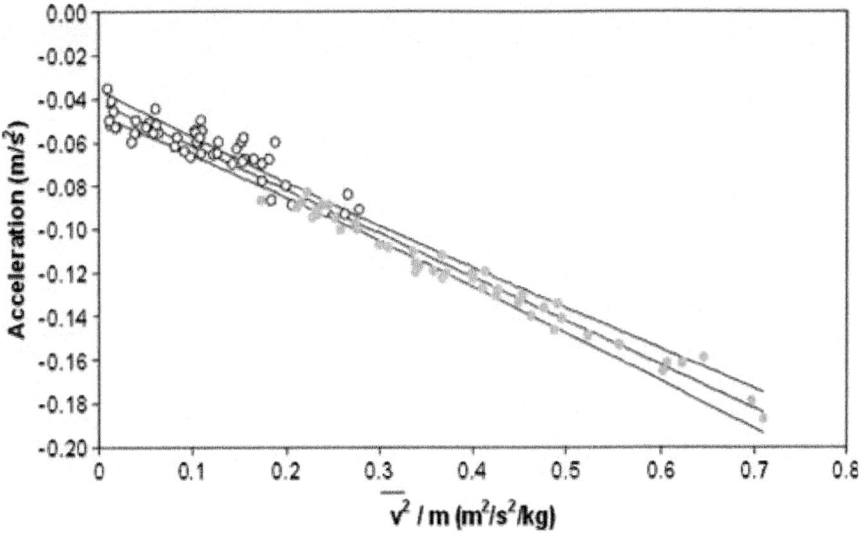

Fig. 7.22 Mean net acceleration $\bar{a}^*$ vs. $\bar{V}^2/m$: linear regression with confidence interval curves and experimental data, [16]

coefficient $\zeta = 0.2 \pm 0.02$ kg/m, while the y-axis intercept was used to identify the friction coefficient $\mu = 0.0042 \pm 0.0009$. Estimated friction coefficient is considerably lower than values obtained through laboratory experiments, probably due to the different testing conditions.

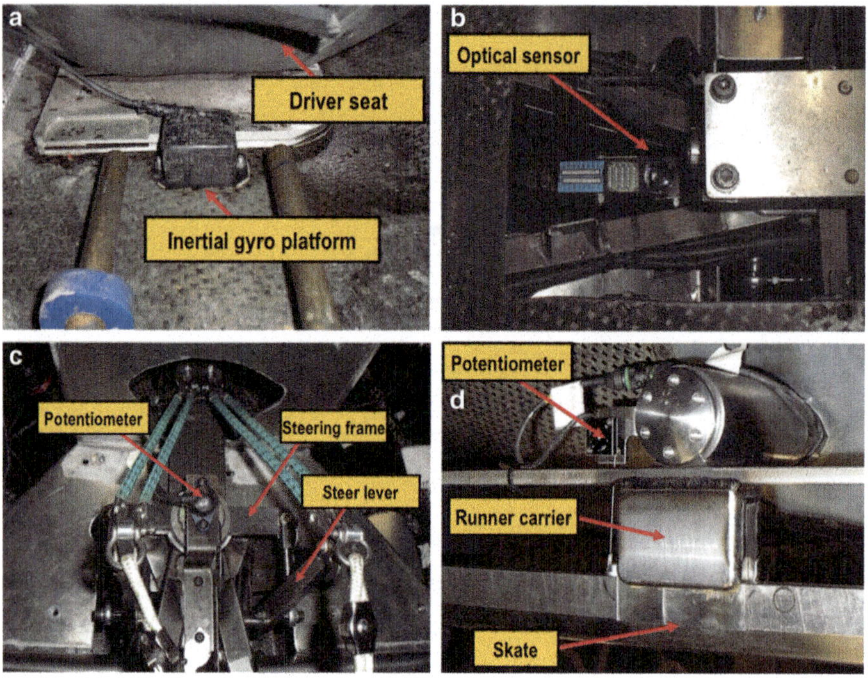

Fig. 7.23 Sensors: (**a**) inertial gyroscopic platform; (**b**) optical sensor; (**c**) rotary potentiometer; (**d**) linear potentiometers. Adapted from [17]

7.3.4.2 Full Scale Measurements with Instrumented Bobsled

An experimental setup for measuring bobsled dynamics and ice-runners contact forces during timed runs performed by professional athletes was described in [17].

In order to assess the dynamics of a two-men bobsled during timed runs on a whole track, the vehicle was instrumented with:

- one inertial gyroscopic platform to measure the longitudinal, lateral, and vertical accelerations of the bobsled and its angular rates of roll, pitch, and yaw (see Fig. 7.23a);
- one optical device to measure bobsled speed and sideslip angle (see Fig. 7.23b);
- one rotary potentiometer to measure the steer angle (see Fig. 7.23c);
- four linear potentiometers to measure the relative displacement between the runner carrier and the skate (see Fig. 7.23d);
- two dynamometric axles in order to measure the ice-skate contact forces (which are described later on).

The position of the inertial gyroscopic platform and of the optical sensor with respect of the bobsled center of gravity (cog) is shown in Fig. 7.24a, while the reference system is shown in Fig. 7.24b.

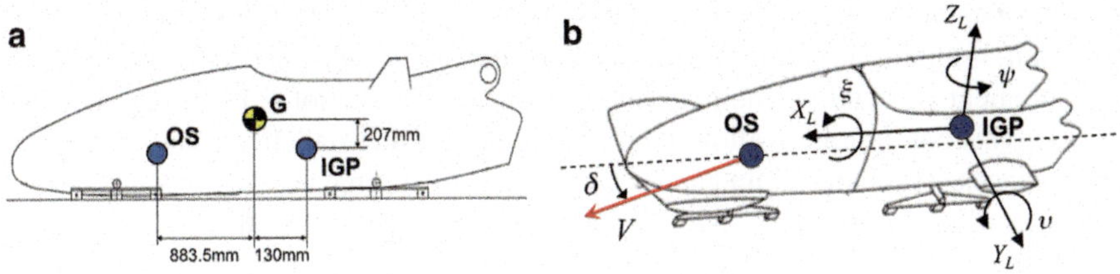

Fig. 7.24 (**a**) Position of the inertial gyroscopic platform (IGP) and of the optical sensor (OS) with respect of the bobsled cog (G); (**b**) moving reference system

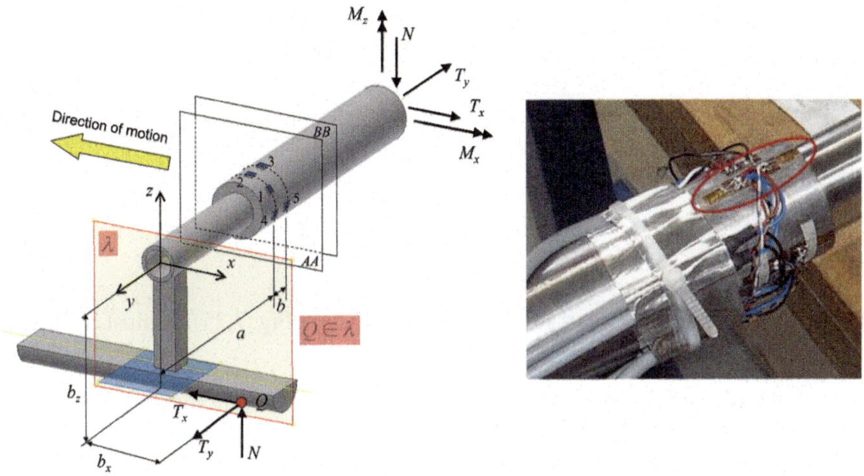

Fig. 7.25 Ice-skate contact forces (*left*) and strain gauges position (*right*). Adapted from [17]

Both power supply and acquiring systems were placed on board the bobsled. 12 V lead acid batteries were used to power supply all the instruments and the acquisition system. Measured signals were recorded on an MTA Digitek Cobra data-logger also placed on board the vehicle and subsequently transferred to a PC for post-processing via an Ethernet link. All the signals were acquired at a frequency of 100 Hz and low-pass filtered at 10 Hz.

Dynamometric axles were designed and developed to assess ice-runners contact forces. Specifically, reconstruction of contact forces was based on the measurement of runners-carrier deformations induced by contact forces themselves. Ice-runner contact forces and position of strain gauges on the axle are represented in Fig. 7.25. Number and position of strain gauges was bounded by the limited space available between the connections of the axle with the runners and the bobsled frame (about 100 mm). Strain gauges indicated with 1 in Fig. 7.25, which have two measuring grids perpendicular one to the other, were connected to form a full bridge allowing axial force (T_y) measurement and compensation of thermal effects and bending.

The pair of strain gauges indicated with 2–3 in Fig. 7.25 (which is connected in a half bridge configuration balanced by means of two resistors) allowed the measurement of bending moment about the x axis in sections AA (M_{xAA}) and BB (M_{xBB}). This configuration allowed compensating for thermal effects and axial force disturbances. The same strain gauges layout (pair of strain gauges 4–5) was selected for measuring the bending moment about the z axis in sections AA (M_{zAA}) and BB (M_{zBB}).

Several laboratory tests [17] were performed to experimentally define the calibration matrix $[C]$ relating the output signals of the strain gauge bridges $\mathbf{v}$ and the measurement vector $\mathbf{f} = \left[T_y \; M_{xAA} \; M_{xBB} \; M_{zAA} \; M_{zBB}\right]^T$:

$$\mathbf{f} = [C]\mathbf{v} \tag{7.9}$$

The lateral component of the ice-skate contact force is already included into vector $\mathbf{f}$, while components N and T_x can be determined by applying the following relations:

$$N = -\frac{M_{xBB} - M_{xAA}}{b}; \qquad T_x = -\frac{M_{zBB} - M_{zAA}}{b} \tag{7.10}$$

being b the distance between sections AA and BB (see Fig. 7.25). It must be in fact considered that:

- the lateral force T_y provides a constant contribution to bending moments about both x and z axes;
- the vertical force N originates a bending moment about the x axis, which increases proportionally to the distance from the runner (triangular distribution, Fig. 7.26a);
- the longitudinal force T_x originates a bending moment about the z axis, which increases proportionally to the distance from the runner (triangular distribution, Fig. 7.26b).

The dynamometric axle developed in [17] thus allows decoupling the measurement of contact force components T_x, T_y, N. Moreover it also theoretically permits the estimation of the position of the application point of ice-runners contact force, once known the contact forces themselves. The z (b_z) and x (b_x) position of application point Q can be in fact determined according to:

$$b_x = -\frac{M_{zBB} - T_x a}{T_y}; \qquad b_z = \frac{Na - M_{xAA}}{T_y} \tag{7.11}$$

It is to point out that b_z is also known by measurement, since it is expected to be the same as the relative displacement between the runners-carrier and the skate.

Both front and rear axles were instrumented in order to measure contact forces in correspondence of each of the four runners.

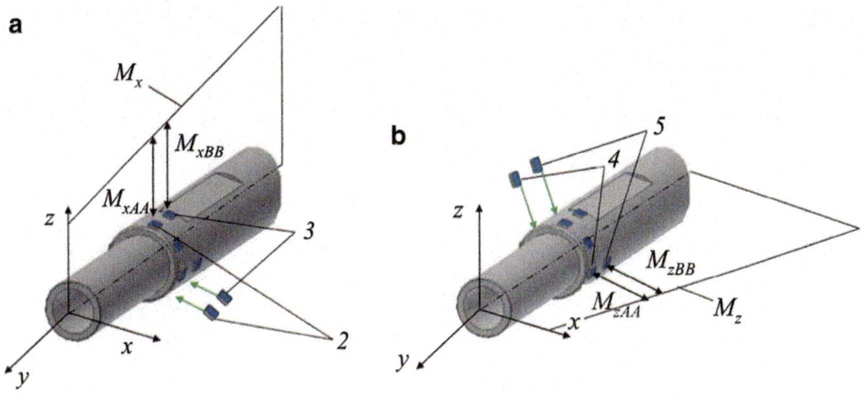

Fig. 7.26 Bending moments about the *x* axis (**a**) and the *z* axis (**b**) [17]

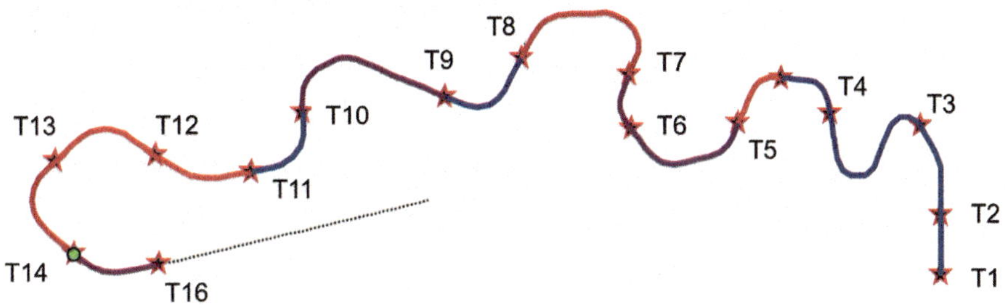

Fig. 7.27 Cesana Pariol Olympic Track [17]

Several runs were carried out with the previously described instrumented bobsled by professional athletes on the Cesana Pariol Olympic Track. The track is 1411 m long and it is characterized by 19 curves, a drop in height of 144 m and an average gradient of 8.08 % (Fig. 7.27).

Figure 7.28 shows the measurements of steer angle, bobsled speed, bobsled sideslip, and roll angles as a function of the traveled distance along the track during one of the performed runs. As it can be seen, bobsled speed increases along the track till reaching a maximum value higher than 130 km/h. Peak values of roll angle (obtained by integrating the time history of measured roll rate) exceed 90° in several turns, reaching its maximum value of 94° in curve 2. Comparing roll and steer angles, it can be noticed that, despite the fact their peaks occur almost simultaneously, a poor correlation between them is visible. This suggests that synchronization of peaks is only due to track geometry: when a turn begins, the steer angle is suddenly increased by the driver and a roll angle is developed as a consequence of the centrifugal force. The maximum (absolute) value of the roll angle is reached with an almost fixed time delay with respect to the corresponding peak of the steer angle (due to bobsled dynamics) and a sudden rise of steer angle is always anticipated by a slower increase of the roll angle. The sideslip angle is of

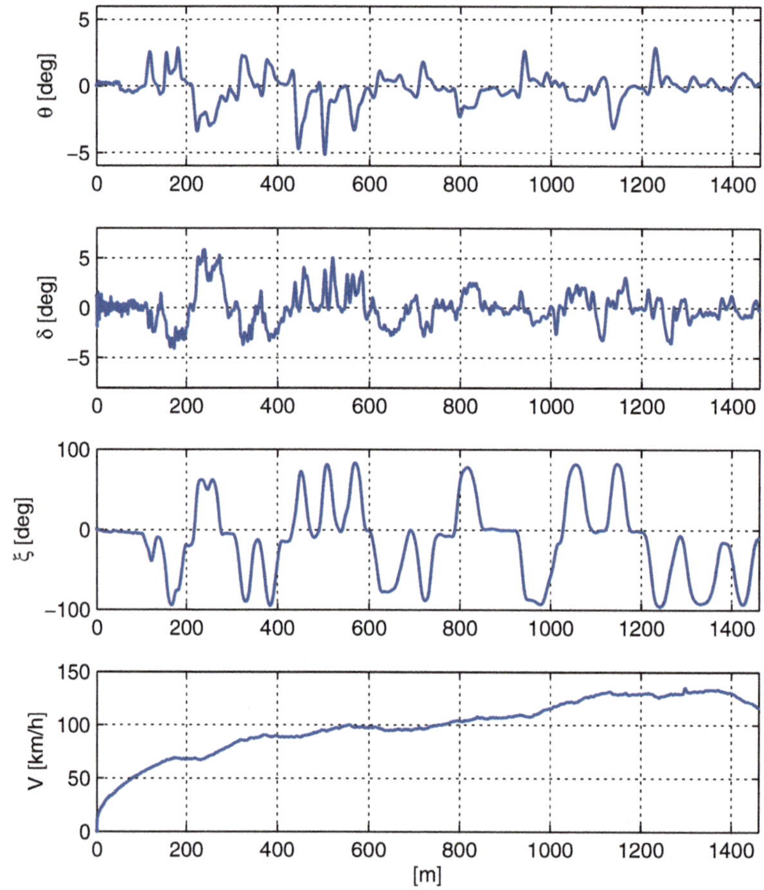

Fig. 7.28 Measurements of steer angle (θ), bobsled sideslip angle (δ), bobsled roll angle (ξ), and bobsled speed (V) vs. traveled distance

the same order of magnitude of the steer angle (it remains in fact bounded between $\pm 5°$). It can be noticed that during the straight sections connecting two curves, the peak of sideslip angle always anticipates the one of the steer angle.

Contact forces measured in correspondence of the four skates in the vertical, lateral, and longitudinal directions vs. the travelled distance are, respectively, shown in Figs. 7.29, 7.30, and 7.31. Contact force components are expressed in a reference system moving with the bobsled (Fig. 7.24b).

Figures 7.32, 7.33, and 7.34 compare the sum of contact forces on the runners divided by the bobsled mass (dashed lines) with the vertical, lateral, and net longitudinal (i.e., longitudinal acceleration compensated for gravitational contribution and aerodynamic drag) accelerations measured by the inertial gyroscopic platform (solid lines) for confirm of ice-runner contact force measurements. A good agreement can be generally noticed between force and acceleration measurements, especially

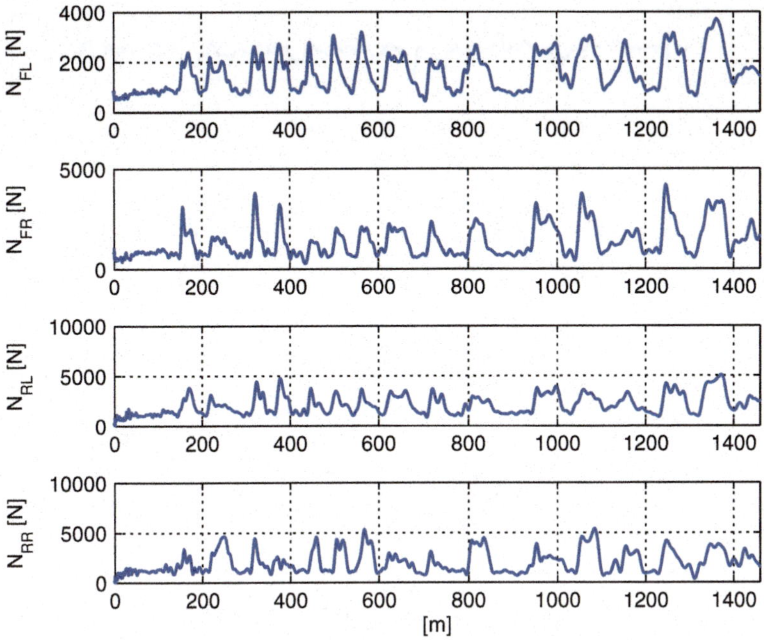

Fig. 7.29 Measured ice-skate contact forces along z axis

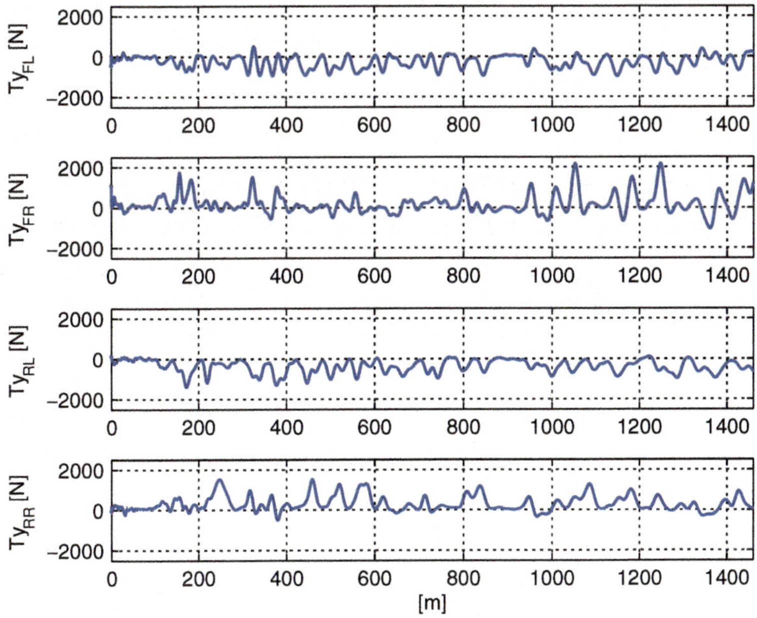

Fig. 7.30 Measured ice-skate contact forces along y axis

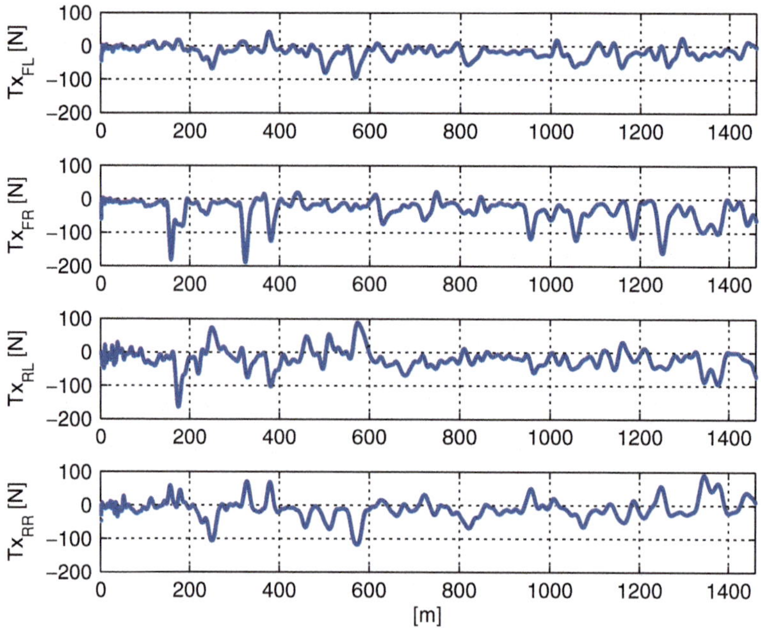

Fig. 7.31 Measured ice-skate contact forces along x axis

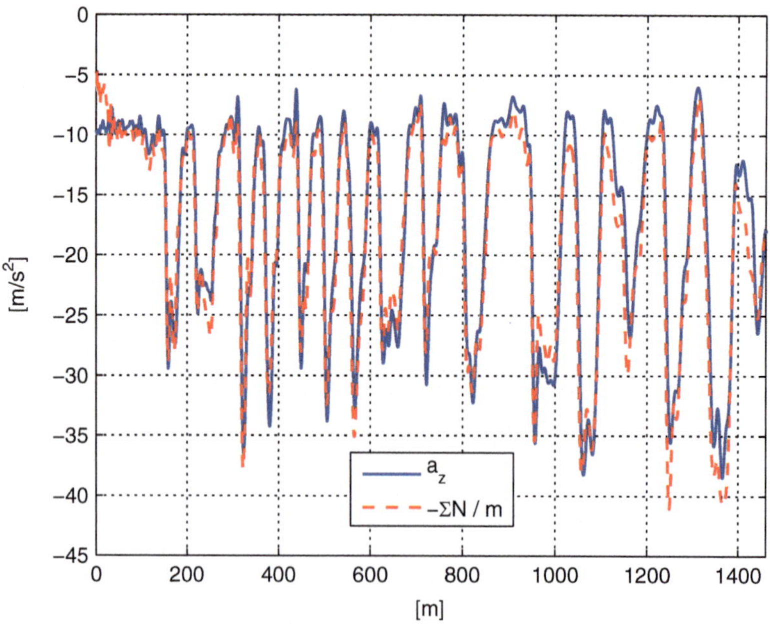

Fig. 7.32 Comparison of equilibrium along z axis

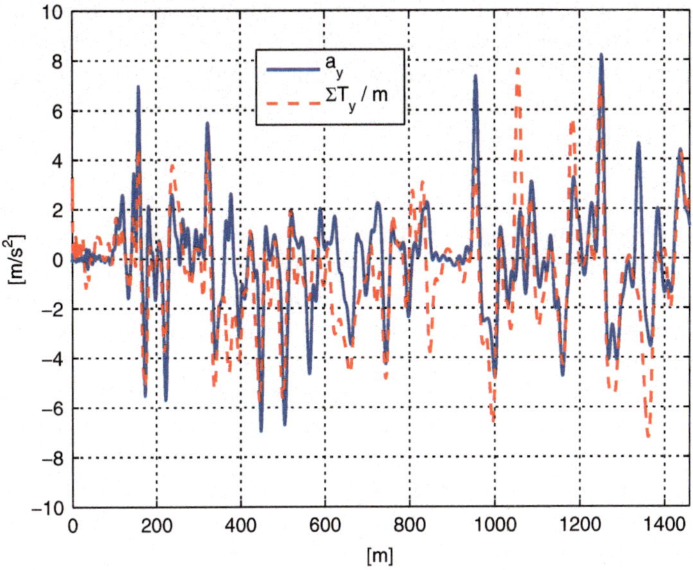

Fig. 7.33 Comparison of equilibrium along y axis

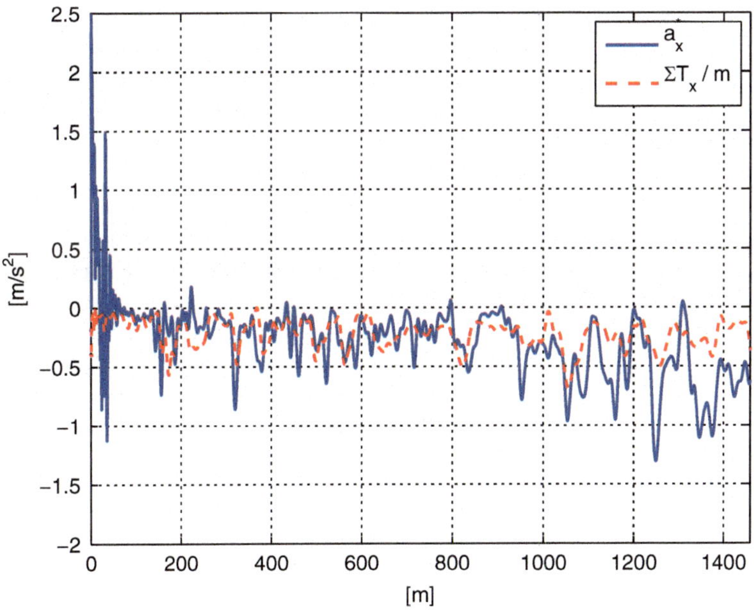

Fig. 7.34 Comparison of equilibrium along x axis

for the vertical direction (Fig. 7.32). Vertical acceleration and contact forces clearly identify curves of the track. Compressions higher than $3.5g$ are reached during most of the curves, coherently with the measurements provided in [10] on a skeleton sled.

Peak values of lateral contact forces (and lateral acceleration) are reached at curves' enter/exit (Fig. 7.33).

Looking at the net longitudinal acceleration of the sled (Fig. 7.34), push phase is clearly visible. It is characterized by a positive value of net acceleration, which ends after about 50 m. Boarding of the crew is distinctly highlighted by a spike in the longitudinal acceleration signal. After that point, the value of net longitudinal acceleration gets slightly negative (its mean value is about 0.015–0.02 $\mathrm{m/s^2}$) becoming coherent with ice-runner longitudinal contact forces. Peaks are present both in acceleration and force measurements in correspondence of curves' enter/exit. Net longitudinal acceleration however shows a decreasing trend along the track (which may indicate an increase of ice-runner friction with speed), which is not captured by measured longitudinal contact forces.

A rough estimation of the longitudinal (μ_x) and lateral (μ_y) ice-skate friction coefficient was performed comparing measured longitudinal and lateral contact forces with the vertical ones:

$$\mu_x = \frac{T_x}{N}; \qquad \mu_y = \frac{T_y}{N} \tag{7.12}$$

It must be underlined that experimental scatter of force measurements (especially in longitudinal direction, where mean force values are in the order of 25–30 N) can be significant, this making estimation of friction coefficient quite uncertain. A value of $\mu_x = 0.018 \pm 0.008$ was however estimated for the longitudinal friction coefficient considering the straight sections of the track, averaging 10 consecutive acquired samples and including all the four runners into the analysis.

Obtained value of friction coefficient is consistent with the literature [13–15] but considerably higher than the value obtained in [16]. In addition to previous comments, it must be however noted the considerable difference of maximum speed reached during the experiments. A maximum speed of 10 m/s was in fact reached in [16] and according to [13, 14] friction coefficient is expected to increase with speed.

A value of $\mu_y = 0.2 \pm 0.08$ was instead estimated for the lateral friction coefficient considering the 15 curves of the track, averaging 10 consecutive acquired samples and including all the four runners into the analysis. This value is considerably higher than the longitudinal one and no data can be found in the literature to confirm the obtained result, since no comparable tests were performed. Obtained results can however be explained considering the different contact geometry during longitudinal and lateral sliding of runners.

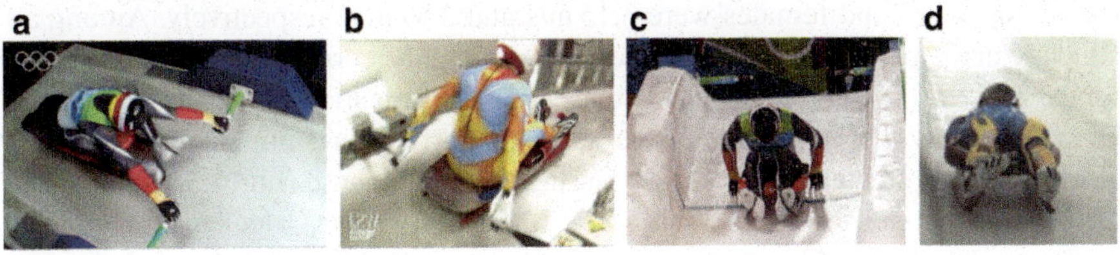

Fig. 7.35 Phases of luge start: (**a**) block; (**b**) drive; (**c**) arm-paddling; (**d**) settling

7.3.5 Full Scale Tests on Luge

As already mentioned, a significant correlation between start time and overall performance exists in luge. Although held in the same tracks as skeleton, luge is a very different sport: athletes begin their run in a seated position, using their arms to propel themselves forwards down a steep tower start. Being luge sleds more responsive than skeleton ones, race times are more tightly grouped. With the smaller margins of time separating finish positions, luge athletes understand the advantage that can be gained in the start. Consequently pull start has always been a focus of experimental studies concerned with luge.

Luge start can be divided into four subsequent phases: block, drive, arm-paddling, and settling (see Fig. 7.35). During the block phase, the athlete grasps the handles and rocks back and forth to load shoulder, trunk, and back extensor musculature prior to the drive phase. At the end of this phase, the athlete briefly stops the backward motion of the sled (Fig. 7.35a). Then the drive phase begins, during which the sled is propelled in forward direction (Fig. 7.35c). Before the athlete settles back into the sled for the run, three or even more arm strokes (arm-paddling phase, Fig. 7.35c) are performed, depending on the start ramp and the physical abilities of the athlete. The paddle motion consists of striking the ice, pulling and finally pushing away from the ice. After the last paddle, maximum start velocity is reached and then the athlete settles in the race position (Fig. 7.35d).

In [18], dynamometric handles able to measure horizontal and vertical forces were used to analyze the first two phases of luge start (block and drive phases, see Fig. 7.35). Force data were used to quantify temporal phases, determine the magnitudes and relative effectiveness of forces and impulses applied by the athletes, and estimate bilateral asymmetries which might have negative effects on the initial sled trajectory. Twelve senior members of US elite luge team took part to the tests, which were carried out on the ice rink in Lake Placid. Both males and females were highly symmetric in the application of forces throughout the luge start, especially during the drive phase. Bilateral asymmetries averaged 8.89 % throughout the entire start sequence. Kinetic effectiveness, defined as the ratio between horizontal and vertical force applied on the handles, indicated that only about 75 % of the force applied during the drive phase contributes to the forward propulsion of sled. Male athletes demonstrated significantly higher absolute impulse and force magnitudes during the start which resulted into higher initial sled speeds. Average initial sled speeds for males and females were 3.15 m/s and 2.87 m/s, respectively. Among all athletes, drive phase kinetic effectiveness was highly correlated with sled speed. Linear regression analysis revealed that 77 % of the variability in luge start can be explained by the magnitude of the horizontal drive phase impulse alone. It was also found that greater kinetic output during the block phase (Fig. 7.35a) enhanced the forward propulsion of the sled during the drive phase. Block phase impulse was strongly associated with initial sled speed, accounting for 45 % of sled speed variation. The associations between initial sled speed and drive phase kinetics were readily explained by the proportional relationship between speed and force (impulse) in the idealized impulse-momentum equation which was used in the calculation of sled speed. Analogous associations between block phase kinetics were explained by a more effective utilization of stored elastic energy during fast start performances. The results of this study suggest that fast luge start performances are a function of both an explosive eccentric loading of shoulder, trunk, and hip musculature during the preparatory block phase and a vigorous forward propulsion of sled during the drive phase.

Sagittal video recordings of the athlete were synchronized with force measurements in [19] to correlate movements of the athlete with the forces applied to the handles. Obtained results are consistent with the ones reported in [18].

High-speed video recordings were also used in [20] to analyze all the different phases of luge start during several tests on race sliding tracks and start training ramps. It was found that generally there exist a difference in duration of lugers' start phases between race sliding tracks and start training ramps. Specifically, it was found that generally athletes tend to decrease duration of arm-paddling when starting on luge race tracks, probably due to the necessity to assume the riding position in time. Only best athletes are able to transfer their typical timing patterns from training to competitive facilities. Most of the athletes instead maintain the same duration of start jerk (i.e., the time interval from the beginning of the sled' forward motion until the moment the athlete releases the start handles) on both starting facilities.

A training tool (called *Speedpaddler*) was developed in [21, 22] to assess and train arm-paddle technique of high performance luge athletes. Arm-paddles in fact are estimate to contribute for 23 % of the total starting performance [22]. The training tool is constituted of an aluminum alloy framework with customary belt conveyor system, which is driven by two synchronized servo-motors. Training with constant speeds up to 12 m/s are allowed by the test bench. Speedpaddler facilitates variations in the inclination and speed of the conveyor belts and thereby resistance and movement speed. If the athlete accelerates the conveyor belts during arm-paddling, the torque of the motors decreases. Torque measurements and high-speed video recordings are provided to evaluate the effectiveness of arm-paddling technique. Tests with 18 luge athletes (Austrian national team and juniors) were performed on luge race tracks and Speedpaddler. Arm-paddle cycle was assessed by means of accelerometers fixed under the athletes' race gloves. Arm-paddle cycle duration on Speedpaddler was found to be consistent with the one measured on ice.

7.4 The Sled

7.4.1 Dynamic Modelling and Optimization

As seen in Sect. 7.2, the rules approved by the Federation for a race bobsled [3] allow to set a number of design parameters (dimensions, weight, suspension stiffness, etc.) within given ranges. Though limited, these ranges may lead to different design choices resulting in different bobsled layouts.

Even if on-track tests represent a basic step for the final tuning of a bobsleigh, experimentation does seem the most suitable tool to objectively determine the influence of the design parameters on the overall performance. In fact, several factors like different start times, driving errors, change in ice characteristics during the day can easily mask their effect. A huge amount of tests would be needed to achieve statistical evidence, but limited availability of tracks and costs of experimentation make this approach hard to follow.

Athletes feedback constitutes an alternative source of indications for bobsled design, but usually leads only to small modifications with marginal effect on the overall performance.

For these reasons, a mathematical model of the bobsled may be regarded as an interesting tool to provide guidelines for the designers. Numerical simulation offers two important advantages: cheapness and repeatability. A huge amount of "virtual" tests could be performed excluding the effect of external variables that usually affects on-track tests; if one structural parameter of the bobsled is changed, its influence on the final time will be clearly quantified. But, even more interesting, several design parameters could be changed at the same time, analyzing their mutual effect on the overall performance. This means that, considering the minimum final time as target, an optimization process could be performed identifying the optimal combination of structural parameters leading to the best performance.

Drawbacks of this approach are all related to modelling approximations or uncertainties. Typical critical elements can be found in the modelling of the skate–ice interaction and of the driver's response; both the elements have a significant impact on the final time and inadequate modelling may lead to improper assessment of the effect of other parameters. But once a good level of modelling is achieved or when limits of models are well understood, numerical simulations undoubtedly constitute a useful aid to bobsled design.

Models of different complexity were presented in literature, ranging from single particle models to 3D multi-body models. The following paragraphs will present three models with increasing degree of complexity showing advantages and limits of the approaches.

7.4.1.1 Single Degree of Freedom Particle Model

A numerical model aiming at reproducing the bobsled dynamics has to include a model for the sled itself, a model for the track, a model describing the interaction between the two and a model for the driver.

A point mass represents the simplest model that can be used to approximate the motion of the sled. A point mass is also known as *particle model* and has just 3 degrees of freedom in the space. In principle, the same model could be used to skeleton and luge too. If one assumes that the point mass only moves on the track surface and no local deformation occurs at the contact interface, its position can actually be described with only two independent variables identifying the position of a point on a two-dimension surface. Under an even more simple hypothesis, if the trajectory of the sled over a given track is known, a single degree of freedom (i.e., the curved abscissa) will be enough to set the bobsled position.

Though very simple, this model was adopted in literature [23] and can be used to analyze the effect of different parameters on the final time. The hypothesis at the basis of this approach is that the reference trajectory of the bobsled along the track is somehow *optimal* and would not be affected by changing the parameters considered in the analysis. Since the trajectory is kept fixed, few information are required to model the track geometry and a driver model is not needed. Interactions between bobsled and track could be computed in a rather easy way, as shown in the following.

The trajectory of the point-mass in 3D space can be expressed in parametric form as reported in (7.13): vector $\mathbf{P}(s)$ represents the position of a point belonging to the selected path as function of the curved abscissa s; the components of $\mathbf{P}(s)$ are the coordinates $x(s)$, $y(s)$, and $z(s)$ of a point of the trajectory in an absolute reference.

$$\mathbf{P}(s) = \begin{Bmatrix} x(s) \\ y(s) \\ z(s) \end{Bmatrix} \tag{7.13}$$

The curved abscissa identifies the position of the point mass along the trajectory and therefore represents the degree of freedom of the model. Curved abscissa is in turn function of time, since the point mass can run along the trajectory with different velocity profiles; the velocity of the point mass $\mathbf{V}$ can be easily obtained as:

$$\mathbf{V} = \frac{d\mathbf{P}(s)}{dt} = \frac{d\mathbf{P}(s)}{ds}\frac{ds}{dt} = \mathbf{t}(s)\dot{s} \tag{7.14}$$

The first derivative of $\mathbf{P}(s)$ with respect to the curved abscissa allows definition of the unit vector $\mathbf{t}(s)$ tangent to the trajectory. Acceleration of the point mass $\mathbf{a}$ is given by:

$$\mathbf{a} = \frac{d\mathbf{V}}{dt} = \mathbf{t}(s)\ddot{s} + \frac{d\mathbf{t}(s)}{dt}\dot{s} = \mathbf{t}(s)\ddot{s} + \frac{d\mathbf{t}(s)}{ds}\dot{s}^2 = \mathbf{t}(s)\ddot{s} + \frac{\mathbf{n}(s)}{R}\dot{s}^2 \tag{7.15}$$

The first derivative of $\mathbf{t}(s)$ with respect to the curved abscissa represents the ratio between the unit vector $\mathbf{n}(s)$, directed toward the trajectory center of curvature, and the local curvature radius R. Vector $\mathbf{n}(s)$ defines the direction of the centripetal

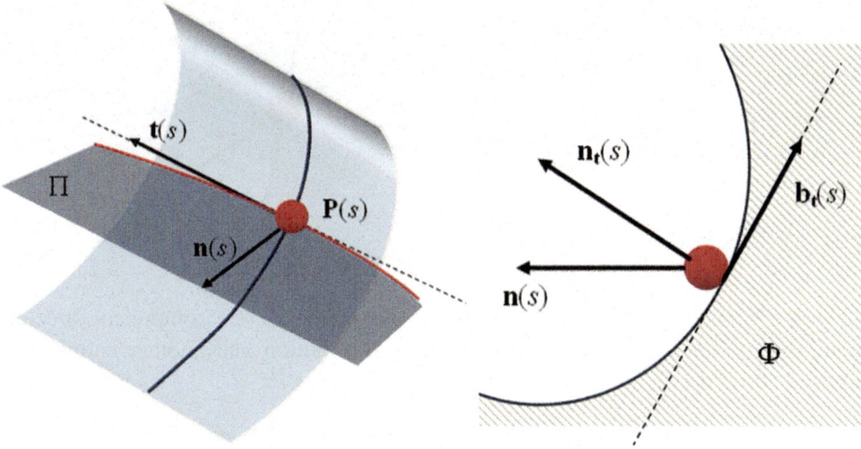

Fig. 7.36 Direction of vectors $\mathbf{t}(s)$, $\mathbf{n}(s)$, $\mathbf{n_t}(s)$, and $\mathbf{b_t}(s)$

acceleration experienced by the point mass when the trajectory presents a curvature. Vectors $\mathbf{t}(s)$ and $\mathbf{n}(s)$ are orthogonal and identify the so-called osculating plane, marked with Π in Fig. 7.36

As shown in the right part of Fig. 7.36, direction of $\mathbf{n}(s)$ is in general different from the direction of a vector normal to the track surface directed outward which hereafter will be indicated with $\mathbf{n_t}(s)$. Under the assumption of infinitely rigid track, the motion of the point mass takes place in a plane which is locally tangent to the track itself. This means that also $\mathbf{n_t}(s)$ is normal to $\mathbf{t}(s)$ so $\mathbf{n_t}(s)$ and $\mathbf{n}(s)$ belong to the same plane Φ. This last contains also the binormal vector $\mathbf{b_t}(s)$ which is instead tangent to the track surface and is defined as:

$$\mathbf{b_t}(s) = \mathbf{t}(s) \times \mathbf{n_t}(s) \tag{7.16}$$

Applying Newton's second law to the point mass, the following vector equation can be obtained:

$$m\left(\mathbf{t}(s)\ddot{s} + \frac{\mathbf{n}(s)}{R}\dot{s}^2\right) = \mathbf{F_g} + \mathbf{F_a} + \mathbf{F_t} \tag{7.17}$$

In (7.17), m is the sled mass, while $\mathbf{F_g}$, $\mathbf{F_a}$, $\mathbf{F_t}$ represent the external forces acting on the sled itself. In particular, $\mathbf{F_g}$ is the gravity force, $\mathbf{F_a}$ is the aerodynamic force while $\mathbf{F_t}$ is the force developed by the track at the contact interface. Gravity force is defined as:

$$\mathbf{F_g} = m\mathbf{g} = m\begin{Bmatrix} 0 \\ 0 \\ -g \end{Bmatrix} \tag{7.18}$$

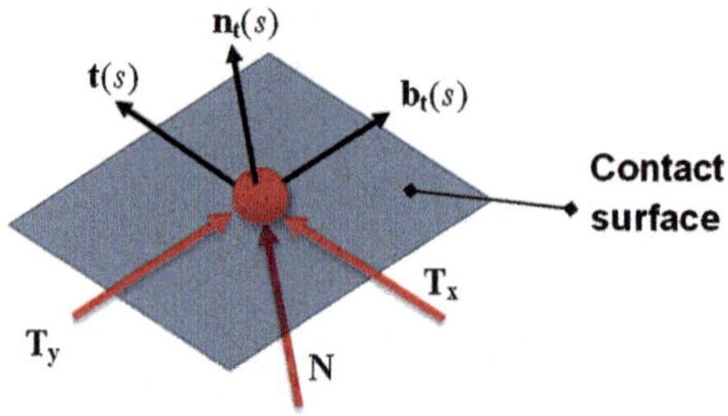

Fig. 7.37 Components T_x, T_y of the contact force $\mathbf{F_t}$

Vector $\mathbf{F_a}$ associated with aerodynamic forces presents one component directed along the motion direction $\mathbf{t}(s)$ due to drag contribution and another component directed along $\mathbf{n_t}(s)$ generated by lift.

$$\mathbf{F_a} = -\frac{1}{2}\rho AC_D\dot{s}^2 \cdot \mathbf{t}(s) + \frac{1}{2}\rho AC_L\dot{s}^2 \cdot \mathbf{n_t}(s) \tag{7.19}$$

Aerodynamic forces are computed considering a front surface A, an air density ρ, lift and drag coefficients, respectively, equal to C_L and C_D.

The force developed at the contact interface is made up of three components, as shown in Fig. 7.37: the component $\mathbf{N}$ is normal to the track surface and therefore is directed along $\mathbf{n_t}(s)$. The tangential contact force can be divided into two components: the first one, named $\mathbf{T_x}$ is associated with longitudinal resistance due to friction and is directed along the motion direction (i.e., $\mathbf{t}(s)$); according to Coulomb's model of sliding friction, $\mathbf{T_x} = -\mu\,|\mathbf{N}| \cdot \mathbf{t}(s)$. The second component of tangential force, $\mathbf{T_y}$, is directed along the binormal direction $\mathbf{b_t}(s)$ and accounts for the skidding resistance developed in lateral direction. Altogether, the contact force $\mathbf{F_t}$ can be written as:

$$\mathbf{F_t} = \mathbf{N} - \mu\,|\mathbf{N}| \cdot \mathbf{t}(s) + \mathbf{T_y} \tag{7.20}$$

Combining (7.18), (7.19), and (7.20), Eq. (7.17) can be formulated as follows:

$$m\left(\mathbf{t}(s)\ddot{s} + \frac{\mathbf{n}(s)}{R}\dot{s}^2\right) = m\mathbf{g} - \frac{1}{2}\rho AC_D\dot{s}^2 \cdot \mathbf{t}(s) + \frac{1}{2}\rho AC_L\dot{s}^2 \cdot \mathbf{n_t}(s) + \mathbf{N} - \mu\,|\mathbf{N}| \cdot \mathbf{t}(s) + \mathbf{T_y} \tag{7.21}$$

Equation (7.21) is in fact a vector equation and can be projected along three directions; projecting (7.21) along a direction normal to the track surface allows computing the normal reaction force N:

$$N = m\frac{\dot{s}^2}{R}\,\mathbf{n}(s) \cdot \mathbf{n_t}(s) - m\,\mathbf{g} \cdot \mathbf{n_t}(s) - \frac{1}{2}\rho AC_L\dot{s}^2 \tag{7.22}$$

Since the motion of the point mass is assumed to take place on a given trajectory, Eq. (7.22) should be used to check for the sign of N; negative values of N could in fact be possible with this simple mathematical model but obviously do not have physical meaning. Besides this, the result of (7.22) allows obtaining a pure motion equation when projecting (7.21) along the direction $\mathbf{t}(s)$:

$$m\ddot{s} = m\mathbf{g} \cdot \mathbf{t}(s) - \mu\, |\mathbf{N}| - \frac{1}{2}\rho A C_D \dot{s}^2 \tag{7.23}$$

As last, when (7.21) is projected along the binormal direction, the lateral contact force required to follow a given trajectory can be computed.

$$T_y = -m\,\mathbf{g} \cdot \mathbf{b}_\mathbf{t}(s) + m\frac{\dot{s}^2}{R}\,\mathbf{n}(s) \cdot \mathbf{b}_\mathbf{t}(s) \tag{7.24}$$

This information is not needed to solve Eq. (7.23) but allows performing an additional check for results reliability: in particular, values obtained for T_y should be compared to the maximum lateral force the bobsled can develop through skates. Assuming a positive value for N and combining Eq. (7.22) with (7.23), bobsled acceleration can be expressed as:

$$\ddot{s} = \mathbf{g} \cdot \mathbf{t}(s) - \mu\left(\frac{\dot{s}^2}{R}\,\mathbf{n}(s) \cdot \mathbf{n}_\mathbf{t}(s) - \mathbf{g} \cdot \mathbf{n}_\mathbf{t}(s) - \frac{1}{2m}\rho A C_L \dot{s}^2\right) - \frac{1}{2m}\rho A C_D \dot{s}^2 \tag{7.25}$$

Equation (7.25) could be formulated in state form introducing the state vector $\mathbf{z}$:

$$\mathbf{z} = \begin{Bmatrix} s \\ \dot{s} \end{Bmatrix} \tag{7.26}$$

Deriving $\mathbf{z}$ with respect to time leads to a set of two first order differential equations; the problem can be thus formulated as:

$$\dot{\mathbf{z}} = f(\mathbf{z}, t) \tag{7.27}$$

Equation (7.27) can be solved numerically assuming an initial position $s(0)$ and an initial speed $\dot{s}(0)$ resulting from the sprint phase which is not modelled.

Equation (7.25) represents the most significant result of the model: the role of motion resistances is clearly evidenced as well as the one of aerodynamic forces. It is noteworthy that increasing the total mass reduces the effect of aerodynamic lift and drag; high drag coefficients could thus be partly compensated adopting mass values close to the upper allowed limit.

Implementation of the proposed model is rather simple and requires few data relevant to track geometry; the reference trajectory could be determined on the basis of data recorded through GPS during a test run and conveniently interpolated. Actually, measurement of vector $\mathbf{n}_\mathbf{t}(s)$ is rather complex but its direction could be approximated knowing the local slope of the track and its transverse orientation [23].

Once implemented, the model could be used to estimate the effect of a number of parameters on the final time: bobsled mass, aerodynamic coefficients, and also efficiency of the sprint phase which can be analyzed by changing the initial conditions. For instance, the model allows quantification of the benefits of a new body with lower drag considering a worsening of the sprint phase due to reduced cockpit accessibility.

The limits of this model are strictly related with its simplicity: according to (7.25), a positive lift coefficient could reduce the motion resistance associated with friction and this suggests the adoption of high positive lift coefficients to improve performance. Actually this would mean that, when a bobsled enters a corner at high speed, its directionality and stability could be jeopardized since also lateral contact forces will decrease. The limited number of parameters considered in the model does not allow a comprehensive analysis of the dynamics of a bobsled; moreover, the hypothesis of a fixed trajectory should be removed to consider high order effects (a change of the speed profile will almost certainly result in a different trajectory) and the influence of driver inputs.

7.4.1.2 Two Degrees of Freedom Particle Model

Allowing the point mass to move over the entire track surface rather than on a fixed trajectory adds little complexity to the model and significantly increases its capabilities. With this approach the problem is no longer monodimensional and two degrees of freedom are required to define the position of the sled over the 2D track surface. The whole track geometry should be modelled: following the method described in Sect. 7.5, this can be done considering a series of track sections interpolated by means of B-splines. Trajectory followed by the point mass along the track will result from the forces acting on it; in particular, lateral contact forces developed through the skates should be adequately modeled. A driver model will also be needed to determine the amount of lateral force required to control the bobsled dynamics and follow a desired path. The following discourse is based on the approach presented in [24]. As shown in Fig. 7.38, the position of a point on the track surface can be identified with two coordinates: α representing the distance from the start line and β representing the distance from the track left side.

Therefore, the position (x, y, z) of a point belonging to the track surface is function of the two coordinates α and β:

$$\mathbf{P}(\alpha,\beta) = \begin{Bmatrix} x(\alpha,\beta) \\ y(\alpha,\beta) \\ z(\alpha,\beta) \end{Bmatrix} \tag{7.28}$$

The sled velocity can be obtained from (7.28).

$$\mathbf{V} = \frac{d\mathbf{P}}{dt} = \frac{d\mathbf{P}}{d\alpha}\dot{\alpha} + \frac{d\mathbf{P}}{d\beta}\dot{\beta} = \mathbf{P}_{\alpha}\dot{\alpha} + \mathbf{P}_{\beta}\dot{\beta} \tag{7.29}$$

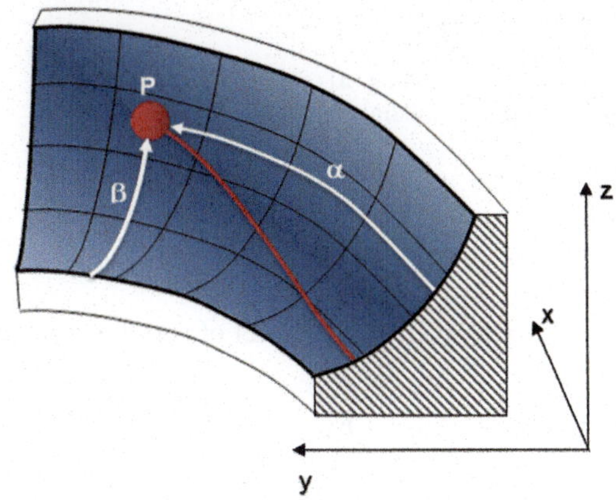

Fig. 7.38 Components of the contact force $\mathbf{F_t}$

Vectors $\mathbf{P}_\alpha$ and $\mathbf{P}_\beta$ identify a plane tangent to the track surface in **P**. Vector **V** lies in this plane and its direction is defined by the unit vector $\mathbf{t} = \mathbf{V}/|\mathbf{V}|$. The unit vector $\mathbf{n_t}$ normal to the track surface in **P** can be instead defined as:

$$\mathbf{n_t} = \frac{\mathbf{P}_\alpha \times \mathbf{P}_\beta}{|\mathbf{P}_\alpha \times \mathbf{P}_\beta|} \tag{7.30}$$

As explained in Sect. 7.4.1.1, the unit vector identifying the lateral direction is called *binormal* vector and is defined as:

$$\mathbf{b_t} = \mathbf{t} \times \mathbf{n_t} \tag{7.31}$$

Starting from Eq. (7.27), sled acceleration can be computed as shown in (7.32).

$$\mathbf{a} = \frac{d\mathbf{V}}{dt} = \mathbf{P}_\alpha\ddot{\alpha} + \mathbf{P}_\beta\ddot{\beta} + \mathbf{P}_{\alpha\alpha}\dot{\alpha}^2 + \mathbf{P}_{\beta\beta}\dot{\beta}^2 + 2\mathbf{P}_{\alpha\beta}\dot{\alpha}\dot{\beta} \tag{7.32}$$

In (7.32) $\mathbf{P}_{\alpha\alpha} = \frac{d^2\mathbf{P}}{d\alpha^2}$; $\mathbf{P}_{\beta\beta} = \frac{d^2\mathbf{P}}{d\beta^2}$ and $\mathbf{P}_{\alpha\beta} = \frac{d^2\mathbf{P}}{d\alpha d\beta}$. Considering Eq. (7.29) and the previous discussion, the term $= \mathbf{P}_\alpha\ddot{\alpha} + \mathbf{P}_\beta\ddot{\beta}$ lies in the plane tangent to the track surface in **P**.

Application of Newton's second law leads to:

$$m(\mathbf{P}_\alpha\ddot{\alpha} + \mathbf{P}_\beta\ddot{\beta} + \mathbf{P}_{\alpha\alpha}\dot{\alpha}^2 + \mathbf{P}_{\beta\beta}\dot{\beta}^2 + 2\mathbf{P}_{\alpha\beta}\dot{\alpha}\dot{\beta}) = \mathbf{F_g} + \mathbf{F_a} + \mathbf{F_t} \tag{7.33}$$

where m represents is the sled mass, while vectors $\mathbf{F_g}$, $\mathbf{F_a}$, and $\mathbf{F_t}$ accounts for the effect of gravity force, aerodynamic force, and contact force, respectively. Similarly to Sect. 7.4.1.1, vector $\mathbf{F_g}$ can be defined as:

$$\mathbf{F_g} = m\,\mathbf{g} = m\begin{Bmatrix} 0 \\ 0 \\ -g \end{Bmatrix} \tag{7.34}$$

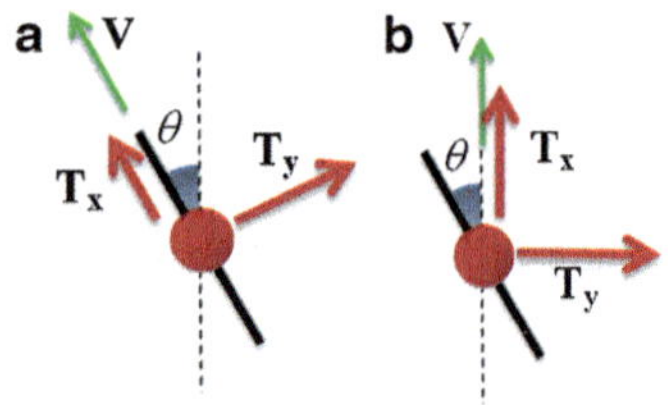

Fig. 7.39 Directions of the tangential contact forces without lateral skidding (**a**) and with lateral skidding (**b**)

Vector $\mathbf{F_a}$ collects the effect of aerodynamic drag and lift and is given by:

$$\mathbf{F_a} = -\frac{1}{2}\rho A C_D |\mathbf{V}|^2 \cdot \mathbf{t} + \frac{1}{2}\rho A C_L |\mathbf{V}|^2 \cdot \mathbf{n_t} \tag{7.35}$$

Vector $\mathbf{F_t}$ can be divided into three components: a reaction force $\mathbf{N}$ directed along $\mathbf{n_t}$, a motion resistance $\mathbf{T_x}$ directed along $\mathbf{t}$, and a skid resistance $\mathbf{T_y}$ associated both with Coulomb friction force and ice shear force and directed along $\mathbf{b_t}$. Both $\mathbf{T_x}$ and $\mathbf{T_y}$ are in fact function of the skate-ice friction coefficient and of the angle between the skates and velocity vector $\mathbf{V}$.

$$\mathbf{F_t} = \mathbf{N} + \mathbf{T_x} + \mathbf{T_y} \tag{7.36}$$

According to the model proposed in [24–27], as long as the steering angle θ is below a given threshold θ_m, the angle between the velocity vector $\mathbf{V}$ and the skates is zero, i.e. no lateral skidding occurs. A friction force rises along the direction of travel while a side shear force normal to the side of the blade is developed. This condition is schematically depicted in Fig. 7.39a. Tangential contact forces are expressed as:

$$\begin{cases} \mathbf{T_x} = -\mu_f |\mathbf{N}| \cdot \mathbf{t} \\ \mathbf{T_y} = -\dfrac{\theta}{\theta_m} \mu_t |\mathbf{N}| \cdot \mathbf{b_t} \end{cases} \tag{7.37}$$

In (7.37), μ_f is the ice-skate friction coefficient in longitudinal direction, while $\mu_t|\mathbf{N}|$ is the maximum shear force. The model assumes that lateral shear force increases linearly with the steering angle.

When the steering angle crosses the threshold θ_m, lateral skidding occurs (Fig. 7.39b). The maximum allowable shear force $\mu_t|\mathbf{N}|$ is developed and still acts perpendicularly to the blade side. Due to lateral skidding, the shear force presents a projection along the travel direction which increases the motion resistance. In addition the friction coefficient of the skate travelling in lateral direction named μ_s is usually different and higher than μ_f. The contact forces become:

$$\begin{cases} \mathbf{T_x} = -(\mu_s + \mu_t \sin\theta)|\mathbf{N}| \cdot \mathbf{t} \\ \mathbf{T_y} = -\mu_t \cos\theta |\mathbf{N}| \cdot \mathbf{b_t} \end{cases} \tag{7.38}$$

These forces can be computed once the reaction force **N** and the steering angle θ are known; steering angle can be regarded as an input for the model, determined on the basis of a driver model (see Sect. 7.6). Normal reaction **N** can instead be computed projecting (7.33) along $\mathbf{n_t}$ direction:

$$\mathbf{N} = m(\mathbf{P}_{\alpha\alpha}\dot{\alpha}^2 + \mathbf{P}_{\beta\beta}\dot{\beta}^2 + 2\mathbf{P}_{\alpha\beta}\dot{\alpha}\dot{\beta}) \cdot \mathbf{n_t} - \mathbf{F_g} \cdot \mathbf{n_t} - \frac{1}{2}\rho AC_L|\mathbf{V}|^2 \tag{7.39}$$

Two additional scalar equations can be obtained projecting (7.33) along two directions normal to ($\mathbf{n_t}$); in particular, the terms $\ddot{\alpha}$ and $\ddot{\beta}$ can be made explicit as follows:

$$\begin{cases} \ddot{\alpha} = \dfrac{1}{m}\mathbf{Q} \cdot \dfrac{|\mathbf{P}_\beta|^2\mathbf{P}_\alpha - (\mathbf{P}_\beta \cdot \mathbf{P}_\alpha) \cdot \mathbf{P}_\beta}{|\mathbf{P}_\alpha|^2|\mathbf{P}_\beta|^2 - (\mathbf{P}_\alpha \cdot \mathbf{P}_\beta)^2} \\ \ddot{\beta} = \dfrac{1}{m}\mathbf{Q} \cdot \dfrac{|\mathbf{P}_\alpha|^2\mathbf{P}_\beta - (\mathbf{P}_\beta \cdot \mathbf{P}_\alpha) \cdot \mathbf{P}_\alpha}{|\mathbf{P}_\alpha|^2|\mathbf{P}_\beta|^2 - (\mathbf{P}_\alpha \cdot \mathbf{P}_\beta)^2} \end{cases} \tag{7.40}$$

In (7.40) **Q** is given by:

$$\mathbf{Q} = -m(\mathbf{P}_{\alpha\alpha}\dot{\alpha}^2 + \mathbf{P}_{\beta\beta}\dot{\beta}^2 + 2\mathbf{P}_{\alpha\beta}\dot{\alpha}\dot{\beta}) + \mathbf{F_g} - \frac{1}{2}\rho AC_D|\mathbf{V}|^2 \cdot \mathbf{t} + \mathbf{T_x} + \mathbf{T_y} \tag{7.41}$$

Equation (7.40) can be formulated in the state form $\dot{\mathbf{z}} = f(\mathbf{z}, t)$ introducing the state vector **z**:

$$\mathbf{z} = \begin{Bmatrix} \alpha \\ \beta \\ \dot{\alpha} \\ \dot{\beta} \end{Bmatrix} \tag{7.42}$$

The opportunity of travelling along different trajectories depending on bobsled parameters, ice-runners contact forces and driver's response, represents the most interesting aspect of this relatively simple 2 dof model. All these elements mutually interact resulting in different bobsled responses and final times. The proposed model allows performing a more global optimization of bobsled parameters, taking into account their effects also on lateral dynamics, including models for side contact forces and for driver's reactions.

Effect of different trajectories on final time (i.e., identification of best trajectory) can also be investigated, always considering the actual physical limits of the sled dynamics. The best combination of structural parameters will be the one allowing the driver to lead the sled along the best possible path with the highest speed.

Besides these considerations, numerical integration of (7.40) requires little computational effort so that real-time simulations are possible; this model was also implemented in a bobsled simulator (see Sect. 7.7) to train professional drivers.

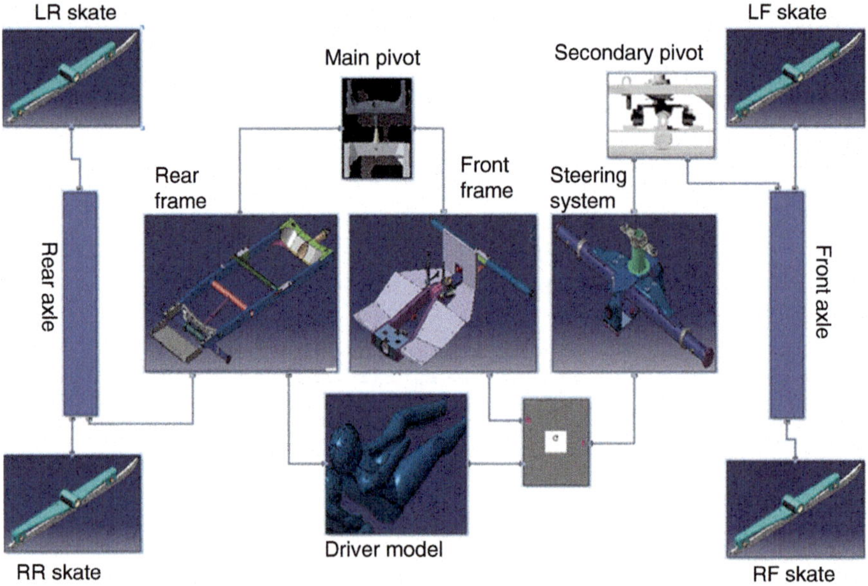

Fig. 7.40 Schematic of the multi-body model proposed in [28]

7.4.1.3 Multi-Body Model

The particle model described in the previous paragraph represents a useful tool to identify the best trajectory on a track and to determine the best combination of a number of structural parameters leading to maximum performance. However, the same particle model is somehow limited since parameters like wheelbase, longitudinal position and height of center of gravity, suspensions stiffness, inclination of main pivot and others cannot be included in the optimization process. To do that, the hypothesis of particle model should be removed, and the bobsled should be modelled at least as one rigid body. Different degrees of complexity are possible; in the following, the multi-body model proposed in [28, 56, 62] will be presented.

The model was developed through the commercial multi-body code Matlab/SimMechanics© and consists of a sled model, a track model, a contact model, and a driver model; this last is described in Sect. 7.5. A schematic of the sled model is reported in Fig. 7.40; the model is made up of seven rigid bodies and its motion is described with 15 degrees of freedom (dof):

- six dof for the 3D motion of the rear frame;
- three dof for the relative motion of the front frame with respect to the rear frame (relative roll plus vertical and lateral displacement);
- one dof for the relative rotation (steer angle) of the steering system with respect to the front frame;

- one dof for the relative roll motion of the front axle with respect to the steering system;
- four dof for the four relative vertical displacement of the skates with respect to the front/rear axles.

Referring to the schematic of Fig. 7.40, rear and front frames are linked through the *main pivot* which allows relative roll motion as well as lateral and vertical relative displacements due to the joint compliances. A nonlinear roll stiffness is introduced to describe the effect of the torsion bar inside the joint and to reproduce the action of the bumpstops that limit the relative roll rotation. An ideal cylindrical joint is adopted instead to describe the relative motion (steer rotation) between the front frame and the steering system. On the basis of the dynamic response of the rear frame (and of the position of the sled along the track), the driver model actuates the steer axle. The *secondary pivot* allows the relative roll motion of the front axle with respect to the steering system; a nonlinear roll stiffness is introduced to include the effects of torsion bar and bumpstops.

The skates are introduced as a rigid bodies linked to the respective axles by means of prismatic joints that allows the relative vertical motion between the skate and the axle. Linear spring-damper elements are introduced to reproduce the elastic behavior of the runner carriers and the damping characteristics of the rubber elements between runners and carriers. The model includes also the presence of front and rear lateral bumpers.

The track is modeled considering a series of transversal sections equally distributed along the track. Data relevant to track geometry are obtained from technical graphs and measurements on the real track. Distance between two subsequent sections is equal to 0.5 m and each section is described through 60 points. Bi-linear interpolation is used to reproduce the 3D geometry of the track.

As far as the contact model is concerned, a single contact is assumed for the skates (perpendicularly below the corresponding axle). Multiple contacts that may rise on skates due to high normal loads are thus neglected as well as the distribution of normal and tangential stresses along the them. Contrarily to the contact models described for the particle model, the hypothesis of indefinitely rigid track is abandoned: both skates actually are assumed to penetrate into the ice and the normal force $\mathbf{F_{zi}}$ developed along the direction normal to the contact surface $\mathbf{n_{ti}}$ in the ith contact point is function of the penetration Δz_i.

$$\begin{cases} \mathbf{F_{zi}} = (k_1 \Delta z_i^2 + k_2 \dot{\Delta z_i})\mathbf{n_{ti}} & \Delta z_i > 0 \\ \mathbf{F_{zi}} = \mathbf{0} & \Delta z_i \leq 0 \end{cases} \tag{7.43}$$

The proposed contact model is thus monolateral and, as a consequence, it allows the skates to detach from the contact surface. In (7.43), a suitable value for k_1 obtained through finite element analysis is 4e8 N/m^2 while parameter k_2, introduced to avoid numerical instabilities, should be kept as small as possible ($\approx$100 Ns/m).

Once the normal contact force is known, tangential contact forces can be computed. Figure 7.41 refers to a plane tangent to the track surface in the ith

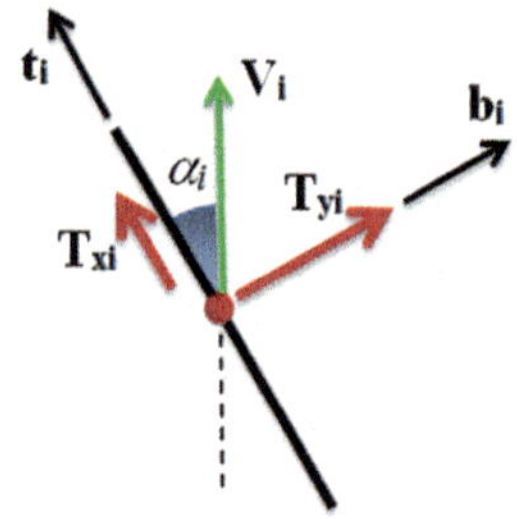

Fig. 7.41 Directions of the tangential contact forces

contact point; direction of vector $\mathbf{V_i}$, representing the velocity of the skate center, forms a slip angle α_i with respect to the longitudinal direction of the skate $\mathbf{t_i}$. The longitudinal contact force $\mathbf{T_{xi}}$ along the skate direction is computed as follows:

$$\mathbf{T_{xi}} = -\mu_x |\mathbf{N_i}| \mathbf{t_i} \tag{7.44}$$

In (7.44), μ_x represents the friction coefficient along the skate direction. Coefficient μ_x is in general function of both normal load and sliding speed. Tests on Cesana Pariol Olympic track led to estimate a value close to 0.02 for this parameter.

Tangential contact force $\mathbf{T_{yi}}$ in a direction normal to the skate side ($\mathbf{b_i}$ in Fig. 7.41) is assumed to be equal to:

$$\mathbf{T_{yi}} = \mu_y |\mathbf{N_i}| \frac{2}{\pi} \arctan(k_3 \alpha_i)\, \mathbf{b_i} \tag{7.45}$$

where α_i represents the slip angle of the considered skate, k_3 is a scaling parameter (= 50 rad^{-1}), and μ_y is the lateral friction coefficient; this last is a function of both normal load and sliding speed and much higher than μ_x due to additional skidding resistance associated with lateral sliding. Due to the lack of specific experimental data and the limited literature available, μ_y was determined on the basis of the full-scale experimental tests carried out on Cesana Pariol Olympic track: a value of 0.2 was identified (see [17]). For what concerns the contact between bumpers and ice, computation of normal contact force and tangential contact force in longitudinal direction is based on the same model described in (7.43) and (7.44).

The proposed model was validated by comparison with experimental data; in particular, as described in Sect. 7.3, a bobsled running on Cesana Pariol track was equipped with several measuring devices so that accelerations, roll rates and sideslip angle of the sled, steer angle and contact forces could be recorded. Figure 7.42 reports a comparison between experimental and numerical results relevant to the normal contact force on the front right skate; model results present a good agreement with measurements as far as left-hand curves are tackled, while overestimations are present in right-hand curves. Figure 7.43 shows another comparison referred to sideslip angle of the rear frame: this quantity was measured through an optical device and represents the relative angle between the velocity vector (usually referred to the cog) and the longitudinal direction of the frame. Sideslip angle is a measure

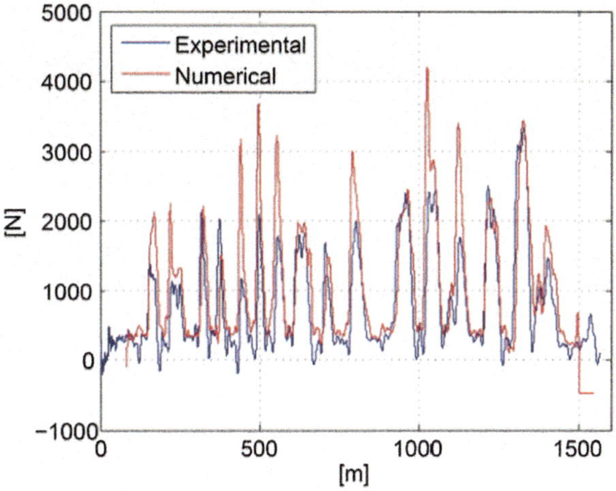

Fig. 7.42 Numerical–Experimental comparison: normal contact force on right front skate during a run at Cesana Pariol track

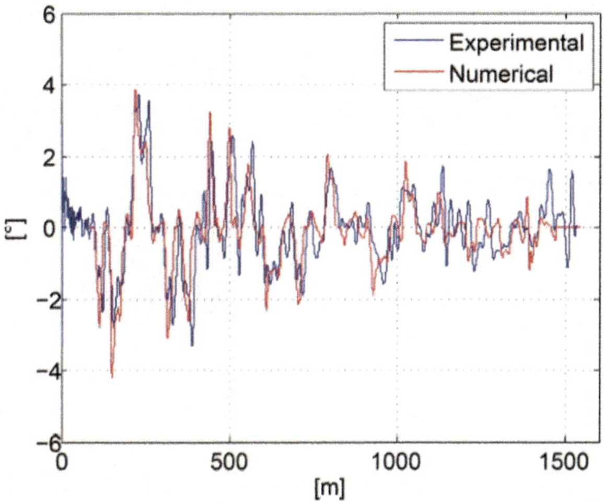

Fig. 7.43 Numerical-Experimental comparison: sideslip angle during a run at Cesana Pariol track

of the global lateral skidding of the sled. Results of comparison can be regarded as good, also considering that values of sideslip angle are quite small.

The complexity of the proposed model is significantly higher than the one of a single particle model, but allows including the effect of numerous parameters in the performance analysis.

One of the most interesting results of the model is relevant to the effect of cog height on final time; the chart in Fig. 7.44 displays the difference in final time on Cesana track as function of variation of cog height with respect to a reference value: a decreasing trend is clearly evidenced. The same result is obtained also in other tracks and can be intuitively explained considering the scheme of Fig. 7.45 which reports a front view of the bobsled during a transition while entering a curve. In this phase the bobsled is changing its roll angle rotating around a point **K**

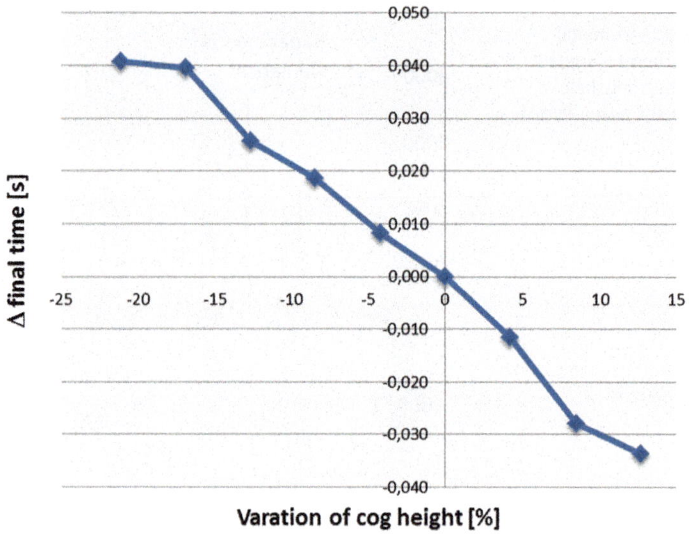

Fig. 7.44 Variation of estimated final time (Cesana-Pariol track) vs. change of cog height with respect to nominal configuration

Fig. 7.45 Variation of roll angle between straight track (*A*) and corner (*B*)

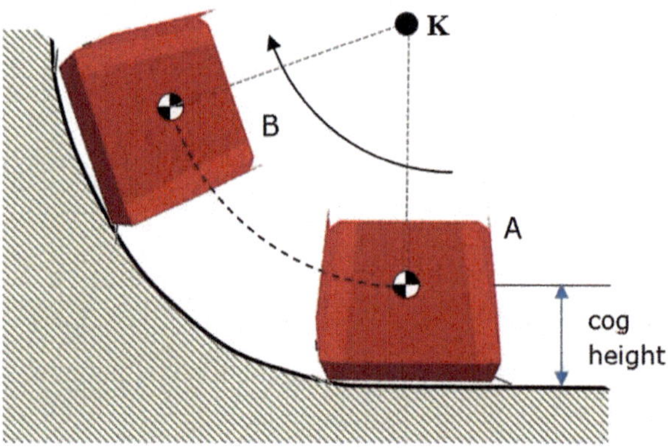

whose position depends on the transversal profile of the track. Increasing cog height actually decreases the distance between cog and **K** thus lowering the moment of inertia of the bobsled around the rotation axis. As a consequence, lower tangential forces (and dissipations) are needed to change bobsled roll angle during transitions at the beginning and at the end of curves. Besides being limited by the physical possibility of distributing mass in the upper part of the sled, increase of cog height may also affect roll-over risk, as could be evidenced by the model analyzing the values of normal loads on runners.

Other interesting results pointed out by the multi-body model are relevant to the effect of wheelbase (Fig. 7.46) and longitudinal position of the cog (Fig. 7.47).

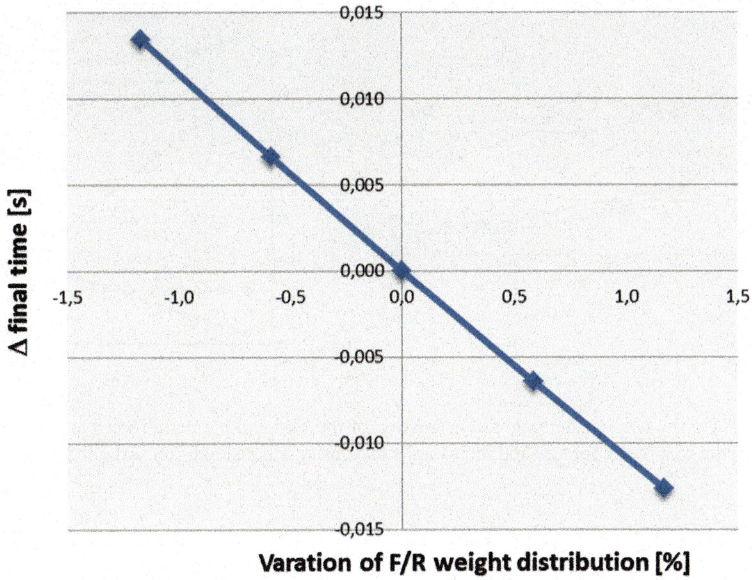

Fig. 7.46 Variation of estimated final time (Cesana-Pariol track) vs. change of wheelbase with respect to nominal configuration

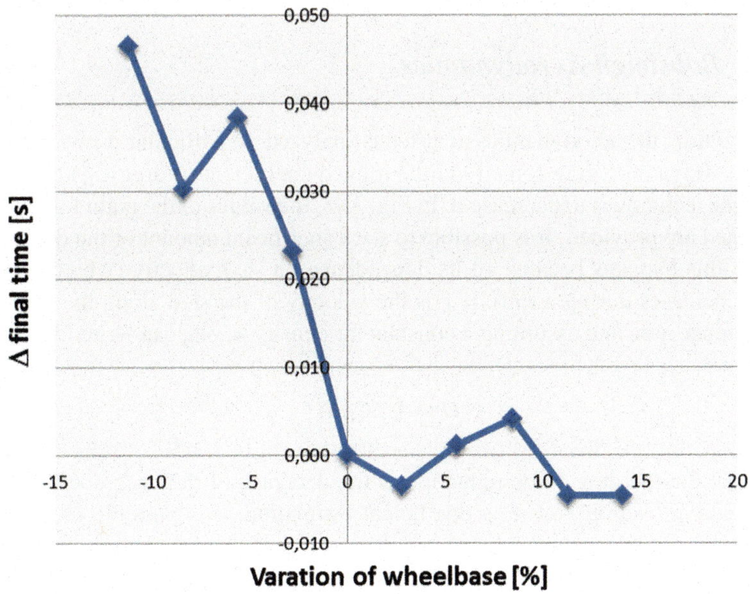

Fig. 7.47 Variation of estimated final time (Cesana-Pariol track) vs. change of front/rear weight distribution with respect to nominal configuration

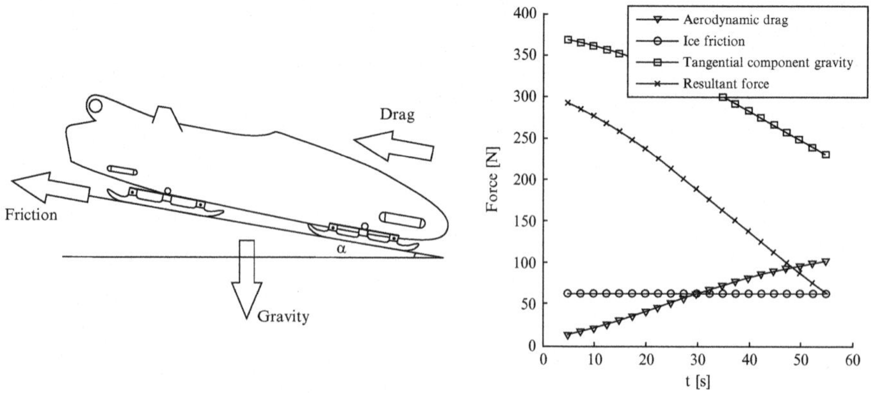

Fig. 7.48 On the *left*, a schematic representation of the sled and the main forces acting on it. On the *right*, the plot of the forces (and their total sum) during a bobsleigh run. Adapted from [30]

Increasing wheelbase improves bobsled stability thus reducing lateral skidding and power dissipation while entering and exiting curves. Changing longitudinal position of the cog influences the distribution of normal loads between front and rear axle optimizing understeer/oversteer response; the optimal value of this parameter was found to depend on the mean slope of the track.

7.4.2 Bobsleigh Aerodynamics

In this section, the aerodynamics of a sled is analyzed. In particular we want to focus on the aerodynamic drag force (and the related coefficient of drag C_D) on the sled and on the techniques to decrease it. In Fig. 7.48, the values of the main forces acting on the sled are provided. It is possible to see a significant amount of the drag during the run: this happens because of its dependence on sled velocity (which normally always increases during a run). If v is the velocity of the sled along the track, A is the reference area and assuming a constant air density ρ, one can write [30]:

$$D = \frac{1}{2}\rho C_D A v^2 \tag{7.46}$$

A goal for the drag force reduction is the decrease of the drag coefficient C_D. To provide an example of drag coefficient estimation, it is possible to separately analyze the components of the sled and then to sum their contribution. Four are the main elements which contribute to the final C_D coefficient: the *main cowling*, the *bumpers* (see Fig. 7.49), the *crew* and then *axles, runners, and runner carries* [30]. The estimation is made at 35 km/h.

- Starting to analyze the main cowling, we can identify three main components: the skin friction drag (due to air viscosity), the fore-body pressure drag (due to

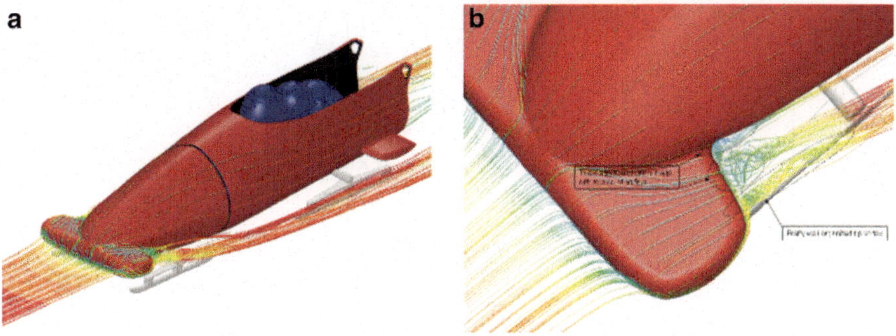

Fig. 7.49 (**a**) The flow-field acting on the front bumper; (**b**) detail of the flow leaving the front bumper

the shape), and the base pressure drag (due to exhaust and wake). In [30] the skin friction drag, that is the downstream resultant of all shear (viscous) forces experienced by the fore body, is assumed to behave like on a flat plate. The skin friction coefficient depends on the flow characteristics, then assuming s the coordinate along the body, for laminar boundary layer it is [31]:

$$C_f = \frac{0.664}{\sqrt{Re(s)}} \tag{7.47}$$

For a turbulent flat plate:

$$C_f = \frac{0.027}{Re(s)^{1/7}} \tag{7.48}$$

Along the profile one can find a transition point in which the thickness of the turbulent boundary layer needs to be matched to the laminar boundary layer. This gives a kind of starting point (s_{vo}) for the turbulent boundary layer. The drag coefficient can be calculated by integrating the value of the skin friction coefficient over the area of the body and dividing by the reference area, so considering the s_{vo} point [30]:

$$C_D = \frac{2\pi}{A}\left(\int_0^{s_{\text{transition}}} C_{f\text{laminar}}\, r\, ds + \int_{s_{\text{transition}}-s_{vo}}^{s_{\text{end}}-s_{vo}} C_{f\text{turbulent}}\, r\, ds\right) \tag{7.49}$$

where r is the radius of the reference area and s_{v0} is a virtual starting point for the turbulent boundary layer. The position where the transition occurs can approximately exhibit a critical Reynolds number of 5×10^5 [30]. In Fig. 7.50, the relation between C_D and the point of transition and the one between the transition point and the velocity are plotted. Moving the transition backward, the skin friction drag can be reduced. An estimation of base drag contribution (resulting from the wake or "dead air" region behind the body) and fore-body

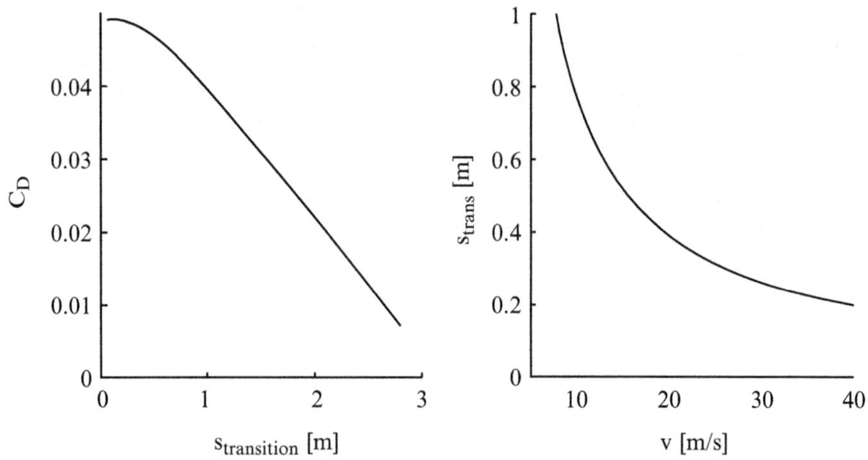

Fig. 7.50 Influence of transition point on drag coefficient (*left*) and influence of free-stream velocity on transition point (*right*) [30]. The *little cross in the left plot* represents the point with velocity of 35 m/s and a transition at $s = 0.22$ m: the resulting drag coefficient is 0.047

drag contribution (the resultant of all the pressure forces on the fore-body) is provided in [30] and is equal to 0.10.

- An estimation of bumpers effect is possible, using simple 2D shapes. In [30], using the data estimated in [32], a total drag coefficient for four bumpers (two front and two rear ones) becomes 0.016, assuming the bumper shape like a rectangle with rounded edges and streamlined back.
- Referring to the crew, the most part of the drag coefficient is due to the driver helmet, because the body of the driver is inside the cowling and the other members of the crew are behind the driver (we'll se later in this section the influence of brakeman position and posture). It is possible to approximate the frontal area of the helmet with the one of a sphere (so $A = \pi r^2$), getting a C_D for the total crew equal to 0.015 [30].
- It is possible to provide a drag coefficient for axles, runners, and runners carries approximating them with the reference shapes calculated in [32]. In [30], a value of 0.147 is evaluated.

At the end of this estimation, a total drag coefficient (at 35 km/h) of 0.322 is found. The cowling pressure drag, the runners and the axles are the most important contributors.

In [33], some of the most important areas which contribute to the total drag are treated. The formation of the wake inside the bobsleigh cavity and the brakeman position are two contributions which most are analyzed and studied in literature. Also the shape of the nose of the sled is considered by Motallebi et al. [33] a source of drag, because of its influence on the flow over the rest of the sled. In the same paper, two modifications are tested on a scaled model: flared sidewalls and shark nose. The goal of the first proposal is to avoid that the flow is diverted and sucked inside the cavity, but in the manned case and at high Reynold's numbers (comparable to the real ones) its effect disappears. The second modification aims for pressure drag reduction, decreasing the area of the separated flow: this technique works well both for manned and unmanned case at low Reynold's numbers, but again loses its effects at higher Reynold's numbers.

It is straightforward, but necessary to underline, that all the aerodynamic modification must respect the rules stated by the Federation (the reader may refer to Sect. 7.2 and to [3]): for example, the use of trailing edge notches or splitter plates to break big vortices, even though really useful and efficient, are not allowed by the rules [30].

It is now interesting to analyze the main hints provided in literature to study, modify, and optimize the aerodynamics of a bobsled. First of all, two leading approaches to this kind of study are the CAD-CFD simulations and the experimental tests, usually developed in wind tunnel. Depending on the goal of the study, the approaches and the parameters for these two methods can be really different.

The CAD model moves from really simple configurations to more complex ones. In [34] a complete model is used, in which front and rear bumpers, axles, runners, the cowling and the pilot's helmet are simulated; only some shape details are neglected, like screws, nuts, the break cavity and the rubber joint between front and rear boat. The first step was the identification of governing shape parameters: thanks to them, it was possible to implement a wire frame model (using the software CATIA V5) from which a surface model was created. In this paper the shape of components was really important, because the impact of shape changing in drag reduction was studied. In [35] a simpler model for the bobsled was implemented: only the internal cavity was represented, without details and with an artificial extent of the rear part to avoid numerical effects due to the outflow conditions. The external part of the bobsled was not considered, and so the edges of the cavity were attached to a solid plane. The goal of [35] was the study of athletes' positions relative to each other and to the sled's cavity: in particular the priority was to determine an optimal seating position for the driver and for the brakeman (or brakewoman), who has no active role during the run and must assume a position that minimizes aerodynamic resistance and perturbations. In Fig. 7.51 the two models are compared.

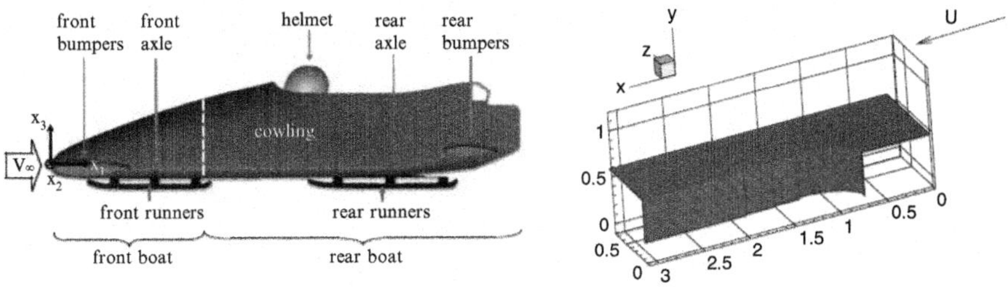

Fig. 7.51 (**a**) A sophisticated CAD bobsled model used in [34]. (**b**) An isometric view of the simpler model developed by Dabnichki and Avital [35]

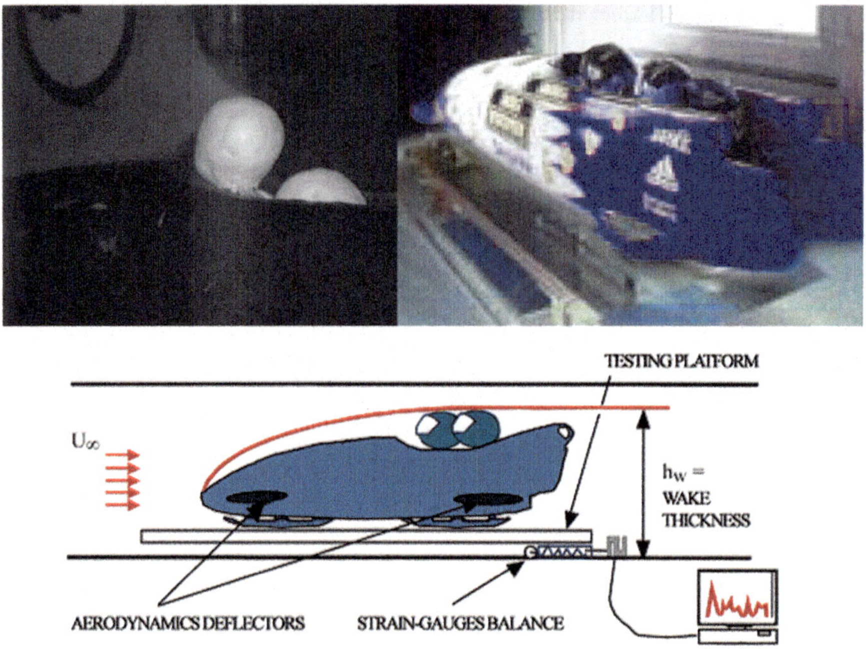

Fig. 7.52 The picture on the left refers to a mock crew, used in [35]. The other two are adapted from [38]: they show the French Olympic four-athletes crew during a wind tunnel experiment and the corresponding schematics of the experimental test

About the fluid dynamic calculation, generally the flow is assumed to be incompressible, viscous and turbulent (see [34–36], in which it is specified that the flow is steady). Due to the high Reynold's number, the Reynold averaged Navier-Stokes (RANS) approach was used and the turbulence was modelled using the k-ω approach (for further information, one can see [31, 37]).

Regarding the crew, different approaches are again followed (in general we can also say that the results found for a two-athlete bobsled are also valid for a four-athlete one). In Fig. 7.52 two different examples of crew during a wind tunnel experiment are shown. These two examples refer to wind tunnel tests, but there are models also for the CFD computation. An interesting model for the athletes bodies is used in [36] (see Fig. 7.53): it is based on the biomechanical properties of the spine and the arms, and the spine is simulated connecting lumbar spine joint, thoracic spine joint, cervical spine joint, and spine-skull connection by splines. To modify the posture the joint angles ϕ_L, ϕ_T and ϕ_C (see Fig. 7.53b) can be changed; the parameters ϕ_E and ϕ_S are used for the bending of the arms. Another parameter is the distance x_{BM} of the brakeman from the front part of the sled. The aim was to minimize the aerodynamic drag area $C_D \cdot A$ optimizing the parameter values. It was easy to prove that the bent area led to a lower area with respect to the stretched arms configuration, posture that is also the conventional one adopted: so it was possible

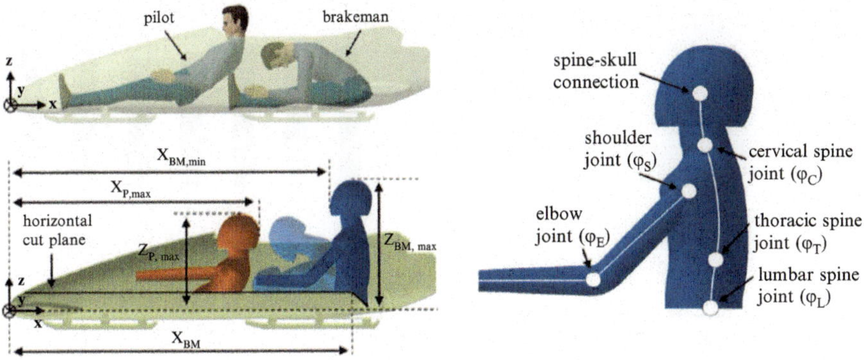

Fig. 7.53 (**a**) Typical position and posture of a two-man crew (*top*) and the crew model with different brakeman postures (*bottom*). (**b**) Main parameters for the biomechanical body model [36]

to abandon the arm joint angle variables, and keep on minimizing the function $C_D \cdot A = f(\phi_L, \phi_T, \phi_C, x_{\mathrm{BM}})$. Also the pressure coefficient was considered:

$$C_p = \frac{p - p_\infty}{q_\infty} \tag{7.50}$$

where p is the local pressure, p_∞ the ambient pressure, and q_∞ the stagnation pressure. The flow over the cowling edge cause a delta-wing vortex evolving from the edge. The vortex presents high rotational velocities and consequentially low pressures: this is a source of pressure drag (so variation of the C_p coefficient, see [36]). It was shown that the presence of the brakeman reduces the suction on the rear side of the pilot: he (or she) alters the strength of the vortices, moving outward the vortex centers and decreasing the pressure drop. As a conclusion, a maximum ducking of the brakeman does not result in the minimum aerodynamic drag, while in contrast an appropriate posture and position can decrease the drag considerably.

We can now list some other considerations and hints coming from the literature on the aerodynamic aspects of the bobsled.

- In [34] a potential for drag reduction is detected in the rear part: in fact together with the floor it forms a diffuser, which leads to a decelerated flow in this region. In Fig. 7.54, the results of a wind tunnel experiment about the influence of the γ angle on the drag coefficient are plotted. Bulky and slim indicate two different contours for the diffuser. With $\gamma = 6.7°$ and bulky contours, the C_D is minimized.

  No significant results were obtained changing the angle of incidence of the bumpers, which is equal to zero.

  Finally, also the yaw angle (Ψ) as influence in the aerodynamic drag. During a perfect run the bobsleigh's central axis must point the direction of motion, resulting in $\Psi = 0$. If the bobsled starts yawing because of pilot's failure, the ice causes an increasing blockage and so a larger drag. For a $\Psi = 10°$, the increase of the C_D is roughly equal to 16 %.

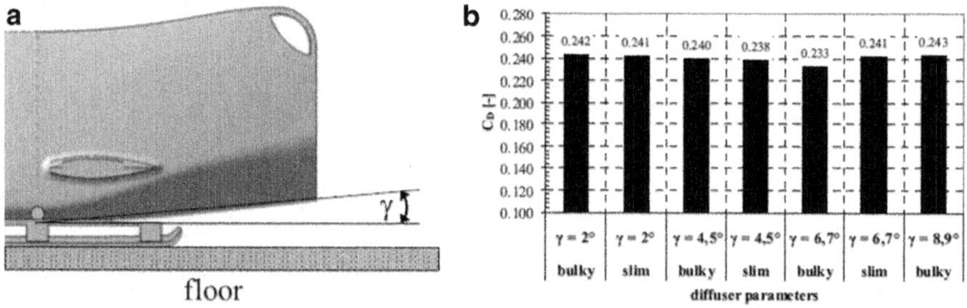

Fig. 7.54 (**a**) Side view and γ angle of the diffuser; (**b**) diffuser optimization with different configurations at $Re = 8.07 \cdot 10^5$ [34]

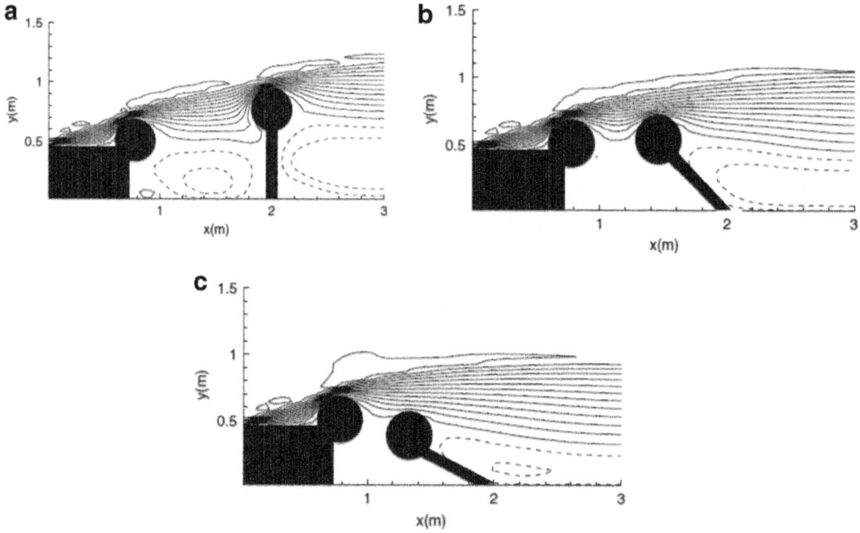

Fig. 7.55 Contour plots of the axial velocity in the mid-plane of the cavity, with 15 contour levels; *solid lines* are positive and dashed negative values. (**a**) Brakeman sits straight; (**b**) bends forward 45° and (**c**) bends forward 60°. Adapted from [35]

- Combined CFD and experimental study is performed in [35] on the influence of the crew position (see Fig. 7.55). Bending forward for the brakeman (or brakewoman, in this paper the study was on female athletes) is beneficial for the pressure drag reduction. In particular the secondary flow between the athletes is reduced, as also the turbulence due to the helmet, and the flow has a better attachment onto the back of the last athlete. Moreover, it was proved that an increase of the cavity width had a negative effect on the drag, because of its influence on the turbulence on the secondary flow.
- A purely experimental approach (using a wind tunnel) was followed in [38]. The purpose of this study was to get a performance improvement of a bobsleigh four-men team, expressing it as a function of bobsleigh shape and relative position of

the athletes inside. The results were obtained after iterative tests and step-by-step improvements. Three specific parameters were found. First of all the position of the athletes inside the sled: keeping the same driver for all the tests, the relative position of the other crew members was sequentially changed, finding an optimal configuration (that is, the one with the lower $C_D{\cdot}A$ parameter). Then the addiction of deflectors (abiding by the rules, see Sect. 7.2) was investigated, proceeding in a way really similar to the one just discussed. Finally the position and postures of the brakeman were studied, getting results close to the ones above described. It was also proved that lowering the position of the brakeman, there is a thinner near wake just behind the bobsleigh which results in a significant drag coefficient reduction.

7.4.3 Skeleton Aerodynamics

The shape of a skeleton sled is totally different with respect to a bobsled (see, for example, Sect. 7.2): this fact implies severe differences also from the aerodynamic point of view. As one can argue, in this case the leading contribution to the drag coefficient is given by the athlete and its position. In [39] it is demonstrated via CFD simulation that, between all the variables that could influence the drag coefficient, the skeleton sled profile has only the 9 % of the total drag, and it means that the main focus for the optimization must rely on the athletes.

One can use a very simple equation to describe the skeleton sled motion neglecting the start sprinting phase (see [23]):

$$\frac{d^2x}{dt^2} = g(\sin(\theta) - \mu\cos(\theta)\cos(\beta)) - \left(\frac{\mu}{r}\sin(\beta) + \frac{\rho C_D A}{2m}\right)\left(\frac{dx}{dt}\right)^2 \tag{7.51}$$

which is valid both for straight descent and banked and curved sections of the track. μ is the friction coefficient of ice, ρ is air density, m is the sum of sled and athlete masses, C_D is the drag coefficient, A is a non-dimensional area, θ is the local track gradient, β is the transverse orientation of the track and r is local radius of curvature during curved sections. In [23] it is proposed to consider the drag coefficient as the sum of two contributions: the normal surface pressure and the tangential shear stress, that is the skin friction. In the same paper it is also proposed an empirical relation for the skin friction coefficient[4]:

$$C_f = \frac{0.075}{(\log_{10} Re - 2)^2} \tag{7.52}$$

[4]Just to have an idea of the Reynold's number value, one can use the values proposed in [10] (or [39]), where for $Re = \frac{\rho v L}{\mu}$ it is assumed: air density $\rho = 1.22\,\text{kg/m}^3$, air dynamic viscosity $\mu = 1.79 \times 10^{-5}\,\text{Pa s}$, velocity $v = 35\,\text{m/s}$ and the athlete length $L = 1.75\,\text{m}$, resulting in $Re = 4 \times 10^6$. This value of the Reynold's number identify a turbulent air regime around the athlete.

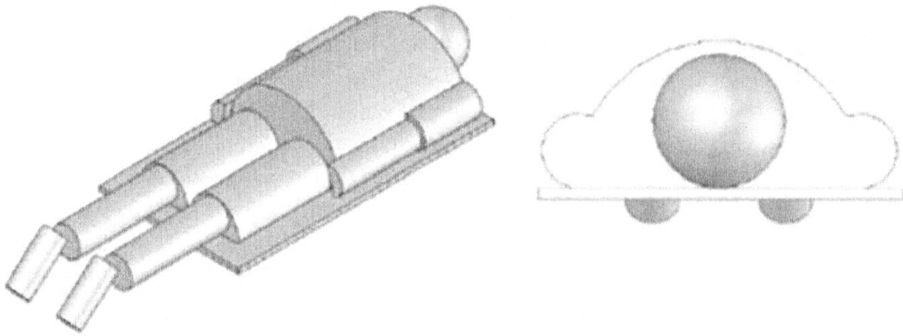

Fig. 7.56 A simple geometric representation of athlete and skeleton bob [23]

With the contributions of [23, 32, 40, 41], one can estimate a total coefficient of drag equal to 0.38.

A simple representation of the body, really useful for aerodynamic simulation and studies (as shown also in Ice Skating Chapter), is based on canonical geometries like cylinders to recreate a simplified human body (see Fig. 7.56). The reader can refer to [42] in which statistical data on proportions and density of total body mass, weighting each part referring to the total height.

A general conclusion achieved in [23] is that for a constant sled weight, heavy and tall athletes, due to an increasing total mass and a reduction of the frontal area, are favorite. By the way, if the mass sled is equalized with the prescriptions of Skeleton FIBT Rules (see Sect. 7.2), lighter (but even taller) athletes have benefits: this is due to the mass equalization which nullify the increased body mass, while the height contributes to reduce the frontal area and to lower the aerodynamic drag.

Nowadays growing attention is devoted to tracksuit materials, achieving a kind of optimization of the fluid flow around the athlete: this is related to vortices and turbulences created by the skin–fluid interaction. An interested reader may refer to [43–46] for further information.

As a final consideration about skeleton sled aerodynamics, in [10, 39] it is stated that the start velocity and the kinetic friction of ice are more critical than the aerodynamic drag, having the kinetic friction 6 times more influence.

7.5 The Track

As underlined in Sect. 7.2.10, the track construction for a bobsled competition (used also for luge runs) is not an easy task.

Except for a really small number of niche competitions performed on "natural tracks",[5] the construction of tracks is highly engineering and only few companies in the word are equipped to do it.

[5]Only France, Austria, Switzerland, and Germany take part in these competitions.

As one can argue, one of the first requirements for a track of this kind is to guarantee the safety of athletes and spectators. Further, despite a correct and precise track design, it always needs changes during the construction phase in order to respect limitations imposed by orography and geology. Finally if one considers the costs of refrigeration and the general (or extraordinary) maintenance of structures, it is clear that a plant of this kind is really expensive, with typical cost around \$15–30 m (see [47]).

A list of existing (and former) tracks is present in [48]. The shortest track is at Oberhof, Germany, with a length of roughly 1070 m, while the longest one is at Sochi, Russia, used for the Winter Olympic Games of 2014, with a length of 1814 m. The vertical drop moves between 96 and 152 m, whereas the number of turns varies, depending both on the specialty (bobsled/skeleton or luge) and on athletes gender.

7.5.1 Design and Verification

An important restriction on track construction is established by the rules about the maximum centrifugal force which the sled (and so the athletes) can receive (see Sect. 7.2.10): the highest allowed peak centrifugal acceleration is $5g$ and must not stand for more than 2 s, and also accelerations of $4g$ must not last for more than 3 s. Neglecting the fact that this rule is not completely clear regarding accelerations and forces,[6] it is possible to create a sort of very easy design rule to determine curve radii (see [49]): having a rough mean value for the bobsled speed v at a specific position of the track and standing the limitation on the maximum centripetal acceleration a_y, one can say:

$$R = \frac{v^2}{a_y} \tag{7.53}$$

This approach presented in [49] is purely stationary (so it cannot account the transition phase between straight and curved track) and the speed is function of the bobsled dynamics and so only approximately known. This fact implies that a model of the bobsled, together with a numerical model of track and driver, are useful to correctly design or verify a track. For the bobsled model and the complete dynamical behavior the reader can find along the chapter some information, here we want to focus on track modeling and verification.

A fundamental starting point for this process is an available 3D CAD model of the track, but it is not so trivial to reconstruct with precise measurements a track that is usually more than 1–1.5 km long. If instead the track is already built, one can directly measures the normal accelerations during a run: it is easier but not always

[6] In fact, due to the roll motion of the bobsled approaching a turn, the athletes perceive a compression force, that is normal to the ground.

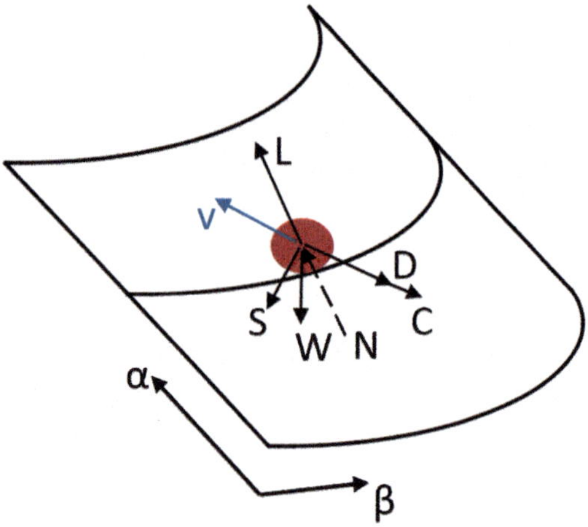

Fig. 7.57 In this picture, adapted from [27], the main forces affecting the motion of the sled are represented. The velocity v lies in the tangent plane at point (α, β), and has components in both directions. W: weight, N: normal force, L: lift, D: aerodynamic drag, C: Coulomb ice friction, S: steering force

possible, and it is important to remember that these data are functions of a specific sled, a team and certain ice/environmental conditions, fact that can strongly affect final results.

In general it is possible to give a mathematical representation of the track. In [27] it is suggested to take a Cartesian earth fixed reference system, with typically a x, y, and z representing east, north and locally vertical directions. If these three coordinates can be computed for every point on the surface we have a exact knowledge of the surface. To locate ourselves on the surface we need a map in which every point corresponds to a point on the surface. We can create a parametrization of the surface:

$$x = x(\alpha, \beta) \quad y = y(\alpha, \beta) \quad z = z(\alpha, \beta) \tag{7.54}$$

In a more compact fashion:

$$\mathbf{r} = \mathbf{r}(\alpha, \beta) \tag{7.55}$$

defining $\mathbf{r} = (x, y, z)^T$. The parameters α and β are dimensionless and increase in the along-track and cross-track direction, respectively (see Fig. 7.57): what we want is a set of differential equations $\alpha(t)$, $\beta(t)$, describing the motion of the sled. In real word the $\mathbf{r}(\alpha, \beta)$ is never exactly known, so practically one works with mathematical models that are good approximations of the surface and are generated with standard modeling techniques.

In [27] it is used an interpolation scheme based on a set of points $\mathbf{p}_{ij}$ whose coordinates (x, y, z) are exactly known for each grid point $(\alpha(t)_i, \beta(t)_j)$ on the map: as general feature one can say that the coarser the grid, the larger the errors between real and approximate surface.

A convenient form for the x, y, and z functions are polynomials in α and β, normally providing, through the use of bicubic surfaces, a C^2 continuity to the shape

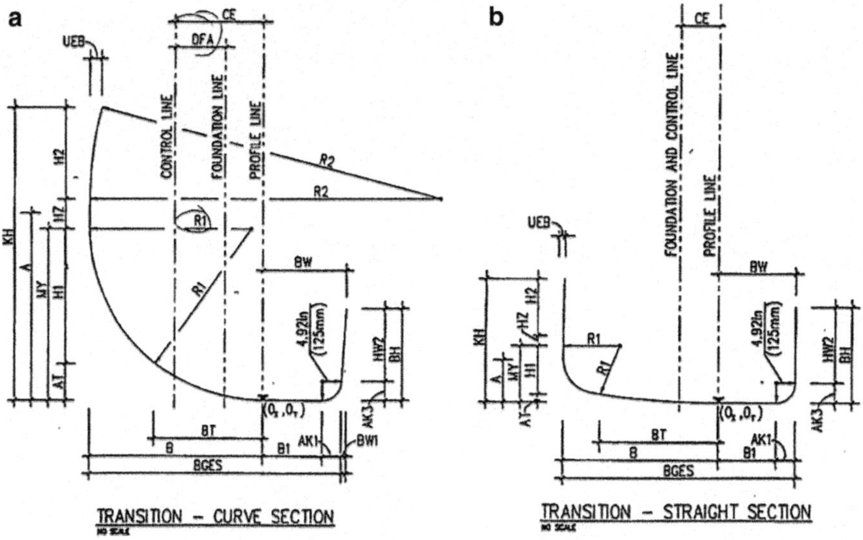

Fig. 7.58 Two example cross sections from Calgary track. (**a**) Near the middle of the turn; (**b**) near the turn exit where the cross section has become nearly rectangular (see [50])

(that is also its first and second partial derivatives respect continuity). This condition can be relaxed, in fact some subsequent cross sections are continuous in slope but have not a smooth transition in curvature (e.g., the Calgary track, see [27]).

A slightly different way to proceed in order to directly recreate the 3D geometry of the track is described in [49]. It is based on three steps:

1. digitalizing pictures;
2. re-sampling;
3. assembling digital pictures.

As first step, a sequence of pictures for several track sections (like in Fig. 7.58) is necessary, with the corresponding curvilinear coordinate to correctly reconstruct the surface. Sections are more dense when relevant track changes occur, and for each section a reasonable position of the sled is shown, together with one or more vertical reference lines.

Then the pictures are converted into a binary matrix and thus into a matrix of polar coordinates, placing the origin in the middle of the image and converting the xy coordinates into polar ones. Between each section it is necessary to guarantee a smooth behavior to perform a simulation. It is demonstrated in [49] that although track description is quite rough (like a section each 50 m), a re-sampling using cubic spline interpolation is sufficient to have a smooth behavior (with a more dense fitting one can use also a linear approximation). After this procedure the sections are converted again in Cartesian coordinates and are placed in an absolute reference system. Usually it is also necessary to do a conversion from curvilinear to Cartesian coordinates, but it can be easily done using [49]:

$$ds = Rd\theta \ \Rightarrow \ \theta(s) = \theta_0 + \int_{s_0}^{s} \frac{du}{R(u)} \tag{7.56}$$

$$\begin{aligned} dx &= ds\cos(\theta) \\ dy &= ds\sin(\theta) \end{aligned} \tag{7.57}$$

$$\begin{aligned} x(s) &= x_0 + \int_{s_0}^{s} \cos(\theta(u))du \\ y(s) &= y_0 + \int_{s_0}^{s} \sin(\theta(u))du \end{aligned} \tag{7.58}$$

being R and θ the polar coordinates describing the geometry of the considered section.

Further, if we know the curvature as a function of the curvilinear abscissa s ($C(s) = 1/R(s)$), it is possible to get the plan of the mean track line in an absolute reference system.

This approach was used in [49] to make some studies on two different tracks: Cesana Patriol Olympic Park and Whistler Sliding Center. The trajectory of the sled was the optimal one, i.e. the trajectory that minimizes the steering action and so the dissipation at skate–ice interface, which is in general not the best nor the fastest possible one.[7] Whistler track is more severe both for a higher mean slope and speeds up to 150 km/h. If we consider the "centrifugal force," it is possible to verify that in one turn the acceleration is equal to almost $5.5g$ (above the rules limitations), while with a simplified approach (see Eq. (7.53)) a value of $4.4g$ would have been estimated, that is an underestimation of almost 20 %.

7.5.2 *Notes on Optimization and Safety*

As written above, the track construction process is long, expensive, and really complicated. First tracks were built without the aid of computers and of a strict mathematical approach, fact that gave rise to a lot of errors (sometimes spectacular and really dangerous) and consequent modifications of the track itself. Nowadays simulations and mathematical models help to reduce this time-consuming process, but this cannot completely avoid errors (at the end of Sect. 7.5.1 it is shown how calculation methods, but also different physical parameters, can strongly affect final results) and thus modifications to the track after the building process.

[7] In [24] a similar concept of optimal trajectory is defined, with a more sophisticated mathematical approach. In [24] it is proved that this kind of optimality does not necessary correspond with best performances. In fact each driver has a totally different way of drive, less or more aggressive with respect to the steering action, and the best athletes are able to change their style during the run, adapting it to the track and the velocity of the sled.

Taking, for example, the Report [51] about Whistler Sliding Center track of 2010, it is clearly written that many difficulties arose during construction and certification phase (also in this case it was necessary also to identify a reference trajectory to perform the study): "...a detailed topographic survey of the ice surface and associated structures integral to the Track, including the roofs, walls and the concrete surface beneath the ice was produced. A system of 3 Dimensional laser scanners, 3-dimensional robotic total stations, and precise digital levels were used to scan the Track with ice in, as well as the concrete surface after the ice melted."

As above cited, the technology behind this process is really complex and the measurements are performed in different seasons to check both the ice condition and the integrity of the structure itself. Further it was necessary to perform more measurements to have a sort of redundancy to overcome errors like differences in the extraction of targets by the scanning system or between the surveyed targets and the scanned target locations, potential changes over time due to freezing, thawing, settling,.. and accumulative error from the environment and tolerances. This process, together with trial runs, was used to determine the critic points of the track and then to correct them in order to abide by the rules and guarantee athletes safety.

Safety is a really critical point: high velocities and low protections (skeleton and luge in particular, where also a kind of external frame is absent) contribute to increase the risks associated with these disciplines. We all have in mind the accident at the Vancouver Olympics in 2010: a tragic crash in the 6th and last training run claimed the life of Georgian luge slider Nodar Kumaritashvili. Few hours after the accident the FIL (International Luge Federation) declared that "there was no indication that the accident was caused by deficiencies in the track," but anyways some modifications were immediately made to improve the safety of the track. The dynamic of the accident was not so clear and a first official report [52] classified the dynamics of the crash as "unknown and unpredictable." After few months a Coroner's report [53] stated that "the collision with the inner wall caused the sled's right runner to ride up onto the wall, causing ... the sled to be launched into the air." Several other reports followed, and in [50] a precise mathematical reconstruction of the accident was made. As a conclusion it stated that "The interaction of the right runner with the fillet at the bottom of the inner wall resulted in the vertical velocity necessary for, and thus was the cause of the ejection of Nodar Kumaritashvili." This ice *fillet* is a common way of connection between inside vertical walls in turns to the flat track bottom (the Rules state that "The transition between the sidewall and the base of the track must be provided with a channel. In the iced state its radius must be 10 cm"). This fillet is not required nor precisely normed in the regulations, and the conclusion of [50] was that it must be not allowed in future track design (and removed where existing in previous tracks).

This tragedy must teach us how the safety component must have a leading role not only during the building and certification process but also in the very beginning phase of rules writing. There must be no lacks on the normative, which can provoke misunderstandings and potential risks. It is not a simple task, that's way norms must be continuously updated using data and results from new competitions and studies.

7.6 The Athletes

Minimum race time is the aim of bobledding competitions. For a given sled and a given track, the minimum race time is determined by push phase, boarding and control of the bobsled. Push phase and boarding of the crew determine initial conditions for the motion of the bobsled down the track. When these initial conditions are determined (i.e., after the crew jump into the sled), the only way to reach high race achievement is to control the sled better than the other teams in most difficult sections of the track. Bobsled control is executed by changing the steer angle. Hence it depends on driver skills in defining the best trajectory and to follow it as close as possible. It is thus straightforward to understand how models able to identify optimal path and steering control strategy would be vital to improve training of crew members and development of new facilities and equipments. These models could in fact allow:

- to determine the minimum race time for a given sled on a given track under given conditions; this could point out the goal for teams to reach and show performance potential;
- to improve training through comparison between the optimal control path and real driving paths;
- to optimize the structural characteristics of the sled;
- to provide on-line help and suggestions in real-time simulator training.

In the following attention will be focused on two different approaches for the implementation of a driver model, computation of optimal path and driver controls. The first approach applies optimum control theory to identify the steering action minimizing the travel time either on a single curve or on the whole track (Sect. 7.6.1). The second approach instead divides the design of the driver model into two tasks: identification of the optimal trajectory and implementation of a controller able to follow it as close as possible (Sect. 7.6.2). Specifically, the optimal trajectory was identified in a pre-processing phase based on a particle model of the bobsled on the purpose of minimizing energy dissipated by longitudinal forces as an index to minimize the run time on the entire track. The controller was instead designed with the aim of reproducing the steering behavior of a real driver and it was developed and tested using a multi-body model of the bobsled. In the following the two approaches are described.

7.6.1 Optimum Steering Control Approach

In [24] optimum control is applied to identify the steering action allowing to minimize the travel time either on a single turn or on the entire track. Optimal steering control strategy design is based on the two-degrees-of-freedom particle model described in Sect. 7.4.1.2. On the purpose, equations of motion of the model can be re-written in state-space form as:

$$\begin{cases} \dot{\mathbf{z}}(t) = f(\mathbf{z}(t)) + g_1(\mathbf{z}(t)) \cdot h_1(u(t)) + g_2(\mathbf{z}(t)) \cdot h_2(u(t)) \\ \mathbf{z}(t_0) = \mathbf{z_0} \end{cases} \tag{7.59}$$

where the state vector $\mathbf{z}$ (see Eq. (7.42)) contains coordinates α (which represents the distance travelled by the particle along the track) and β (which represents the distance of the particle from the left side of the track) and their time derivatives

$$\mathbf{z} = \begin{Bmatrix} \alpha \\ \beta \\ \dot{\alpha} \\ \dot{\beta} \end{Bmatrix} \tag{7.60}$$

$\mathbf{z_0}$ is the vector collecting the initial conditions (i.e., speed and position of the sled after the crew boarded consequently to the push phase), $u(t)$ is the steer angle normalized by threshold value θ_m:

$$u(t) = \frac{\theta}{\theta_m} \tag{7.61}$$

Taking into account Eqs. (7.37)–(7.41), functions $f(\mathbf{z}(t))$, $g_1(\mathbf{z}(t))$, $g_2(\mathbf{z}(t))$, $h_1(\mathbf{z}(t))$, and $h_2(\mathbf{z}(t))$ are defined as:

$$f(\mathbf{z}) = \begin{Bmatrix} \dot{\alpha} \\ \dot{\beta} \\ \frac{1}{m}\mathbf{e} \cdot \frac{|\mathbf{P}_\beta|^2 \mathbf{P}_\alpha - (\mathbf{P}_\beta \cdot \mathbf{P}_\alpha) \cdot \mathbf{P}_\beta}{|\mathbf{P}_\alpha|^2 |\mathbf{P}_\beta|^2 - (\mathbf{P}_\alpha \cdot \mathbf{P}_\beta)^2} \\ \frac{1}{m}\mathbf{e} \cdot \frac{|\mathbf{P}_\alpha|^2 \mathbf{P}_\beta - (\mathbf{P}_\beta \cdot \mathbf{P}_\alpha) \cdot \mathbf{P}_\alpha}{|\mathbf{P}_\alpha|^2 |\mathbf{P}_\beta|^2 - (\mathbf{P}_\alpha \cdot \mathbf{P}_\beta)^2} \end{Bmatrix} \tag{7.62}$$

$$g_1(\mathbf{z}) = \begin{Bmatrix} 0 \\ 0 \\ \frac{1}{m}\mathbf{b_t} \cdot \frac{|\mathbf{P}_\beta|^2 \mathbf{P}_\alpha - (\mathbf{P}_\beta \cdot \mathbf{P}_\alpha) \cdot \mathbf{P}_\beta}{|\mathbf{P}_\alpha|^2 |\mathbf{P}_\beta|^2 - (\mathbf{P}_\alpha \cdot \mathbf{P}_\beta)^2} \cdot \mathbf{N} \\ \frac{1}{m}\mathbf{b_t} \cdot \frac{|\mathbf{P}_\alpha|^2 \mathbf{P}_\beta - (\mathbf{P}_\beta \cdot \mathbf{P}_\alpha) \cdot \mathbf{P}_\alpha}{|\mathbf{P}_\alpha|^2 |\mathbf{P}_\beta|^2 - (\mathbf{P}_\alpha \cdot \mathbf{P}_\beta)^2} \cdot \mathbf{N} \end{Bmatrix} \tag{7.63}$$

$$g_2(\mathbf{z}) = \begin{Bmatrix} 0 \\ 0 \\ \frac{1}{m}\mathbf{t} \cdot \frac{|\mathbf{P}_\beta|^2 \mathbf{P}_\alpha - (\mathbf{P}_\beta \cdot \mathbf{P}_\alpha) \cdot \mathbf{P}_\beta}{|\mathbf{P}_\alpha|^2 |\mathbf{P}_\beta|^2 - (\mathbf{P}_\alpha \cdot \mathbf{P}_\beta)^2} \cdot \mathbf{N} \\ \frac{1}{m}\mathbf{t} \cdot \frac{|\mathbf{P}_\alpha|^2 \mathbf{P}_\beta - (\mathbf{P}_\beta \cdot \mathbf{P}_\alpha) \cdot \mathbf{P}_\alpha}{|\mathbf{P}_\alpha|^2 |\mathbf{P}_\beta|^2 - (\mathbf{P}_\alpha \cdot \mathbf{P}_\beta)^2} \cdot \mathbf{N} \end{Bmatrix} \tag{7.64}$$

$$h_1(u) = \mu_t \cos(u\theta_m) \tag{7.65}$$

$$h_2(u) = -(\mu_s + \mu_t \sin(u\theta_m)) \tag{7.66}$$

being:

$$\mathbf{e} = -m(\mathbf{P}_{\alpha\alpha}\dot{\alpha}^2 + \mathbf{P}_{\beta\beta}\dot{\beta}^2 + 2\mathbf{P}_{\alpha\beta}\dot{\alpha}\dot{\beta}) + \mathbf{F_g} - \frac{1}{2}\rho A C_D |\mathbf{V}|^2 \cdot \mathbf{t} \tag{7.67}$$

Target for optimal control is identifying the steer angle (i.e., the value of $u(t)$) able to minimize the terminal time t_f under given initial conditions $\mathbf{z_0}$ at time t_0 and specified terminal conditions. This can be achieved minimizing a proper performance index:

$$J = S[\alpha_f - \alpha(t_f)] + \frac{L}{\dot{\alpha}(t_f)} + \frac{Tt_f^\sigma}{\sigma} + \frac{1}{2}\int_{t_0}^{t_f} Ru^2(t)\,dt \tag{7.68}$$

where $S > 0, L > 0, T > 0, R > 0$ and $\sigma \geq 1$.

The first term in the performance index is to make the sled arrive at the terminal position of the curve/track α_f. The second term is to increase the final along-track speed of the sled. This contribution is particularity important for the single-curve optimal problem. A higher terminal along-track speed will in fact increase the sled speed in the remaining part of the track. General value of the parameter L is in the order of the remaining distance to go.

The third term is to reduce the final time, which is the goal of optimization. To increase the penalty for the final time, a value of σ larger than 1 can be chosen.

Finally the last term is to penalize an excessive control action and to smooth it. It must be in fact pointed out that any larger steering angle than θ_m does not increase the bobsled control authority, but increases the resistance in the direction of motion. Then the optimal control problem can be solved subject to the constraint:

$$|u(t)| \leq 1 \tag{7.69}$$

but this, in absence of the last term of Eq. (7.68), produces a typical minimum time control problem whose solution is given by a bang-bang control strategy. In practice, however, a bang-bang control of the steering angle is unrealistic. The steering angle can in fact only be changed continuously by the driver since he has a limited bandwidth. Therefore it is reasonable to consider a trade-off between the constraint condition on the control variable and the performance, which can be achieved by adding the last term of Eq. (7.68).

Previously described optimal control strategy was applied to identify the steering angle along individual curves of the Lillehammer Olympic Track in [24]. The main achievements can be summarized as follows:

- increasing the friction coefficient strengthens the control effects and this means that better driving skills are more important in lower-atmospheric temperature conditions;
- increasing speed reduces sled controllability;
- optimal control of the sled in early low-speed stages (when sled controllability is higher) during a competition is extremely important.

In [24], a piecewise minimum-time optimal control algorithm was also developed to calculate the optimal solution for the whole track and applied to the Lillehammer sled track. According to the developed algorithm, to find the piecewise optimal algorithm, the Lillehammer sled track was divided into 16 curves and an optimal solution was calculated for each curve separately. Each curve takes the optimal terminal conditions of the previous curve as initial conditions for its optimal calculation. Moreover the along-track distance at the terminal time is specified as boundary condition ($\alpha(t_f) = \alpha_f$). Initial condition $\mathbf{z}_0$ of curve 1 is determined by how fast the crew pushes the bobsled at the beginning of the race. In this way, the whole track piecewise optimal solution is obtained through sequential calculation from curve 1 to curve 16.

The time difference between the piecewise optimal path and the free-travelling path (i.e., the path the bobsled would follow in absence of a driver model) is shown in Fig. 7.59a, while the speed difference is depicted in Fig. 7.59b. Although piecewise optimal solution for the whole track is only a suboptimal solution, it shortens the travel time of more than 1 s and increases the sled final speed by almost 2 m/s. Thus, single-curve optimum is consistent with whole-track optimum, although one-curve optimal solution may not be the optimal solution of the whole-track problem in that curve. This means that the entire track performance can be improved if performance in each individual curve is enhanced.

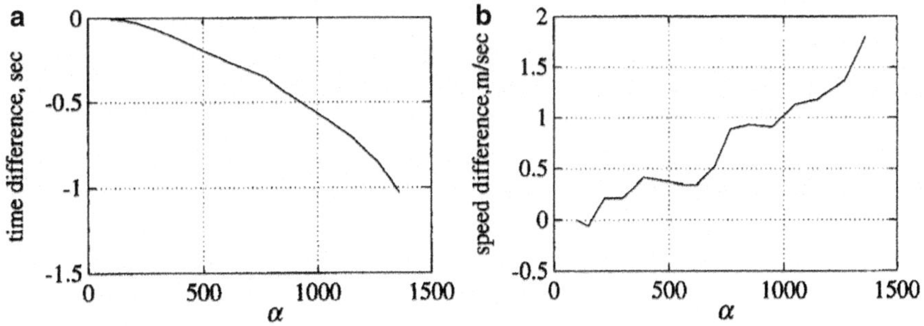

Fig. 7.59 Time difference (**a**) and speed difference (**b**) of whole-track piecewise minimum-time control for the Lillehammer Olympic Track, [24]

7.6.2 Trajectory Planning and Steering Control Approach

According to the approach described in [54], the design of the driver model can be split into two tasks:

- to determine the trajectory that allows to achieve the minimum run time and
- to follow it as close as possible by acting on the steering axis.

The development of the driver model is therefore divided into two phases: trajectory planning and controller design. Note that the trajectory planning phase takes into account the whole track and not just single curves in order to guarantee a continuous and smooth trajectory.

Controller design aims at reproducing the real behavior of professional drivers, even if this may not lead to the optimal solution represented by the minimum possible run time.

7.6.2.1 Trajectory Planning

As already mentioned, improving performance with a given sled, a given track and given initial conditions consists in reducing the run time. Trajectory must thus be planned on this purpose. In [54, 63] different strategies (minimum-length path and minimum-dissipated-energy path) were compared to achieve this target and it was found that performance can be improved (i.e., the final time is reduced) if energy dissipated by longitudinal forces is minimized.

According to [54], in order to identify the trajectory able to minimize energy dissipated by longitudinal forces, the bobsled is modelled as a three degrees of freedom particle and its motion is integrated in the time domain. The position of the particle in space is identified by vector $\mathbf{x_G}$ whose components are expressed in a global reference system $\mathbf{x} - \mathbf{y} - \mathbf{z}$. At the generic integration step, the track surface around $\mathbf{x_G}$, described through a set of discrete points, is approximated by its tangent contact plane τ as shown in Fig. 7.60.

Thus, the unit vector $\mathbf{n}$ normal to the contact surface is determined. The motion of the particle is assumed to take place in the plane τ. A second unit vector $\mathbf{t}$, normal to $\mathbf{n}$ and tangent to the trajectory, i.e. parallel to the velocity vector $\mathbf{V}$, can therefore be defined. Finally, the bi-normal vector $\mathbf{b}$ is equal to the cross product of $\mathbf{t}$ and $\mathbf{n}$. Matrix $[\Lambda]$, that allows expressing a vector according to the reference system identified by $\mathbf{t} - \mathbf{n} - \mathbf{b}$, is thus given by:

$$[\Lambda] = \begin{bmatrix} t_x & t_y & t_z \\ n_x & n_y & n_z \\ b_x & b_y & b_z \end{bmatrix} \tag{7.70}$$

The rows of $[\Lambda]$ are the three components of the unit vectors $\mathbf{t}$, $\mathbf{n}$ and $\mathbf{b}$ expressed in the global reference system. The equations of motion of the particle expressed in the global reference system can therefore be written in matrix form as:

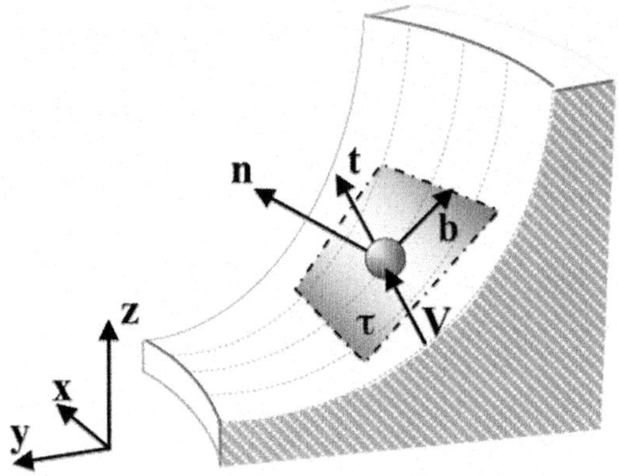

Fig. 7.60 Global ($\mathbf{x}-\mathbf{y}-\mathbf{z}$) and local ($\mathbf{t}-\mathbf{n}-\mathbf{b}$) reference systems [54]

$$m\mathbf{a_G} = m\begin{Bmatrix}\ddot{x}_G\\ \ddot{y}_G\\ \ddot{z}_G\end{Bmatrix} = \begin{Bmatrix}0\\ 0\\ mg\end{Bmatrix} + [\Lambda]^T\begin{Bmatrix}-\mu_x F_z - \frac{1}{2}\rho C_D A V^2\\ N\\ \eta\mu_y F_z\end{Bmatrix} \tag{7.71}$$

since the forces acting on the point-mass are the weight mg (directed along the global vertical axis $\mathbf{z}$), the contact forces (in tangential, normal and bi-normal directions), and the aerodynamic drag (also defined in the local reference system). To project the forces defined in the local reference system on the global one, the projection matrix $[\Lambda]^T$ is used. The normal contact force F_z is computed using the same approach implemented in the multi-body model of the bobsled and described in Sect. 7.4.1.3: once the contact surface is approximated through the tangent contact plane, the penetration of the point-mass in the ice surface along direction $\mathbf{n}$ is identified and thus the normal contact force is determined according to Eq. (7.43). As far as the bi-normal (lateral) contact force is concerned, the scaling factor η is adopted to account for the amount of lateral force ($\mu_y F_z$) used to alter the trajectory (η is a linear function of the imposed steering angle and is limited in the range $[-1, 1]$; η=1 corresponds to the maximum clockwise allowed steering angle while $\eta = -1$ corresponds to the maximum counter-clockwise allowed steering angle). The value of η varies along the track and is determined through an optimization algorithm aimed at minimizing the run time. To limit the number of variables to be optimized (η value at each section along the track), η is set to 0 during straight sections and is described through an analytical function inside each turn. Thus, the parameters of such analytical functions become the optimization variables. Referring to Fig. 7.61, five nodal points $S_0, S_1, \ldots, S_4$ are identified along the Track Center Line (TCL) for each curve including entrance and exit sections. The curve is therefore divided into four segments. S_0 and S_4 are placed at the beginning and at the end of the curve while S_2 is placed at its center. S_1 and S_3, instead, are points placed in the first and second half of the turn whose positions are determined during the optimization process (first two optimization variables).

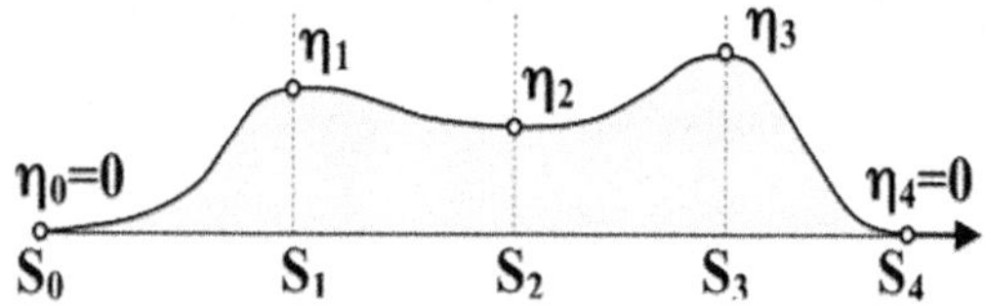

Fig. 7.61 Variation of η inside a turn [54]

For continuity reasons, the value of η is assumed to be equal to 0 in S_0 and S_4 while η_1, η_2, and η_3 are determined during the optimization procedure (additional three optimization variables). The value of η along each segment in which the curve was divided (necessary for determining the lateral contact force at each track section and thus solving the equations of motion (7.71) that allow to evaluate the run time) is determined by using a piecewise cubic polynomial (i.e., cubic spline) that interpolates the five considered nodal points. Five optimization variables S_1, S_3, η_1, η_2, and η_3 are associated with each curve thus leading to $5 \times n_c$ optimization variables on the entire track, being n_c the number of curves of the track. The identification of these variables is carried out simultaneously for all the curves of the track, with the assumption that lateral contact force varies according to a piecewise cubic polynomial along each curve starting from zero at curve entrance and finishing to zero at curve exit. Curve entrance and exit points are assumed to be equal to the corresponding geometric points. The shape of the function shown in Fig. 7.61 gives information about the driving style determined through the optimization procedure: during curve entrance, value of η is increased up to η_1 meaning that the driver has to guide the sled into the curve; then, η is maintained almost constant along the full curve thus meaning that the driver should let the sled slide keeping the angle of attack of the front runners constant; finally, the steer axle is released during curve exit. The length of entering section (in which η is increased) is usually longer than length of exit section (in which η is decreased).

7.6.2.2 Controller Design

Once identified the trajectory, a controller acting on the steer axle was developed in [54] to follow the desired path in a way to mimic the behavior of a professional bobsled driver. While the trajectory was determined through a simplified dynamic analysis (making reference to the motion of a particle along the track), the controller was developed based on the multi-body model of the sled described in Sect. 7.4.1.3, which allows to accurately account for the effect of the steering action on the sled trajectory.

To assess the behavior of a real professional driver, the experimental tests described in Sect. 7.3.4.2 and carried out at the Cesana Pariol Olympic Track with an instrumented bobsled were analyzed [54, 55, 63]. Making reference to Fig. 7.28, results of the performed experimental analysis of driving behavior can be summarized as follows:

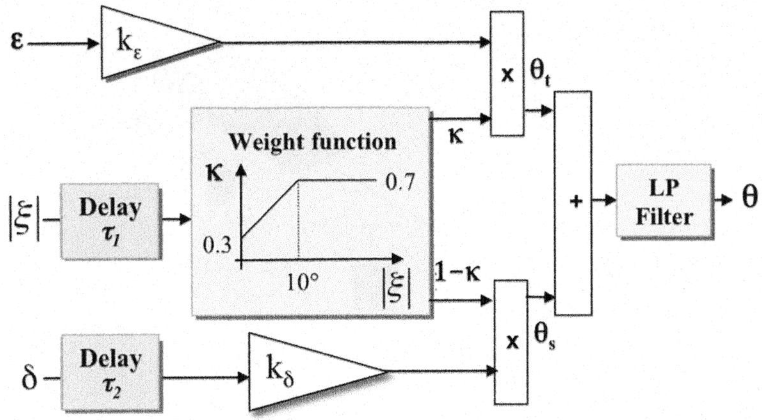

Fig. 7.62 Scheme of the driver model (controller). Adapted from [54]

- the driving style appears to be mainly based on a feedforward contribution, i.e. driver's actions on the steer axle are pre-determined on the basis of driver's experience acquired during previous runs;
- corrections on the steer axle are performed by the driver based on vehicle dynamic response feedback:
  - the roll angle ξ acts as "trigger" warning the driver to start steering at curves entrance;
  - driver's reaction to roll angle is delayed of about 0.25 s (depending on his promptness);
  - the driver reacts to bobsled sideslip angle, δ, during straight tracks, in presence of lateral drifts, or when exiting a curve;
  - driver's reaction to sideslip angle is delayed of about 0.1 s (depending on his promptness).

Based on these experimental evidences, the control action θ was designed as the sum of two contributions, one associated with trajectory following (θ_t that reproduces the feedforward action of the driver) and one associated with the sideslip angle (θ_s that reproduces the feedback action of the driver). Figure 7.62 shows the scheme of the driver model (controller) proposed in [54]. The controller gains were optimized through a series of simulations having as target the reproduction of the actual driving style ("learning phase") of a professional driver.

The trajectory following contribution (upper part of Fig. 7.62) is based on the difference ϵ between the desired path and the actual one. As shown in Fig. 7.63, this difference is determined considering two points, P_r on the desired trajectory and P along the longitudinal axis of the bobsled (called X_L), l^* meters in front of the cog of the sled (marked with G).

The angular deviation ϵ between the desired position and the one the bobsled is directed to quantifies the trajectory error. Note that this error is computed l^* meters

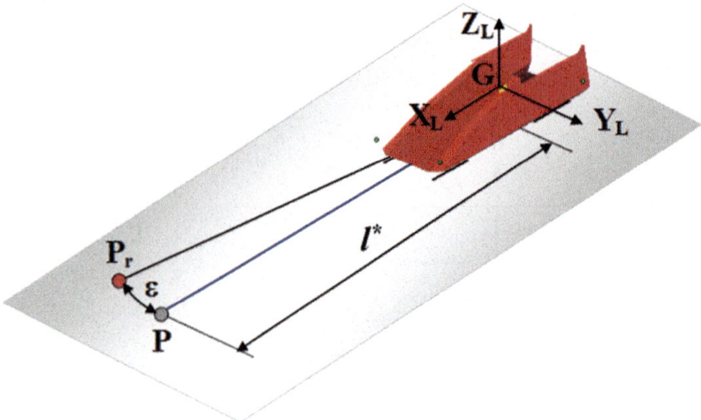

Fig. 7.63 Angular deviation ϵ with respect to the reference trajectory. Adapted from [54]

in front of the cog in order to anticipate track changes (as occurs for real driver). The distance l^* is called "visual distance" and is assumed to be a function of the bobsled's speed V [29]:

$$l^* = Vt^* \tag{7.72}$$

being t^* a constant time interval that has to be determined during the controller learning phase. According to Eq. (7.72), the visual distance increases with vehicle speed. This is in accordance with the fact that at high speeds longer visual distances are necessary to keep the bobsled on track.

This trajectory following contribution to the steer angle (control action) is finally defined as:

$$\theta_t = k_\epsilon \epsilon \tag{7.73}$$

where k_t is the gain of the adopted proportional controller, correlated to the promptness of the driver's reactions. As t^*, also k_t is determined during the learning phase. The sideslip contribution (lower part of Fig. 7.62) aims at zeroing the sled sideslip angle. According to the analysis of the driving style, the amplitude of steer angle is of the same order of magnitude of sideslip angle. Thus the proportional gain k_δ can be assumed equal to 1:

$$\theta_s = -k_\delta \delta(t - \tau_2) = -\delta(t - \tau_2) \tag{7.74}$$

where δ is the sideslip angle of the bobsled cog computed $\tau_2 = 0.1$ s before the actual simulation time. The time delay introduced in l^* is associated with the promptness of the driver, which was identified by the experimental data. In particular, the value of 0.1 s represents the mean measured time delay between the peak of sideslip and steer angles. The two contributions θ_t and θ_s are weighted according to the value of the roll angle ξ: the weight parameter κ increases linearly from 0.3 to 0.7 when the roll angle is below 10° and is then kept constant for higher roll angles. As a consequence, during straight track sections, where the roll angle is close to zero, the sideslip contribution has a higher weight with respect to the trajectory following contribution. The lateral drift that may arise due to driving errors when exiting a curve or due to impacts against the lateral walls can thus be controlled. During curve entrance, instead, the roll angle rises and the trajectory following contribution prevails. According to the analysis of driving style, value of roll angle is delayed by $\tau_1 = 0.25$ s, i.e. the mean measured value of delay between the rising of roll angle and the subsequent rising of steer angle. Thus, even if a trajectory error over the desired trajectory is computed and a contribution θ_t is present, its value will become effective only when a certain value of roll angle is crossed ("trigger effect"). Note that, being the weighting factor in the range of 0.3–0.7, both contributions are present in all track sections. The weighting function moreover smoothes out the tendency of cutting in the curve generated by the visual distance l^*.

Once the control action θ is determined, before being passed to the multi-body model of the bobsled, it is low-pass filtered using a 4th order Butterworth filter with a cut-off frequency of 3 Hz in order to achieve a bandwidth equal to the experimentally measured one.

7.6.2.3 Results

Performance of the above described driver model on the Cesana Pariol Olympic Track is illustrated in [54]. Results were obtained coupling the driver model with the multi-body model of the bobsled described in [28, 56, 62] and reported in Sect. 7.4.1.3.

The steer angle demanded by the driver model and multi-body model response were compared with the experimental data collected during the tests described in 7.3.4.2 at the Cesana Pariol Olympic Track with an instrumented two-men bobsled. Figure 7.64 compares the steer angle demanded by the driver model (green dashed line) and the experimental data (black solid line) along the whole track. As it can be seen, numerical and experimental steer angle appear very similar in terms of magnitude and shape; overestimations of the steer amplitude are present in turn 9 and 15. The steer angle is always directed towards the inner side of each curve, reproducing the response of a real driver.

Capability of the driver model to mimic the behavior of a real driver is confirmed by the comparison between experimental and numerical roll angle of the bobsled during the run (Fig. 7.65). This comparison in fact allows to quantify the difference between the trajectory followed by the bobsled and the one predicted by the numerical model. As it can be seen, although the roll angle is underestimated in some turns, amplitude and shape of numerical and experimental roll angle are very similar. Roll rate during curves enter/exit is also very close to the experimental one.

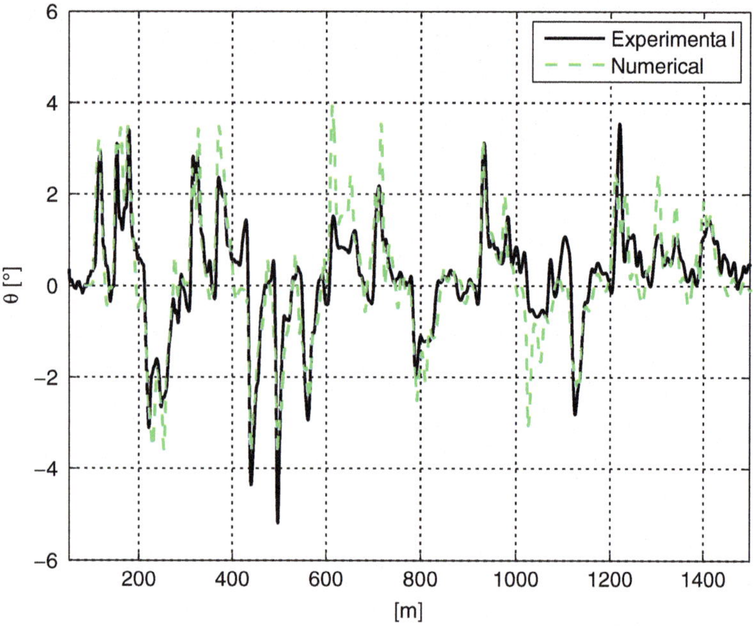

Fig. 7.64 Steer angle vs. travelled distance: comparison between numerical and experimental data. Adapted from [54]

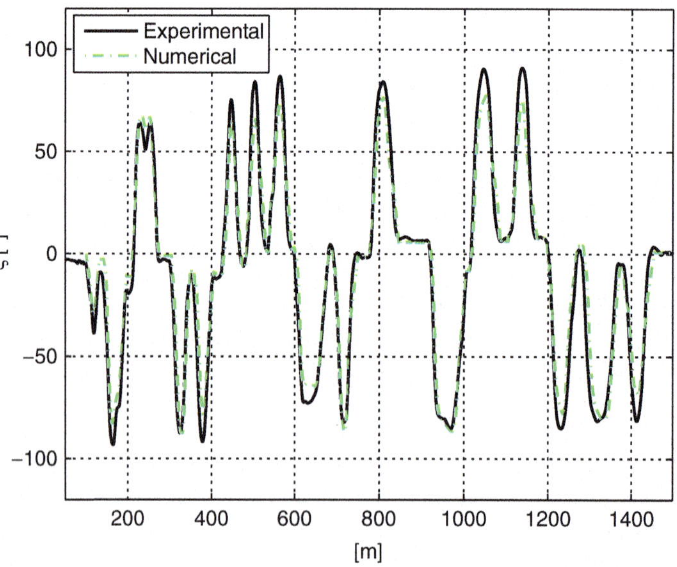

Fig. 7.65 Roll angle vs. travelled distance: comparison between numerical and experimental data. Adapted from [54]

7.7 Simulator

In bobsled, the simulation of a run has a fundamental importance to improve the performance of athletes. The goal of this section is to give a general overview on the main reasons of this choice, and to show different examples of simulators. At the end, some proposals about simulator developments are provided.

7.7.1 Introduction

As in other sports, practice is a key element which can help athletes in lowering the lap time by those fraction of seconds required to win a race. Unfortunately, practising bobsled is neither cheap nor easy. Few internationally ratified tracks exist, mainly set in Europe, and costs associated with international travel can be significant. In addition, tracks are usually available for a limited number of months during the year, resulting in different bobsled teams practising the same day or sharing the facility with athletes from skeleton and luge. As a consequence, runs completed in a day are usually far less than 10, a number largely insufficient to develop an adequate level of muscle and visual memory.

Besides logistic aspects, repeatability is another critical factor for track testing: different start times, changing of track conditions during the day (e.g., ice temperature, smoothness of track surface), fatigue associated with hard pushing at the start and with high accelerations (up to $5g$) experienced during each run, are all factors influencing the overall performance. With only few runs all resulting in similar finish times and affected by a number of variables, it is difficult for a bobsled driver to determine the best trajectories and the most appropriate steering controls which lead to lap time minimization.

From one side, increasing the number of tests and removing the influence of external factors would allow the driver to understand how the choice of different trajectories and steering commands influences the lap time; in this way, a best trajectory could be identified. On the other side, the same factors would allow the driver to achieve familiarity with the sequence of curves and the associated steering inputs; in particular, considering that bobsled speeds largely exceed 100 km/h in the final part of the track (above 130 km/h at Cesana and nearly 140 km/h at Whistler), the driver has to start steering even before seeing the next curve. In other words, since the contact forces developed at the ice–skate interface are low with respect to the system inertia, the driver's inputs are effective only if he anticipates the dynamics of the bobsled. Thus, once the driver identifies the best trajectory to be followed, he has to memorize the sequence of inputs on the steering by repeating a series of almost identical track laps.

As outdoor tests seldom offer this opportunity, a bobsled simulator could represent a valid alternative for athletes training. Reproducing the experience of running with a racing bobsled in a controlled environment can overcome several limits of track tests:

- costs related to travel expenses and use of the track could be significantly reduced if a simulator is designed to reproduce in a virtual environment the geometry of various tracks spread around the world;
- though different athletes would use the simulator, its availability would be certainly greater than that of an Olympic track, thus tests could be repeated as needed;
- effectiveness of training would be improved since influence of external factor would be minimized and the effect of each single action on the finish time could be quantified.

Besides these advantages, a bobsled simulator would also offer the opportunity of training athletes in a safe environment: driver's errors or mechanical failures will not result in injuries for the team members or damages for the bobsled itself.

7.7.2 Characteristics of a Bobsled Simulator

A bobsled simulator should reproduce as much as possible the real experience of running on a racing bobsled down a twisted, banked, icy track. Higher realism will result in a more immediate transfer of the experience gained in the simulator to the real track.

The mathematical model of a bobsled moving on a track is clearly the heart of a simulator; a detailed analysis of these models was presented in Sect. 7.4. When a model is implemented on a simulator, the basic requirement consists in real-time computation; this means that an adequate hardware is required and excessive complexity should be avoided.

As known from scientific literature, most part of the perception of motion (almost 70 %) is related with the visual system. Thus, a bobsled simulator should first of all be equipped with an adequate graphic engine able to generate in real time a detailed 3D representation of the track. Updating frequency is also important and should be higher than 30 Hz. Additional refinement of visual perception are achieved with the use of several monitors surrounding the driver: the track is not just projected on a single flat screen but also on side monitors enhancing the perception of a 3D environment and stimulating the peripheral vision. It is important to avoid the presence of visual inputs different from those generated by the graphic engine; for this purpose the monitors and the driver should be located in a closed and dark environment. As an alternative Head Mounted Displays can be used.

Secondly, motion perception is associated with the vestibular system able to record accelerations and yaw rates. The equipment required to emulate the dynamics of a bobsled would be extremely complex: roll, yaw, and pitch dynamics should be controlled simultaneously with continuous and sudden changes of angular speeds; in addition, all the controls should be applied in perfect synchronization with the graphic engine to avoid delays between the inputs to visual and vestibular systems. The design of the system would be a challenge and its realization would imply costs far beyond reason.

As an alternative, electric or hydraulic actuators could be added to reproduce at least the most significant motions characterizing the bobsled dynamics. Roll rate while entering and exiting curves can exceed $60\,\mathrm{rad/s^2}$ and represents one of the most relevant inputs for the vestibular system; an electric motor is used in the simulator presented in [25] to control the roll motion (in particular the roll rate) of a bobsled cockpit. Accelerations experienced while cornering may exceed $5g$ and undoubtedly activate the vestibular system too. Accelerations higher than $1g$ may be obtained through a linear actuator whose control should be conveniently synchronized with the actuation of roll motion. Actually high-g accelerations lasting for the time needed to complete a corner would require high power actuators and would significantly increase the dimensions of the facility; altogether this solution does not appear practically feasible. High-g may also be obtained through a centrifuge, but the sudden transitions from left-hand to right-hand curves characterising a bobsled run could be hardly obtained with this device. In general, considering the purposes of a simulator, there is no practical way to reproduce the accelerations characterizing a bobsled run. This means that an important input for the vestibular system will be missing.

Some researchers pointed out that, controlling the roll angle of a simulator without the possibility of reproducing the corresponding g-forces would lead to inconsistent feedback. According to this point of view, it would be preferable not to excite the vestibular system at all [57]. Other researchers [25] believe instead that a partial reproduction of some movements of the bobsled can enhance the realism of the simulator.

Another element that is considered important is the feedback from the steering system; in the simulator described in [25] the steering mechanism is controlled through an electric motor to reproduce the moment along the steering axis generated by the ice-skates contact forces.

As last, also the environment surrounding the athletes plays a role. Besides the aforementioned need of removing external visual inputs, the presence of element typical of a bobsled run will enhance the perception of a real experience. Recreating a real cockpit with a real steering system will help a driver to memorize the proper sequence of movements. Also playing sounds that are usually heard during a run will heighten the realism of the simulator and will also limit the effect of external auditory disturbances.

In the following paragraphs different simulators described in technical literature will be presented.

The first one is quite complex from a hardware point of view and consists of a real cockpit which rolls according to the position of the bobsled along the virtual track.

The second one follows a different philosophy, avoiding a complex hardware required to excite vestibular system, and focusing on visual and auditory elements; in particular, this simulator is conceived to run on different platforms (also on a simple PC) with a wide range of optional peripherals. This simulator thus is somehow limited in terms of reproduction of the real experience but allows the athletes to train almost everywhere.

The last section presents a completely different kind of simulator aiming at improving the athletes performance during the start time. Since optimal start time is often a crucial factor to win a race, a simulator based on a treadmill and a dynamometric bobsled handle is used to optimize the position of the athletes to maximize push force and to determine when they should jump into the sled.

7.7.3 US Bobsled Team Simulator

Information around this simulator was directly taken from [25], the paper was published in 2000 and described the 3rd Version of a simulator developed for the US bobsled team. No further versions were presented in technical literature. This simulator uses a complex hardware to recreate a realistic experience of a bosleigh run by activating visual, vestibular, tactile, and auditory systems.

As shown in Fig. 7.66, the simulator externally appears as a fiberglass shell mounted over a base frame. The upper cowling is opened to access the simulator where the athletes seat in a cockpit made from a real bobsled mould. A steering system actuated with bungee chords and resembling the actual one is located in front of the driver. The fiberglass shell houses a monitor used for the graphical representation of the track and audio speakers playing sounds normally heard inside a bobsled during a run.

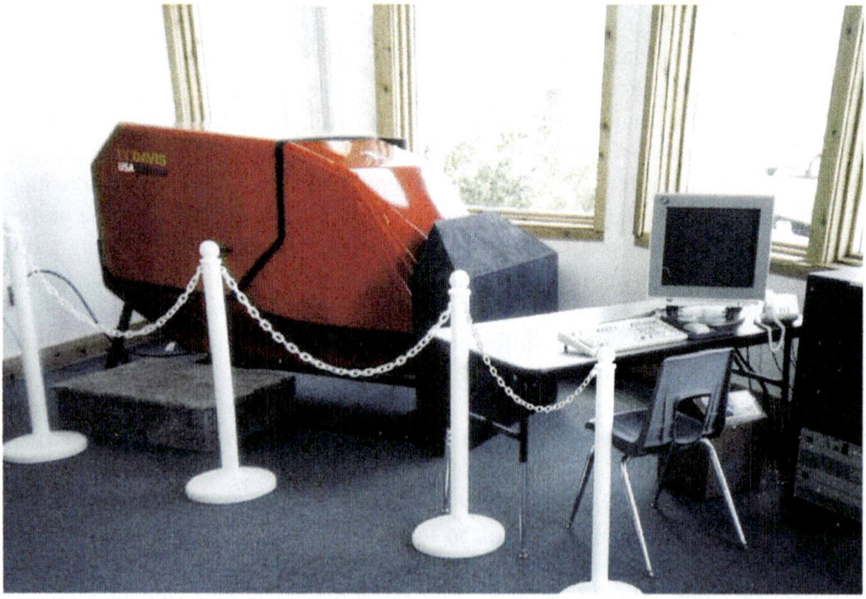

Fig. 7.66 US Team bobsled simulator. Adapted from [57]

The fiberglass shell is mounted on an internal aluminium chassis which supports the weight of the driver and of the other devices (steering system, monitor, etc). The cockpit is also provided with adjustable foot pegs and seat. Once the cowling is closed, the athlete can start training in a realistic environment virtually isolated from optical and auditory disturbances. The cockpit is linked to the base frame by means of two bearings, which support the weight of the structure and allow the simulator to rotate along a longitudinal axis; exploiting this degree of freedom, the motion of the cockpit is thus controlled to reproduce the roll motion of the bobsled during a run.

The core of the simulator is the mathematical model of the bobsled; since the simulation has to be performed in real time, a simple particle model was chosen. The sled is subjected to gravity force, track normal force, track tangential force in terms of ice friction and lateral steering force, aerodynamic lift and drag. All the data required for the modelling of the bobsled (inertial parameters, aerodynamic coefficients) and of the contact forces were directly measured or taken from literature.

The track is modelled as a sequence of transversal sections interpolated by means of cubic polynomials. The bobsled is assumed to be always in contact with the track surface; its position along the track is univocally identified with two coordinates α and β, respectively, representing the distance from the start and from the right edge of the track. Lines of constant α and β are then used to obtain a graphical representation of the track.

The numerical simulation of bobsled dynamics is performed by an IBM workstation.

The most remarkable component of this simulator is represented by the hardware, which allows to include some important elements characterizing a run on a race bobsled. Besides the use of components taken from a real cockpit, the simulator actuates controls over the roll motion and on the steering axle: an electric motor is used to rotate the cockpit along the longitudinal axis while another one provides a force feedback on the steering system. Controls are performed by the so-called motion controller which directly receives signals from the workstation and generates proper voltage output to the amplifiers; these lasts convert the signals to current used to drive the torque provided by the motors.

Controlling the roll motion represents a challenge, due to the high inertia of the system and to the need of rotating the bobsled in close synchronization with the graphic engine. Even though in this simulator the roll angle of the cockpit is scaled 20 % with respect to the one computed by the bobsled model, a powerful motor (4.5 kW) is required to actuate the roll motion updating the position at 30 Hz. Control over the roll motion is performed in proportional-derivative (PD) mode, comparing the reference roll angle outputted by the bobsled model, with the one measured on the cockpit through a motor encoder. Though roll acceleration does not reach values of $60\,\text{rad/s}^2$ normally recorded during a run, roll rates obtained when entering and exiting curves significantly contribute to enhancing the realism of the experience.

The roll motion of the cockpit is controlled also while cornering; in particular, after the rotation at the beginning of a curve, the roll angle is gradually zeroed, with a roll rate that cannot be detected by the vestibular system. This technique, called *washout*, is used to generate the feeling of an external force directed toward the bottom of the cockpit. In fact, during a real run, the combination of high g-forces and roll angles close to 90°, set the direction of the force perceived by the driver perpendicular to the track. Since the simulator is not able to reproduce high g-forces, zeroing the roll angle while cornering allows to exploit gravity acceleration to mimic in part the real condition.

A second interesting control implemented in this simulator is the feedback generated on the steering system. Since the main purpose of the simulator is driver's training, the athletes should be able to memorize a proper sequence of steering inputs and the corresponding forces required to generate them with the correct timing. As mentioned above, the simulator cockpit is equipped with a steering system actuated through bungee cords, mostly similar to the one present on a race bobsled. During a run, the forces developed at the contact interface and the inertia of slave arm, axle, runner mounts and runners, generate a moment along the steering axis which provides an important feedback for the driver. On the simulator, a simple mathematical model is used to estimate the moment generated by the contact forces while an electric motor connected to the steering system reproduces this effect. A disc with proper mass distribution is also linked to the steering system to introduce the moment associated with inertial effects.

Great attention was paid toward safety, considering that the simulator was developed also to serve as tourist attraction. The roll motion of the shell can be locked through a safety pin; the motion can take place only if simulation program, motion controller, and amplifiers are all operating properly. Several safety switches are used, including an optical roll limit switch.

7.7.4 Virtual Reality Simulator

As already mentioned, two approaches are possible for a bobsled or a skeleton simulator. In one case the platform is fixed and the world moves around the athlete. In the other case, the world is fixed and the position of the athlete relative to the horizon changes using a moving platform [57]. The second approach was described in the previous section. In this section, the first approach is analyzed, so the platform does not move and the simulation is purely virtual.

In [57] a project of a virtual reality simulator for bobsled and luge, developed at the University of Calgary, was described. This prototype was composed by four screens (powered by a PC cluster) and the point of view is the one of a driver with a fixed camera attached to the sled (this is an useful point of view, in fact it is common for athletes to use a small camera attached to the sled to tape and then study their performances): the sled appears fixed with the world rotating around the subject.

Fig. 7.67 An example of a PC display view during a luge simulation [57]

The simulation was based on the track of Salt Lake.[8] In a CAD environment, the 3D geometry of the track was reproduced, as well as the bobsled and luge sliders (used, at athlete's discretion, like avatars). These models were then imported into Virtool[9] to realize the simulation: it implements an accurate real-time rendering, assigns physical properties to objects, and uses the Havoc Physics Engine (for an accurate real-time simulations, the effects of forces on the sled must be evaluated at a minimum of 30 fps [57]).

The results of the use of this simulator were discussed with coaches and athletes [57]. First of all the choice of using a fixed viewpoint was appreciated because in a motion platform the center of gravity doesn't change, but in an actual bobsled run the center of gravity is always perpendicular to the track runners. This could provide the athlete with a wrong feedback. Several advantages are then related with the training environment. It is safe (moreover this method has a great portability, in fact this software can be used by athletes also on a common PC or laptop, see Fig. 7.67) and useful to learn the shape of the track. In fact at high speed (like 130 km/h), the athlete must anticipate the turn, often before it can effectively have it in view. This simulator would help also to acclimate to the visual blur of moving at high velocity down a track. Further features are included to help the athletes, for example a line along the track to show the optimum path and a "playback mode" in which the athlete's path is compared with the optimum one. To improve the level of the simulation also the sound level is controlled: it is related to the speed of the sled.

[8]This track is part of the Utah Olympic Park. It was completed in December 1996, the track length is 1680 m roughly, with a vertical drop of 120 m and 15 curves.

[9]One may refer to [58].

7.7.5 Bobsled Start Simulator

In bobsled competition, the difference between a winning and a losing performance is often less than thousandths of a second over the course of two or more usually four, runs down the track with a duration of the order of one minute (see [23]). A phase of the run which plays a great role on the final time is the start, that is the sprint and the jump of the athlete(s) into the sled. As suggested by [59], an adequate model for the acceleration of the bobsled and the pushers is similar to that described for the sprinters by Vaughan. He wrote the equation of motion in the form

$$\frac{dv}{dt} = A - Bv^{\alpha} - Cv^2 \tag{7.75}$$

where v is the sprinter's instantaneous velocity, A, B, C and α are constants (see [60]). An optimal value for α proposed by Vaughan is 0.7, and consequently one may choose the values for the other parameters (which are provided as range of values, see again [60]).

In [61], a "*Bobsled Start Simulator*" is described: its proposal was to give bobsledders a tool to optimize the start variables without having to train on a track. The device was tested and proved successfully in improving not only pushing force but also timing for six US Olympic Team athletes.

The basic system schematic is shown in Fig. 7.68. Two are the main subsystems which compose the simulator: the treadmill and the bobsled handle. The *treadmill*, which can reach approximately 45 km/h and its grade moves from +40° to −10° (even if this parameter is not so critical for this kind of simulation), has a speed control system by computer: the pre-existing motor controls send four I/O signals and an analog voltage signal to the treadmill power box to run the basic functions of the treadmill. The motor controller bypass is controlled by the computer, and

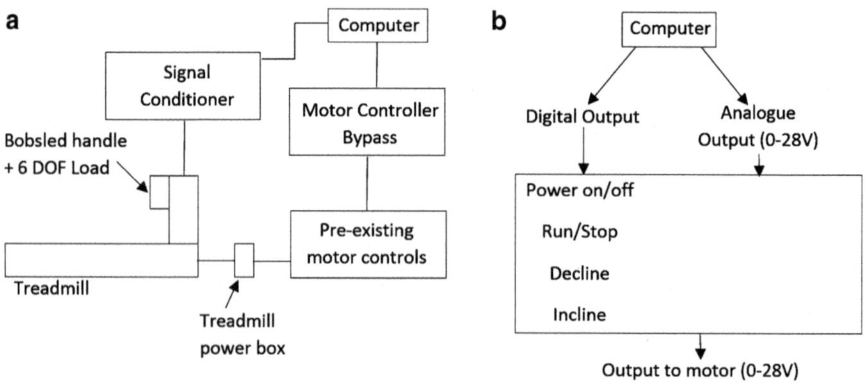

Fig. 7.68 (**a**) A schematic representation of simulator, (**b**) motor controller diagram. Adapted from [61]

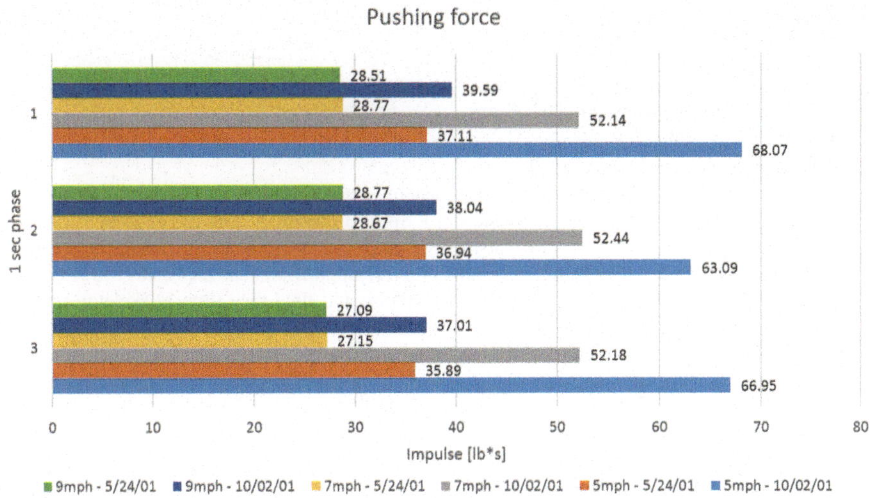

Fig. 7.69 The impulse trend over five months of training with the simulator. These results refer to athlete N.1 and are obtained at three different treadmill speeds (adapted from [61])

simulated these five inputs to the treadmill power box [61]. The *bobsled handle* is attached to a six DOF load cell to measure the force output: these signals are conditioned and passed to the computer A/D board, where they are sampled at a maximum frequency of 1000 samples/s. The load cell outputs 6 ± 10 V analog signals on separate channels, and it can measure roughly up to 450 kg of compression load and 55 Nm of torque. The force forward, which is the force that accelerates the bobsled and the one that athletes are trying to maximize. The force down is particularly important because it is used to adjust the handle to the correct height for the athlete, and thus optimize the athlete's forward force input into the bobsled. It can also indicate that the athlete is using an incorrect form and directing their energy downward instead of into the forward acceleration of the bobsled (see [61]).

The device was tested and proved successfully in improving not only pushing force but also timing for the six US Olympic Team athletes. For specific results, the reader can refer to [61], but to summarize the results we can report the data related to the impulse (that is the pushing force). In Fig. 7.69, the results of the impulse per second measure at three different treadmill velocities over five months of training with the simulator are provided. A drastic increase of the pushing force (in certain cases the value is almost doubled) is evident.

Also, the simulator was used to optimize the height of the push bar and to determine the best moment for a bobsledder to jump inside the sled.

References

1. FIBT, Federation Internationale de Bobsleigh et de Tobogganing (2013). http://www.fibt.com/
2. Wikipedia, Bobsleigh – Wikipedia, the free encyclopedia (2015)
3. International Bobsleigh Rules (2012)
4. International Skeleton Rules (2012)
5. M.M. Morlock, V.M. Zatsiorsky, Factors influencing performance in bobsledding: I: influences of the bobsled crew and the environment. Int. J. Sport Biomech. **5**, 208–211 (1989)
6. G. Bruggemann, M. Morlock, V.M. Zatsiorsky, Analysis of the bobsled and men's luge events at the XVII olympic winter games in Lillehammer. J. Appl. Biomech. **13**(1), 98–108 (1997)
7. C. Zanoletti et al., Relationship between push phase and final race time in skeleton performance. J. Strength Condit. Res. **20**(3), 570–583 (2006)
8. N. Bullock et al., Characteristics of the start in women's world cup skeleton. Sports Biomech. **7**(3), 351–360 (2008)
9. W.A. Sands et al., Anthropometric and physical abilities profiles: US national skeleton team. Sports Biomech. **4**(2), 197–214 (2005)
10. I. Roberts, Skeleton bobsleigh mechanics: athlete-sled interaction. Ph.D. thesis, The University of Edinburgh, 2013
11. F.P. Bowden, T.P. Hughes, The mechanism of sliding on ice and snow. Proc. R. Soc. Lond. A **172**, 280–298 (1939)
12. F.P. Bowden, D. Tabor, The friction and lubrication of solids. Technical report, Oxford University Press, Oxford (1950)
13. D.C.B. Evans, J.F. Nye, K.J. Cheeseman, The kinetic friction of ice. Proc. R. Soc. Lond. A **347**, 493–512 (1976)
14. B.A. Marmo, J.R. Blackford, C.E. Jeffree, Ice friction, wear features and their dependence on sliding velocity and temperature. J. Glaciol. **51**(174), 391–398 (2005)
15. K. Itagaki, G.E. Lemieux, N.P. Huber, Preliminary study of friction between ice and sled runners. J. Phys. (Paris) Colloq. **48**(C3), 297–301 (1987)
16. L. Poirier et al., Experimental analysis of ice friction in the sport of bobsleigh. Sports Eng. **14**(2–4), 67–72 (2011)
17. F. Braghin et al., Experimental assessment of Bobsleigh dynamics and ice-skate contact forces. Top. Modal Anal. II **6**, 487–498 (2012)
18. P.F. Vint et al., Kinetic analysis of the luge start technique. J. Biomech. **27**(5), 620 (1994)
19. P. Stergiou, L. Katz, Performance analysis of the pull start in luge: Integrated and real-time use of video and force measurement technologies, Technical report (2010)
20. V. Fedotova, V. Pilipiv, Comparison of lugers' start elements on a sliding track and an iced start ramp. Port. J. Sport Sci. **11**(Suppl. 2), 223–226 (2011)
21. H.P. Platzera, C. Raschnera, C. Pattersona, Performance-determining physiological factors in the luge start. J. Sports Sci. **27**, 221–226 (2009)
22. S. Lembert, O. Schanchner, C. Raschner, Development of a measurement and feedback training tool for the arm strokes of high-performance luge athletes. J. Sports Sci. **29**, 1593–1601 (2011)
23. J. Roche, S. Turnock, S. Wright, An analysis of the interaction between slider physique and descent time for the bob skeleton. Eng. Sports **7**(2), 101–109 (2008)
24. Y.L. Zhang, M. Hubbard, R.K. Huffman, Optimum control of bobsled steering. J. Optim. Theory Appl **85**(1), 1–19 (1995)
25. A. Kelly, M. Hubbard, Design and construction of a bobsled driver training simulator. Sports Eng. 3.1 (2000), pp. 13–24.
26. M. Hubbard et al., Simulation of vehicle and track performance in the bobsled. Am. Soc. Mech. Eng. Appl. Mech. Div. **98**, 373–376 (1989)
27. M. Hubbard, M. Kallay, P. Rowhani, Three-dimensional bobsled turning dynamics. Int. J. Sport Biomech. **5**, 222–237 (1989)

28. F. Braghin et al., Bobsleigh performance optimization through a multi-body model, in *Multibody Dynamics 2009 ECCOMAS Thematic Conference, Warsav, Poland, 29 June–2 July*, Institute of Aeronautics and applied Mechanics – Warsaw University of Technology (2009), pp. 1–12
29. F. Braghin et al., Race driver model. Comput. Struct. **89**, 1503–1516 (2008)
30. O. Lewis, Aerodynamic analysis of a 2-man bobsleigh. MA thesis, TU Delft, 2006
31. F.M. White, *Viscous Fluid Flow* (McGraw-Hill, New York, 1991)
32. S.F. Hoerner, *Fluid-Dynamic Drag*, Hoerner Fluid Dynamics (Backersfield, 1965)
33. F. Motallebi, P. Dabnichki, D. Luck, Advanced bobsleigh design. Part 2: Aerodynamic modifications to a two-man bobsleigh. Proc. Inst. Mech. Eng. B J. Mater. Des. Appl. **218**(2), 139–144 (2004)
34. A. Winkler, A. Pernpeintner, Improving the performance of a bobsleigh by aerodynamic optimization. Eng. Sport **7**(2), 329–338 (2008)
35. P. Dabnichki, E. Avital, Influence of the postion of crew members on aerodynamics performance of two-man bobsleigh. J. Biomech. **39**, 2733–2742 (2006)
36. A. Winkler, A. Pernpeintner, Automated aerodynamic optimization of the position and posture of a bobsleigh crew. Proc. Eng. **2**(2), 2399–2405 (2010)
37. F.R. Menter, Improved two-equation k-omega turbulence models for aerodynamic flows, NASA Technical Memorandum 103975, 1992
38. E. Berton et al., Aerodynamic optimization of a bobsleigh configuration. Int. J. Appl. Sports Sci. **16**, 1–13 (2004)
39. R. Hastings, The use of computational fluid dynamics to investigate and improve the aerodynamics of bob skeleton racing, University of Edinburgh, 2008
40. C.R. Walpert, R.A. Kyle, Aerodynamics of the human body in sports. J. Biomech, **22**(1), 1096 (1989)
41. K. Bromley, Factors affecting performance of skeleton bobsled. Ph.D. thesis, Mechanical Engineering, University of Nottingham, 1999
42. G.R.L. Dempster, W.T. Gaughran, Properties of body segments based on size and weight. Amer. J. Anatomy **120**, 33–54 (1969)
43. L.W. Brownlie, Aerodynamic characteristics of sports apparel. Ph.D. thesis, Simon Fraser University, 1992
44. Aerodynamic testing methodology for sports garments (2008)
45. H. Chowdhury et al., Design and methodology for evaluating aerodynamic characteristics of sports textiles. Sports Technol. **2**(3–4), 81–86 (2009)
46. H. Chowdhury, F. Alam, A. Subic, Aerodynamic performance evaluation of sports textile. Procedia Eng. **2**(2), 2517–2522 (2010)
47. M. Hubbard, Recreating the cresta run. Phys. World, Feb 21–22. International Olympics Committee "Salt Lake City 2002 Olympic Winter Games Global Television Report" (2002)
48. Wikipedia, List of bobsleigh, luge, and skeleton tracks -Wikipedia, The Free Encyclopedia (2013)
49. F. Braghin et al., Design and verification of bobsleigh track, in *ASME 2010 10th Biennial Conference on Engineering Systems Design and Analysis, ESDA2010*, vol. 4 (2010), pp. 505–512
50. M. Hubbard, Luge track safety. Sports Eng. **16**(2), 123–135 (2013)
51. Sait Polytechnic, Whistler Sliding Center -Sled Trajectory and Track Construction Study, Technical report (2010)
52. International Luge Federation, Official report to the International Olympic Committee on the accident of Georgian athlete, Nodar Kumaritashvili, at the Whistler Sliding Center, Canada, on February 12, 2010, during official luge training for the XXI Olympic Winter Games. Technical report (2010)
53. T. Pawlowski, Coroner's report into the death of Kumaritashvili, Nodar. British Columbia Ministry of Public Safety and Solicitor General. Case No: 2010-0269-0002. Technical report (2010)
54. F. Braghin et al., A driver model of a two-man bobsleigh. Sports Eng. **13**(4), 181–193 (2011)

55. F. Braghin et al., Bobsleigh driver model, in *Proceedings of the Mini Conference on Vehicle System Dynamics, Identification and Anomalies*, 2008, pp. 673–679
56. F. Braghin et al., Multi-body model of a bobsleigh, in *Proceedings of the Mini Conference on Vehicle System Dynamics, Identification and Anoma lies*, 2008, pp. 661–672
57. R.M. Levy, L. Katz, Virtual reality simulation: bobsled and luge, in *IACSS International Symposium Computer Science in Sport*, 2007
58. Virtools (2014). www.virtools.com
59. M. Hubbard, Simulating sensitive dynamic control of a bobsled. Simulation **65**(2), 147–151 (1995)
60. U. Holmlund, von H. Raimo, Model of sprinting based on the Newton's second law of motion and their comparison. Rakenteiden Mekaniikka **30**, 7–16 (1997)
61. M. Wacker et al., Design build and test of a bobsled simulator for olympic athletes. Trans. ASME **1**, 96–102 (2007)
62. F. Braghin, Cheli, M. Donzelli, S. Melzi, E. Sabbioni, Multi-body model of a bobsleigh: comparison with experimental data. J Multibody Syst. Dyn. 25(2), 185–201 (2011)
63. F. Braghin, F. Cheli, S. Melzi, E. Sabbioni, Development of a numerical model of a bobsled driver: trajectory planning, in *Proceedings of 12th Mini Conference on Vehicle System Dynamics, Identification and Anomalies*, 2010, pp. 501–510

Chapter 8
Ice Skating

Edoardo Belloni, Edoardo Sabbioni, and Stefano Melzi

What we are proposing in this chapter is an overview of the ice skating. We are now not considering the figurative ice skating, but only the ice speed-skating. Furthermore, only few hints will be provided about the outdoor skating, focusing then on the indoor competitions. After an historical reconstruction of the origins of this sport, the analysis will be divided into two main blocks: the dynamical model and the aerodynamic analysis of a speed skater (both numerical and experimental). Few notes about the track will be discussed along the chapter.

8.1 Introduction

8.1.1 Historical Notes

As many other sports (for example, the bobsleigh), the origins of this discipline are more related to real and concrete needs with respect to recreation. Dating back one millennium, the first ice skates were used by natives to travel on frozen rivers, canals, and lakes, especially in Northern Europe, Scandinavia, and the Netherlands: they were nothing more than shoes with some bones added on, used to ease the deambulation along these slippery lands [1]. Even if there are some

E. Belloni (✉)
Dipartimento di Meccanica, Politecnico di Milano, Via Giuseppe La Masa, 1, Milano, Italy
e-mail: edoardo.belloni@polimi.it

E. Sabbioni • S. Melzi
Department of Mechanical Engineering, Politecnico di Milano, Milano, Italy
e-mail: edoardo.sabbioni@polimi.it; stefano.melzi@polimi.it

F. Braghin et al. (eds.), *The Engineering Approach to Winter Sports*,
DOI 10.1007/978-1-4939-3020-3_8

previous testimonies,[1] only starting from the sixteenth century people really began to consider skating as fun, and this activity had some developments also outside the above mentioned lands: in 1592 a Scotsman designed a skate with an iron blade. The *Skate Club of Edinburgh*, born in 1642, was the first official skating club, mostly devoted to figure skating. There are some discrepancies with the real foundation date, and the most part of people accept 1742 or 1744 (see [3]), whereas the above-mentioned date is reported only one time in a small book published by the club council in 1865 [4].

The first ice speed skating competition dates from 1763 on the Fens in England and, in the same period, in Netherlands people set up competitions along the waterways connecting the 11 cities of Friesland. Later, during the half of 1800, also North America started to appreciate this activity succeeding in the construction of an all-steel blade, which was both lighter and sharper with respect to the others [1]. The Netherlands arranged the first world championship in 1889 [1] and, subsequently, the ISU (International Skating Union) was born in 1892 (see [5, 6]). Along the nineteenth century lot of competitions arose, with athletes from Northern Europe, United Kingdom, the Netherlands, North America, and also Russia. Furthermore, standard lengths for the competitions were established: 500, 1500, 5000, and 10,000 m.

After the inclusion of the figure skating at the Olympics Games of 1908, the delegates decided to include speed skating in the 1916 Olympics: World War I forced to suspend these Games, so the first official appearance was in 1924 at the Chamonix Olympics Games. First official competitions for women were organized by ISU in 1936 [1].

8.1.2 Rules

We are now referring to the official regulation of the speed skating [7], the most up to date at this moment was approved in 2012, and can be freely consulted on the ISU website [6].

The standardized distances for races are: 100, 500, 1000, 1500 m (called short distances), 3000, 5000, 10,000 m (called long distances). Sometimes also different distances are accepted. There are specific rules for each competition (that is, European, World,... championship), which fixes a set of races with different lengths.

About the equipment, racing suits must not change the natural shape of athlete body, no insertions or attachments are allowed (except for some permanently

---

[1]There is a written Latin text dated back to circa 1190 of William Fitzstephen (sub-deacon to Archbishop Thomas-a-Becket) which states [2]:

> *When the great fen or moor which watereth the walls of the City (London) on the Northside is frozen, many young men play upon the ice [..] Some tie bones to their feet, [..] and shoving themselves by a little picked staff, do slide as swiftly as a bird flyeth.*

attached stripes with strict rules on dimensions): this is made in order to avoid unfair improvements from an aerodynamic point of view. Nowadays a real step in performances for this discipline is gained with an improvement of the apparel textile: we will discuss it in Sect. 8.3.

An obvious problem arises: how to ensure safety and protection for athletes. In fact this lack of authorized equipment has the consequence to not prevent the skaters from injuries, particularly critic if one thinks that sometimes velocities can reach even more than 50 km/h. To protect the most delicate part of the body, that is the head, an helmet is allowed, but it has to respect specific shapes (the head's one) and rules: the reader may refer to [7], Rule 291, paragraph 1a. A pair of glasses is allowed in order to reduce the glare due to the light on the track, to improve the visibility and to protect eyes from ice chips. Further equipments like gloves, shinguard, neck and knee protections are allowed.

The skates are an important part of the equipment: they are a passive mechanical extension of the leg providing the contact of the athlete with the ice, meant to facilitate the gliding technique. As written in [7]: "The skate consists of a blade and a shoe of any material with dimensions which do not change significantly during a competition, and a construction which connects the shoe to the blade. This construction may allow any translation and rotation of the shoe relative to the blade, in order to enable an optimal utilization of the physical capacity of the Skater, as far as it does not jeopardize the safety of the Skaters." An important specification is that heating the blades by means of whichever solution is totally prohibited. The part of the blade which comes in contact with the ice forms a straight line, and the ending part of the blade must be rounded off, with a minimum radius of 10 mm. Generally, blades used by male skater are 42–46 mm long. The longer the blade, the faster the skate up to a point where length would become an obstacle. The underside of the blade is only about 1 mm thick [8].

A regular skate has two rigid connection points between the blade and the boot, but for ice skating, allowed only for long tracks, a different kind of skate can be used, called *clap skate*. In this pair of skates, the blade is attached to the boot by a hinge at the front (see [8, 9]): the blade briefly disconnects from the heel of the boot, thereby keeping the blade on the ice longer and increasing the skater's pushing power. Reaching the blade maximum extension, a spring mechanism (typically mounted on the front of the boot) snaps the blade back, causing the typical sound which gives name to these skates. First developments of this technology are at the end of nineteenth century, with the first patent in 1894 by Karl Hannes. This model was totally re-invented by Gerrit Jan van Ingen Schenau in the 1980s (the reader can see the patent [10]), and it has experienced further changes during recent years (a more modern patent is, for example, [11]). The history of the use of this kind of skates has experienced several phases, but nowadays the use of them is quite common and an effective increase in final result is demonstrated. Research completed in 2001 showed that the speed gain from using the clap skate does not originate in using the calf muscle to stretch the ankle, as was assumed in the creation of the clap skate, but in the fact that the point of rotation is moved from the tip of the skate to the hinge, facilitating the transfer of power to the ice (see [9]).

8.1.3 Track

The construction of a skating arena is not an easy task, and there is a long history behind on construction techniques to reach the actual knowledge and regulation.[2] The related bibliography is quite vast, and a lot of "side aspects" of the construction are also analyzed: examples of rink building [13–17], more specifically on the external structure [18–21], and finally some focus on the refrigerating system [22–26].

A standard (see [7]) speed skating track is an open, covered, or enclosed ice rink with a double-laned competition track. The length of a circuit must be enclosed between a minimum of 333 1/3 and 400 m (which is the official length for World Cup and Olympic Games), with two curves of 180° and a curvature radius included between 25 and 26 m. The length of the track is rigorously measured by a qualified surveyor, with details about the position of all starting and finishing lines. Starting and finishing points must be denoted by colored lines: a pre-start line is placed 2 m before the starting lines and the last 5 m before the end must be clearly marked.

A strict regulation on safety measures is stated by the Rules [7]: before each competition, the organizers must provide a technical report of all planned or still existing safety installations, including relevant data obtained from tests.

The protection mats (padding) must have rectangular or trapezoidal shape and are composed by a superposition of layer made with different densities, to ensure a high level of safety for the athletes, that must be proven through scientific tests. The protection minimum thick is at least 60 cm, with a minimum height of 100 cm (and a maximum one of 120 cm to allow an adequate viewing angle for spectators) and must be placed all along the track.

Also the ice condition is subordinate to a strict analysis and continually monitored. "The ice technical expert of the ice rink must submit to the ISU Ice Commission all information that is relevant to decide the ice preparation schedule and to monitor the ice and racing conditions at all stages during the competition. [...] On in-door tracks the ice technical expert has to provide information also on wind or air-flow conditions at any time during racing, and is responsible that the ventilation system of the track is controlled so as to avoid unequal conditions due to differences in air-flow on the level of the competition track" [7].

8.2 Dynamic Model

8.2.1 Primary Observations

It is not an easy task to get a propulsion on an icy surface: by ordinary experience we will know the risks correlated with a slippery ground. So the way in which an

[2]An interested reader can refer to Martin [12].

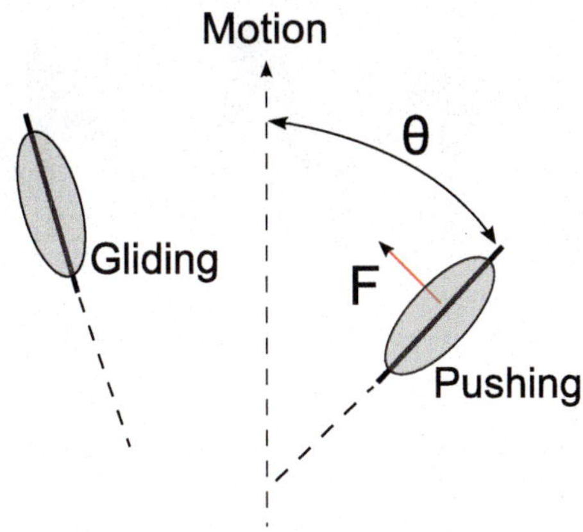

Fig. 8.1 The position of skates during the motion and the pushing force acting on the rear skate (adapted from [27])

ice skater can get a strong and "controlled" motion is something fascinating, which deserves to be at least some investigations.

The motion of an ice skater is basically composed by a gliding on ice and by the use of edges to push off the ice so that to obtain a gain in speed [27]. In order to get a propulsion, the skater cannot directly rely upon the friction between skate blades and ice, which is inherently close to zero, so the only way to propel himself on the ice is to tilt a bit the skate with respect to the forward direction of motion, generating a force perpendicular to the edge (see Fig. 8.1).

While the rear skate is pushing, the other one is gliding on ice (or is raised). The only component of F which can effectively propel the skater is projected on the direction of motion, depending on the skate angle of inclination θ: raising this angle, an athlete increases the useful part of the force, reaching higher accelerations [27]. During the motion, an athlete does not maintain the body upright, but bends/crouch in the direction of motion. An intuitive consequence is a benefit from the aerodynamic point of view: a reduction of the frontal area corresponds to a smaller coefficient of drag. Anyway, this body position is fundamental to maintain the body equilibrium: this prevents the athlete from falling backwards because of the torque generated by force forward component. Moving the center of mass (COM) a counter-torque provides balancing for the athlete [27]. Observing a skater, this particular way of pushing results in a kind of sinusoidal motion of the upper body on the ice (as shown in [28], see also Fig. 8.2).

Another expedient to increase performances is the use of clap skates (briefly described in Sect. 8.1.2), which allow to increase the pushing phase of the skater during stride and consequently the acceleration.

As one can notice, the strides of a skater are wider in the straight part of the track with respect to the turns, because it is easier to steer if the lateral distance between the strides is small [27]. The lean of skaters is remarkable in curve (especially in short tracks, where the radius of curvature is smaller), so they can't do much in curve and must rely on momentum obtained during the straight part.

To have a rough estimation of skater lean in curve, a simple force equilibrium can be used, as shown in [27].

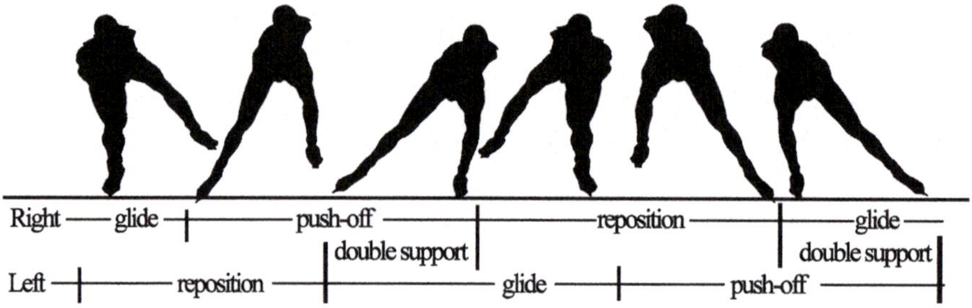

Fig. 8.2 The phases of skater motion (from [28])

Fig. 8.3 A simple scheme of forces acting on a skater during a turn (adapted from [27])

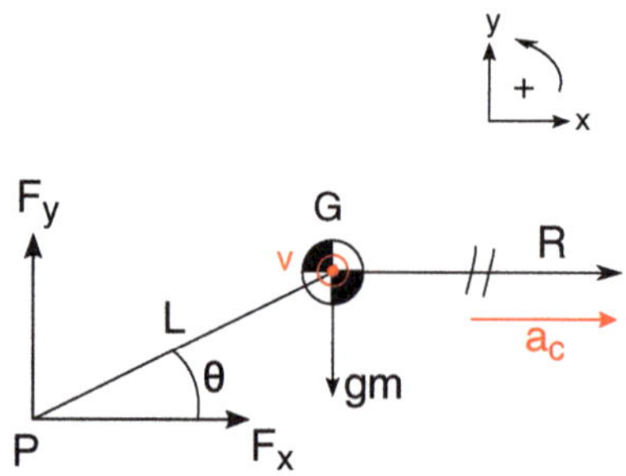

In Fig. 8.3 a scheme of the forces during a turn is represented. We indicate with P the contact point of the blade on ice, F_x and F_y the horizontal and vertical contact forces, respectively, at point P, G the athlete's center of mass (we consider the athlete a rigid body), L the distance between P and G. In Fig. 8.3 R is the radius of turn (measured from the center of the turn to G), a_c the centripetal acceleration (pointing the center of the turn), θ the lean angle, calculated between the ground and the line connecting P and G, v the velocity of point G, pointing out of the page.

Let's now compute the dynamic equilibrium of the body. In vertical direction we have:

$$F_y - mg = 0 \tag{8.1}$$

While in the horizontal one $\left(\text{with the centripetal acceleration equal to } a_c = \frac{v^2}{R}\right)$:

$$F_x = ma_c = m\frac{v^2}{R} \tag{8.2}$$

Considering θ constant, we are in a state of rotational equilibrium, which can be expressed as:

$$F_x \sin\theta \cdot L - F_y \cos\theta \cdot L = 0 \tag{8.3}$$

Using (8.1)–(8.3) we get:

$$\tan\theta = \frac{Rg}{v^2} \tag{8.4}$$

Fig. 8.4 The clap skate used in the 1998 Olympic Winter Games (from [36])

With a radius of $R = 8.5\,\text{m}$ and a velocity equal to $v = 10\,\text{m/s}$, the lean angle is $\theta = 39.8°$. Decreasing the radius of curvature, one can easily notice that the angle becomes smaller and smaller.

8.2.1.1 The Influence of Clap Skates

In Sect. 8.1.2 we have briefly described the clap skates (see Fig. 8.4), pointing out that the performances during the competitions are risen since they were adopted. In particular, as stated in [29], the maximum velocity grew by 4 %, which corresponds to an increase of about 10 % in power output. We have described the motion of a speed skater in Sect. 8.2.1, and the push-off phase of the motion is really constrained (despite the high velocities it is possible to reach, see [30]): the trunk in horizontal position (to decrease the influence of air friction) and the necessity of plantar flexion suppression (to perform a gliding push-off and to prevent the scratching of blade's tip with the ice) limit the motion [29]. The major effects are on the contribution of the ankle plantar flexor, which cannot supply the maximum power, and on an incomplete knee extension at the end of the push-off [31] (that is, the instant when the rear part of the blade disconnects from the ice, even if the blade tip is still on the ice).

A performance comparison between the two kinds of skate was managed by Houdijk et al. [29], involving a group of ten elite speed-skaters (from the Dutch national team). Each athlete performed eight 400-m laps, alternating conventional and clap skates. The kinematics of the body was reconstructed using a vision system based on two 16 mm high speed cameras and specific markers on the suit and the skate: it was possible to calculate the 3D position of the body using ad-hoc methods (see [32, 33]). As it was shown by De Koning et al. [34], the in-plane forward/backward force (y-direction) never exceeded the 10 N, that is almost 1 % of the vertical force (z-direction), so it was chosen to neglect the measurement of it (as also the sideward x-axis component). To measure the force in z-direction, for the traditional skate was used a classical instrumentation shown by Jobse et al. [35], for the clap skate a potentiometer fixed at the level of the hinge was used to register the angular displacement between the skate and the blade (see [29]).

It was immediately possible to notice the rising of average velocity: the net increase was close to ~0.056 % using clap skates, passing from 11.23 to 11.86 m/s and resulting in a lap time equal to 33.7 s and 35.6 s, respectively. Comparing the stroke frequency one gets 1.36 versus 1.30 strokes/s. To go deeper in the analysis, one can refer to [29] which characterizes completely the athlete and its motion, we can now briefly sum them up:

- with both skates (conventional and clap skates) the joint extension and the duration of the push off phase seem to be identical;
- with conventional skates, the knee extension is more complete;
- conventional skates imply higher peak angular velocities of knee and hip joint;
- the mean power output grows up of ~12 % switching from conventional (209 W) to clap skates (234 W);
- through the use of electromyography (attaching pairs of electrodes to the skin of the most relevant muscles), there were found no considerable differences in timing and amplitude of muscle activation between the two configurations.

So finally the increased skating velocity is directly connected with the boost in mechanical power output. Referring to the gain of 25 W, one can attribute around 15 W to the increase of work per stroke (see also [37]) and the other 10 W to the higher stroke frequency[3] (see [29, 36]).

8.2.2 A Two Dimensional Model

In Sect. 8.2.1 we have used a schematic 1 dof model to deal with the dynamic of a skater. In literature several models have been developed: the reader can see [39–41]. Each model has different characteristics and weak points (as recapped in [28]):

- De Koning and van Ingen Schenau model [40] was one of the first ever developed. It analyzed, through a sensitivity approach based on a power balance, the influence of human and environment parameters on performances. A limit of this model is its difficult validation and further it is not robust to the differences in coordination pattern (which is typical of each athlete).
- Otten [41] created a model merging forward and inverse kinematics, using up to 19 rigid bodies and 160 muscles. It is critical because of its manually tuning of the kinematics and the lack of measurements for the validation, but it can potentially simulate skating and provide force/moments in the joints.

[3]This analysis has considered only the skating mechanics in the plane through the ankle, knee, and hip. An exorotation or abduction of the hip can provide some additional work outside this plane (see [29, 38]).

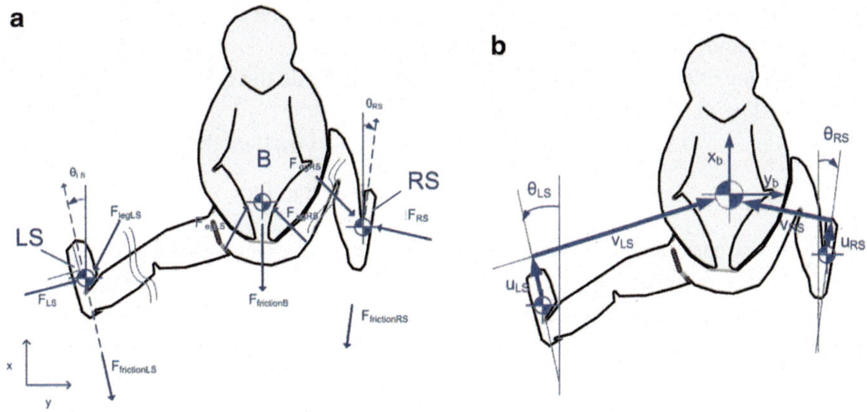

Fig. 8.5 (**a**) Schematic representation of a skater, underlining the most important parameters like COM of body and skates, forces and skate lean angles. The inertial reference system is fixed to the ice rink. (**b**) Velocity vectorial representation. Adapted from [28]

- Allinger and Bogert [39] presented a model based on inverse dynamics, driven by individual strokes. Even if the model is accurate and useful for coordination pattern optimization, it is strongly dependent on a presumed leg function in time and again the validation using real measurements is missing.

We are now presenting a model developed by Fintelman, Braver, and Schwab (the reader refers to [28] for a complete dissertation and results): a two-dimensional inverse dynamic model, which uses three lumped masses and measures the two-dimensional in-plane positions (*x*,*y*) of the skates and the upper body.

Referring to Fig. 8.5, it is possible to write the equations of motion in *x* and *y* direction in a general fashion, considering friction forces and constraint:

$$\begin{aligned} m\ddot{x}_i &= -F_{\text{friction}X_i} + F_{\text{constraints}X_i} \\ m\ddot{y}_i &= -F_{\text{friction}Y_i} + F_{\text{constraints}Y_i} \end{aligned} \tag{8.5}$$

where subscript i can indicate: B, body of the skater, LS, left skate, or RS, right skate. The configuration of the skater is a combination of the upper part motion plus the leg extensions, and can be represented by the following generalized coordinates (see Fig. 8.5b):

$$\mathbf{q} = (x_b, y_b, u_{\text{LS}}, v_{\text{LS}}, \theta_{\text{LS}}, u_{\text{RS}}, v_{\text{RS}}, \theta_{\text{RS}})^T$$

As stated at the beginning of this description, the mass is considered distributed in three positions along the body: one on the COM of the upper part of the body and the other two on the skates. A coefficient α is introduced to represent the mass distribution along these three points. If m is the total mass of the skater, one can define $m_B = (1-\alpha)m$, $m_{\text{LS}} = (\alpha/2)m$ and $m_{\text{RS}} = (\alpha/2)m$ (see [28]).

The constraints are both the leg extensions (which are a connection between the upper body and the skate) and the skates on the ice.

For the leg extension constraints, one can easily write the relations:

$$\begin{aligned} c_1 &= x_{\mathrm{LS}} - x_B + \cos(\theta_{\mathrm{LS}})u_{\mathrm{LS}} + \sin(\theta_{\mathrm{LS}})v_{\mathrm{LS}} = 0 \\ c_2 &= y_{\mathrm{LS}} - y_B - \sin(\theta_{\mathrm{LS}})u_{\mathrm{LS}} + \cos(\theta_{\mathrm{LS}})v_{\mathrm{LS}} = 0 \end{aligned} \tag{8.6}$$

These two equations are referring to the left side, it is exactly symmetrical for the right side (getting c_3 and c_4).

The equations of constraints due to the skates must represent the condition of zero lateral velocity of the skate: this can be expressed by a non-holonomic constraint[4]:

$$\begin{aligned} c_5 &= -\sin(\theta_{\mathrm{LS}})\dot{x}_{\mathrm{LS}} - \cos(\theta_{\mathrm{LS}})\dot{y}_{\mathrm{LS}} = 0 \\ c_6 &= -\sin(\theta_{\mathrm{RS}})\dot{x}_{\mathrm{RS}} + \cos(\theta_{\mathrm{RS}})\dot{y}_{\mathrm{RS}} = 0 \end{aligned} \tag{8.7}$$

for the left and right sides, respectively. The vertical motions are not considered, and the contact condition is easily expressed by an on/off switching of Eq. (8.7).

It remains to define the friction forces acting on the skater, specifically the friction between skate and ice (which is roughly only the 20 % of the total, see [32]) and the air friction.

For the first one, it is possible to express the relation using a Coulomb's friction law (the reader refers to [34]):

$$F_{\mathrm{ice}} = \mu F_N = \mu m g \tag{8.8}$$

where F_N is the skate normal force on ice (here the height of the skater is considered constant, see also [43], and no double stance phases are present, please compare with Fig. 8.2).

The air friction is calculated with the classical formula for the drag force:

$$F_{\mathrm{air}} = \frac{1}{2} A C_d \rho v^2 = k_1 v^2 \tag{8.9}$$

which can be distributed along the three masses multiplying the mass distribution coefficient by the total air drag.

Merging together the above relations, it is possible to write the dynamical model (the reader can refer to [28] for the complete formulation). The model is then validated for four different skaters through experimental tests. The global positions are measured by a radio frequency system, with the transponders placed approximately at the positions of the point masses. A dedicated instrumented skate

[4]A non-holonomic constraint states an algebraic relationship in differential, non-integrable form, usually expressed in form of the time derivative of q: $\Psi(q, \dot{q}, t) = 0$, generally written in the form $A(q, t)\dot{q} = b(q, t)$. See [42].

was used to measure the normal and lateral forces, and to obtain the lean angle (fundamental to make a comparison with the output forces of the model).

It was found a good matching of the model with measured data. The model has less accuracy in forward velocity, $\dot{x}_B$, with respect to sideward velocity $\dot{y}_B$: quantization errors have more influence on forward direction, where the forces are less. Accuracy of lean angle sensor introduces a local normal force error included between ±20 N. Other inaccuracy in left and right force measurements is due to the double stance phase which is neglected in the model (but the sum of these forces corresponds well to the predicted one).

8.2.3 Power Balance Model

To predict average skating performances it is possible to use a power balance model, as demonstrated in [44–47]. De Koning and other researches worked in this directions for some years, and a good model was presented in [48]. In general, it is possible to write a power balance model which considers production and dissipation:

$$P_o = P_f + \frac{dE_{\text{mcb}}}{dt} \tag{8.10}$$

where P_o is the general power output of the skater, P_f is the average power loss due to air and ice friction, and dE_{mcb}/dt represents the change of kinetic, rotational, and potential energy of the body [48]. This change in mechanical energy is determined by the rate of change of the kinetic energy of COM:

$$\frac{dE_{\text{mcb}}}{dt} = \frac{d(1/2mv^2)}{dt} = mv(dv/dt) \tag{8.11}$$

where m is the body mass and v is the average speed.

Seeking a good precision model, it is necessary to have expressions for P_o and P_f. It is possible to obtain an estimation of them indirectly. The aerobic power (P_{aer}) is related to measurements of oxygen uptake during breathing ($\dot{V}_{\text{O}_2}$), modelled as (see [44, 46, 49]):

$$\dot{V}_{\text{O}_2} = \dot{V}_{\text{O2max}} \left(1 - e^{-\lambda(t-t_d)}\right) \tag{8.12}$$

with λ rate constant and t_d time delay, and to measurements of efficiency.

In [44] the anaerobic power is modelled as a first order system:

$$P_{\text{an}} = P_{\text{an-con}} + P_{\text{an-max}}\, e^{-\Gamma t} \tag{8.13}$$

where $P_{\text{an-con}}$ is the mechanical P_{an} at the end of the race, $P_{\text{an-max}}$ is the maximal mechanical P_{an} at $t = 0$ minus the value of $P_{\text{an-con}}$, and Γ is a rate constant (see [44]). In general this parameter is quite critical, and to obtain this formulation the athletes were forced to adopt an "all-out" approach during the lab tests, while other studies (see [50]) allow athletes to adopt a freely chosen pattern of energy distribution. Further it is important to remind that the values of P_{an}, as also the ones of P_{aer}, are estimated in literature for cycling, so it is not straightforward they are also really meaningful for skating.

Finally the power output can be calculated as:

$$P_{\text{an}} = P_o - P_{\text{aer}} \tag{8.14}$$

The power dissipated is basically a sum of effects due to ice and air frictions, where the friction with air is the most relevant (see also Sect. 8.2.2). One can use Eq. (8.9) for the drag force estimation, but studies in wind tunnel have underlined the strong influence of body's position [46, 51], so a different form for the air friction was estimated (see [48, 52]):

$$F_{\text{air}} = 0.02051\text{m}^{1/3}\rho_0 e^{-0.000125h} F(\theta_1)G(\theta_0)H(v)v^2 = kv^2 \tag{8.15}$$

where l is the body height (m), m is the body mass (kg), ρ_0 is air density at sea level (kg/m^3), h is the altitude above sea level (in m), $F(\theta_1)$ and $G(\theta_0)$ are expressions that account for trunk position (θ_1) and knee angle (θ_0), respectively, $H(v)$ is the influence of the velocity on the drag coefficient, and k is the resulting air friction coefficient (kg/m) [48].

It is possible to say that a power balance method can predict well the final time in speed skating competitions [44, 45, 47], but it is not so efficient for the momentary velocity profile estimation.[5] So, at least at this step of model development, it is not relevant for individual athletes rather than for a group behavior prediction.

8.3 Aerodynamics of an Ice Speed Skater

An important aspect of the dynamic in ice speed skating is the aerodynamic interaction of the skater with air. It is already explained in Sect. 8.2 how the aerodynamic contribution enters in the equation of motion: the goal of this section is to explain in a more specific fashion the most relevant factors contributing to the drag force on an athlete.

[5]One of the limitations of this model was to assume a constant body position of the athlete, but the combination of trunk and knee angle during a run involves an increase in k, the air friction coefficient. Further the fatigue of the athlete and its consequences are not considered.

8.3.1 Wind Tunnel Tests

A classical way to get experimental data in the field of aerodynamics passes through some tests in a wind tunnel and also in the case of ice skating it is one of the leading strategies for this kind of measurements. It is important to get results as close as possible to the ones observable during a real run, that is a test model as close as possible to reality is essential. This implies a great similitude both for shapes/dimensions and for environment conditions, apart from other ways of drag reduction using dedicated fabrics (it will be discussed later).

D'Auteuil has developed a 1:1 scale model of a skater (see [53–55]): it respects the body and cross sections dimensions of a typical elite athlete (see Fig. 8.6).

In [53], the mannequin was instrumented with 124 surface pressure taps (allowing local measurements of pressures and, by integration, of forces acting on relevant body positions, see Fig. 8.6) and mounted on a turntable to simulate a rotation, in the range of $\pm 20°$, of the body and measure the consequent drag coefficient variation (see Fig. 8.7).

A set of different tests was made in the wind tunnel: in static configuration, the mannequin was positioned with different yaw angles with respect to the air flow direction, and seven different wind speeds (from 30 to 60 km/h) were simulated.

Fig. 8.6 On the *left*, the mannequin in sidepush position inside the wind tunnel. In the *middle*, the same model with the locations of pressure taps (the *thin circular* "*lines*" one can notice on the mannequin). On the *right*, cross sections on the right side of mannequin [53]

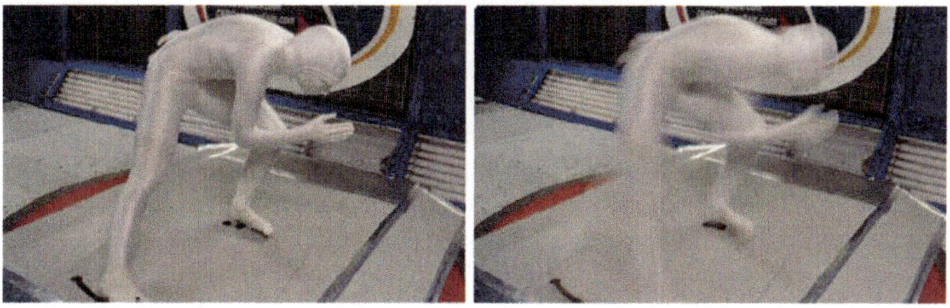

Fig. 8.7 Mannequin in static position at 0° yaw (*left*) and rotating thanks to a turntable from $\pm 20°$ (*right*) [53]

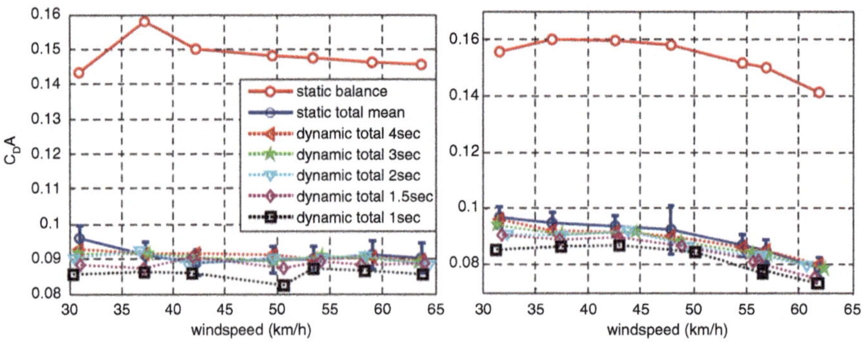

Fig. 8.8 The variation of drag area coefficient for static configuration and six different period of rotations, varying the wind speed. On the *left* with a bare mannequin, on the *right* with a smooth fabric suit [53]

Data were then collected for the model oscillating at different periods for the same range of wind speed variation. Some results of the drag coefficient for the whole body are shown in Fig. 8.8 (in [53] the reader can find the result on each instrumented cross section).

In Fig. 8.8 it can be seen that, while the wind speed is increasing, a reduction of the drag coefficient is present and comparable both for static and dynamic tests. With a bare mannequin in laminar flow this effect is less obvious: a possible explanation can be that the Reynolds number is in the subcritical range for the most part of the body, presenting smaller coefficient variation.

Together with other observations (see [53]), it was possible to state that the reduction of drag coefficient can be done using a static configuration of the mannequin. Further it is possible to do the hypothesis of quasi-steady aerodynamics for speed skating. In the sidepush position, the limit of this hypothesis is given by a rotation period which is lower than 1.5 s (or with a frequency higher than 0.67 Hz), where the trend of drag coefficient is not coincident between static and moving configurations [53].

Standing the fact that a static mannequin is a good model for wind tunnel tests, one can identify the most important parameters for the drag reduction.

Dimensions and proportions of the whole body are a fundamental starting point, that can condition the whole analysis. This contribution is explained well in [54] where, before using the same mannequin described above (see Fig. 8.6), the simplified approximation of cylindrical shapes was investigated (see also [56]). For any case-study, the results did not match properly the 3D flow and its interactions with the limbs of the body. Also the relative position of limbs between themselves exposes the body to different upcoming flow conditions.

A dedicated paragraph (Sect. 8.3.2) will be devoted to the use of technical fabrics and their influence in drag coefficient modification. In Fig. 8.9 an example of how the effect of textile roughness changing influences significantly the pressure distributions along the body.

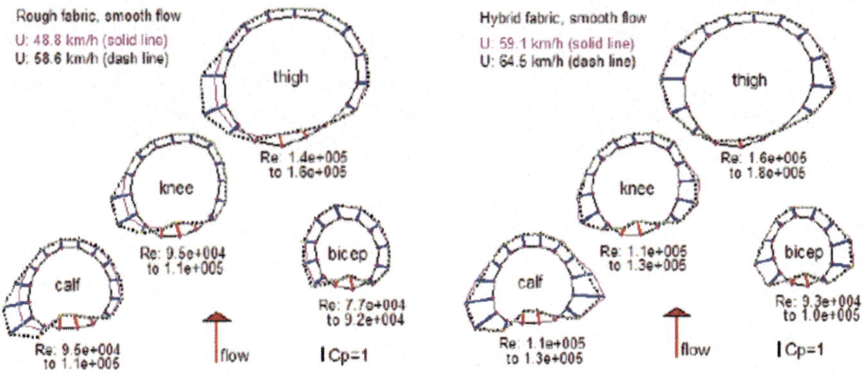

Fig. 8.9 *Left*: Distribution of surface pressure coefficient for four different cross sections for a rough fabric in smooth flow at 48.8 km/h (*solid magenta line*) and 58.6 km/h (*dotted black line*). *Right*: Distribution of surface pressure coefficient for a hybrid skinsuit with smooth/rough fabrics, in smooth flow at 59.1 km/h (*solid magenta line*) and 64.5 km/h (*dotted black line*). From [54]

Another important parameter for the drag area coefficient is the free-stream turbulence [54]. The effects are strongly dependent on the suit fabric and flow regime: for smooth fabric the minimum of the drag coefficient is reached at 60 km/h with a free-stream turbulent flow, corresponding to the end of the 2-bubble regime (TrBL2)[6] on the right leg and bicep. These observations are not still valid with a laminar flow and the same fabric, because in this case the same parts are at the beginning of TrBL1 phase, reaching its minimum value for higher speeds. Similar considerations can be done with a rough fabric (the reader can refer to [54]). Figure 8.10 summarizes the test done by D'Auteuil.

The wind turbulence is characterized by wind speed fluctuations due to mechanical and convective sources: the motion of the athlete generates the main wind flow. Other sources are the environment conditions, obstacles, people around the oval and also the convective turbulence due to temperature gradient on the ice surface. It is not easy to simulate the same conditions in a wind tunnel, and further the simulation must cover a frequency range as wide as possible because it is not straightforward to identify which part of the spectrum affects the aerodynamics the most. In [58] some measurements are done to evaluate a good experimental setup for a wind tunnel simulation (see also Fig. 8.11). In [58], D'Auteuil develops a more complete analysis of local pressures on the mannequin, considering three different (and most relevant) positions of the athlete during the run, going deeply inside the considerations explained few lines above.

[6]It is a regime with a transition in boundary layer. The TrBL regime has a lower Reynolds number bound of 100000–200000 and an upper bound of about three to five million [57]. In Zdravkovich studies, like [56], this flow regime is further sub-divided, and the TrBL2 condition presents two laminar bubbles, while TrBL1 has a laminar bubble on one side of the cylinder and TrBL3 presents a spanwise disruption of bubbles.

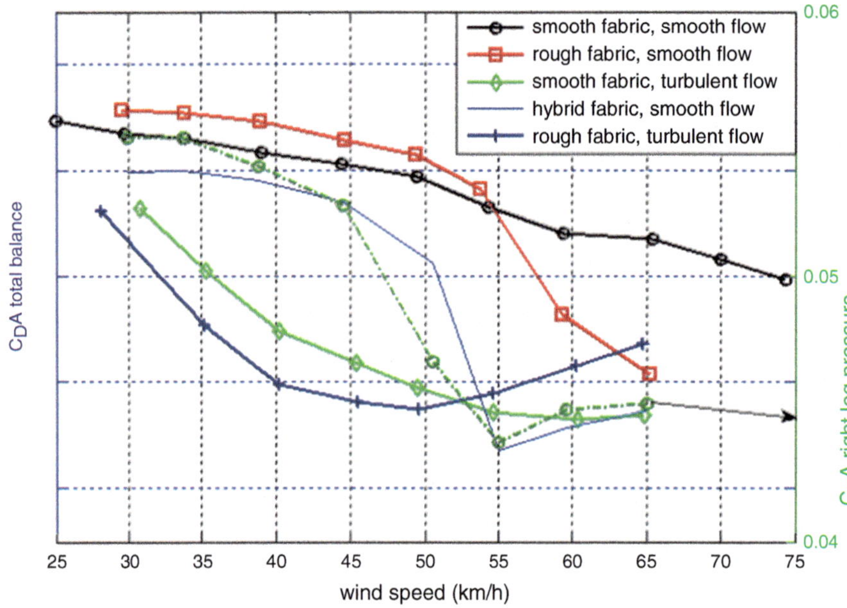

Fig. 8.10 Drag area coefficient measured with the external balance for five different conditions and from integration of surface pressure (*dash line*) of the right leg for the hybrid fabric, smooth flow condition. Adapted from [54]

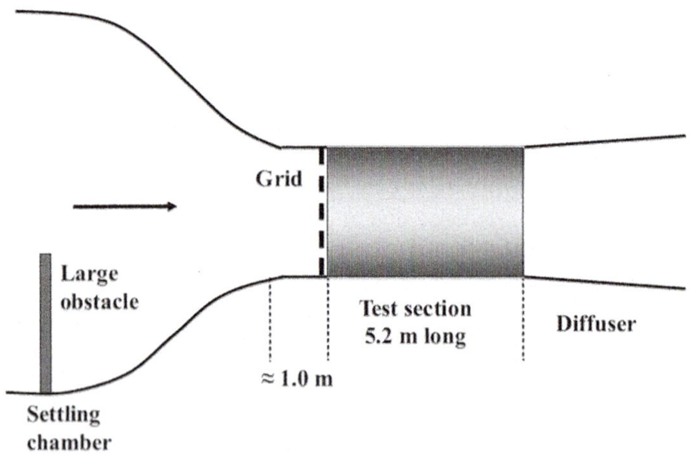

Fig. 8.11 Schematic sideview of a wind tunnel, with the objects used to create the desired turbulence. From [58]

8.3.2 Suits and Fabrics

Generally, in competitions the top athletes have really small differences in performances, which can be influenced by the equipment and the apparel used by the competitors [59]. Numerous clothing developers and sponsors work hard to improve the quality of suits, in particular during Olympic Winter Games the most relevant and newest technologies are presented (see also [60]). To directly have a "weight" of the fabric influence on performances is not easy: it is hard to perform a complete simulation of all the environmental conditions to really test only the effects of the sport apparel. It is indirectly possible to see the impact of this kind of technology looking more in general to the evolution of results in competitions before and after its introduction [61].

Nike, Descente and Hunter, during the Winter Sports of 2002 in Salt Lake City, presented their aerodynamically optimized suits, respectively SWIFTSkin, Vortex C2, and Delta-Flash. In later years, till 2005, all the manufactures started to incorporate SWIFTSkin features (rough textured fabrics for arms and legs, smooth polyurethane on torso and thighs, low friction panels in groin region and gloves with polyurethane coated stretch fabric). The reader can take a look at Fig. 8.12 to figure out the fabric performances for different kinds of material and technology.

Mathematical modeling suggested that the original SWIFTSkin can reduce the race time of approximately 1 %, a result which is not achievable by other brands [61].

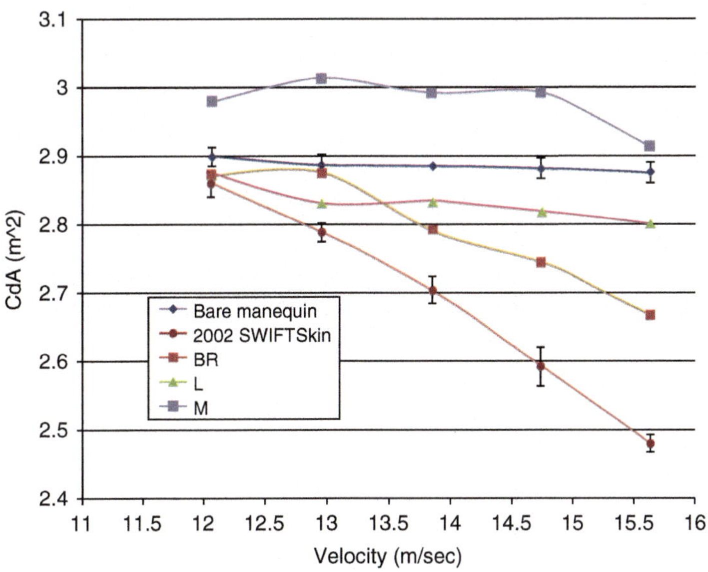

Fig. 8.12 Variation of the drag area coefficient changing the fabric of the suit inside a range of wind velocity. *BR*, *L*, and *M* refer to other suits. Adapted form [61]

8.4 Sliding on Ice

Some water on the floor or a large amount of rain can create dangerous situations for people and cars, respectively: a mobile layer of liquid on a rigid surface makes the surface slippery. Ice slipperiness is then roughly close to ask how a liquid or liquid-like layer can occur on the ice surface in the first place [62]: a lubricating layer of meltwater separates ice and slider surfaces also in case of ice skating. In particular during ice speed skating there is also a friction generation from ploughing a groove in the ice and from the arising of shear stresses in the lubricating Couette-like flow [63]. The mechanism of ice friction and slipperiness has been studied for years, at least starting from the half of the nineteenth century. A precise historical reconstruction is above the scopes of this text: an interested reader can refer, for example, to [34, 62–64].

8.4.1 Fundamental Aspects

Two main theories about ice slippery one can found in literature are the ice melting under pressure and the frictional heating of ice (see, for example, [62]).

Ice melts under the pressure due to the weight of the skater: water has a higher density with respect to ice (about 10 % more), so according to Le Châtelier's principle a pressure rising corresponds to ice melting and sample's volume decreasing. Joly in 1866 was probably the first who correlated the high pressure to the small area of skater's blade edge: he calculated a pressure of 466 atmospheres with a melting point of −3.5 °C, temperature that creates a film of water on which the skater can slide. In [65], the Clausius-Clapeyron equation was used to point out the decrease of the melting point (dT) for a given pressure drop (dp):

$$\frac{dT}{dp} = -\frac{\Delta V}{\Delta S} = -0.0074\,°\mathrm{C}/10^5\,\mathrm{Nm}^{-2} \tag{8.16}$$

where ΔV is the ice-water volume change and ΔS is the ice-water change in entropy. As an example, a 75 kg mass skater with a contact area of 0.5 cm^2 the pressure is 150×10^5 Nm^{-2}: this means, using Eq. (8.16), that the temperature must be not lower than −1.1 °C to form water (see also [34]).

Both the two results above do not explain how is it possible to skate with lower temperatures; further, optimum temperatures for figure skating is −5.5 °C (a "soft" ice, useful for landings) and for hockey −9 °C (an harder, faster ice), whereas skating is also possible in climates as cold as −30 °C (see [62]). Bowden and Hughes in 1939 (see [66]) were the first to show the role of frictional heating on the melting of ice. They measured both static and kinetic friction using wood and metal surfaces: metal skis showed higher friction, and frictional heating was affected by the conductivity of skis. Pressure melting seemed to play a role only near the melting point [62]. Bowden in 1953 (see [67]) suggested that friction generates enough heat to melt some of the ice, and pressure-melting of ice has a role only at temperatures very close to 0 °C and with slow moving surfaces (like glaciers). The formula:

$$P = \mu N v \tag{8.17}$$

characterizes the rate of heat supplied to the rubbing surfaces, where μ represents the coefficient of friction, N is the normal force (that is, the weight of the skater), and v is the sliding velocity (see [34]). Using an ice friction coefficient of 0.005, a normal force of 750 N and a velocity of 10 m s^{-1} the energy flow is equal to 37.5 W: the heat of this plane source is conducted away into the two rubbing bodies (skate and ice, see [67]). One can use the heat given by Eq. (8.17) in the formula presented by Furey in 1971 (see [68]) to get the increase of temperature dT (it was derived for a circular contact area moving on a flat surface with velocity v, and all the heat P is considered flowing into the ice):

$$dT = \frac{0.31P}{ka}\sqrt{\frac{k}{\text{reva}}} \tag{8.18}$$

where $a = 4$ mm is the radius of the circular surface (this value refers to [34], the typical contact area of a skate blade in those years), $k = 2.1$ W m^{-1} °C^{-1} is the thermal conductivity of ice, $r = 2200$ J kg^{-1} °C^{-1} is the specific heat, $c =$ 917 kg m^{-3} and $v = 10$ m s^{-1}. With these values the resulting dT was equal to 7 °C (it might be lower if not all the heat is assumed to be transmitted to the ice, but to the skate too, see [34]).

Another mechanism, that can influence the ice slippery, and can also clarify it even while one is standing still on it, are the liquid-like surface properties of ice. Already in 1859 Faraday proposed the existence of a liquid-like layer on ice surface, and further this layer exists also without the contact of another body, so no friction is required (see [64]). In [69] a synthesis of different experiments on liquid-like ice surface is displayed: in Fig. 8.13 a comparison between different technique is performed, nature and sensitivity of techniques are leading factors to explain the differences. This large scatter of results has brought to different explanations of this phenomenon both for the general nature of the layer and the onset temperature of its formation. Review in the literature is provided by Hobbs [70] and Petrenko and Whitworth [69].

As briefly summed up by Kietzig et al. [64], Fletcher (see [71, 72]) attributed the formation of the liquid-like layer to electrostatic interactions, whereas a minimization of the energy was the explanation given by Lacmann and Stranski (see [73]), and further carried on by Dash [74]. Starting from the idea that the free surface energy of a solid–vapor interface γ_{sv} is higher than solid–liquid (γ_{sl}) and liquid–vapor (γ_{lv}) interface energies, so the system will "lower its free energy by converting a layer of the solid to liquid" [64]:

$$\gamma_{sv} > \gamma_{sl} + \gamma_{lv} \tag{8.19}$$

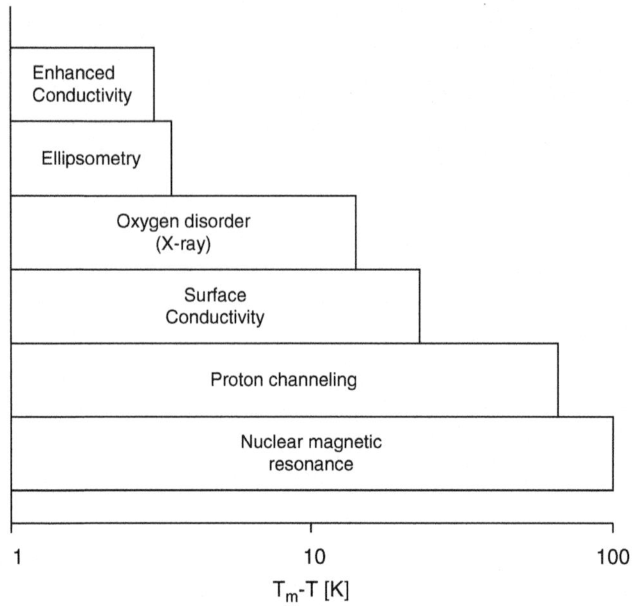

Fig. 8.13 Different techniques show the special characteristics of ice surface with different temperature ranges below the melting point T_m, logarithmic scale (see [64])

It was proved (see, for example, [75]) that this is not strictly valid for "dry" ice [64]. Fukuta [76] justified the liquid-like layer on ice by surface pressure melting, due to the neighbors conditions of the surface molecules, and [75] showed that this pressure was enough to reduce the melting point of about 13 K. Through low-energy electron diffraction (LEED) experiments[7] and molecular dynamics simulations, several authors have better investigated the nature of this surface layer. Kroes (see [77]) has found a partial charge of the surface molecules due to their motion, while Devlin and Buth (see [78]) confirmed the theory of energy minimization for the breakdown of the solid structure. Furukawa and Nada (see [79] explained the disordered configuration of the surface molecules, and Materer et al. (see [80]) found that the LEED cannot detect atoms at a temperature as low as 90 K due to their high vibration [64].

8.4.2 Ice Friction

The ice friction is not so easy to characterize: one can identify several friction regimes, depending on liquidlike layer thickness (see Fig. 8.14).

[7]This technique is based on a collimated beam of low energy electrons (20–200 eV) which bombards the surface of a crystalline material. It is useful to determine both the symmetry of the surface structure and the atomic positions of molecules on the surface.

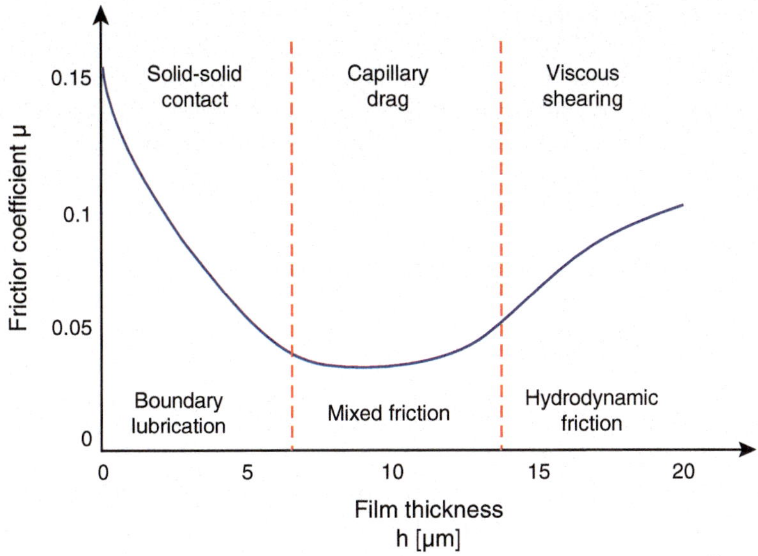

Fig. 8.14 A qualitative diagram of the most relevant friction regimes (except dry friction regime, which is not possible under atmospheric conditions), varying the liquidlike layer thickness on the surface (adapted from [64])

8.4.2.1 Dry Friction

It happens when the sliding between two surfaces is totally without lubrication, so that the asperities of surfaces are interested by physical and chemical adhesive bonds. This means that a relative motion of the two surfaces must break these adhesive links, generating a tangential friction force F_T equal to:

$$F_T = \tau_c A_c \tag{8.20}$$

with τ_c shear strength necessary to shear the asperity contact and A_c area of contact [64]. Considering now the applied load F_N and the softer material's hardness H, Bowden (see [81]) proposed a relation:

$$A_c = \frac{F_N}{H} \tag{8.21}$$

So,

$$\mu = \frac{\tau_c}{H} \tag{8.22}$$

that is, the friction coefficient depends on the applied load and the surfaces' hardness, but not on the sliding velocity (see also [82–84]). This regime is not possible for the ice under atmospheric conditions, also at very low temperatures (layer of few molecules are always present, see [85]).

8.4.2.2 Boundary Friction

This regime is characterized by a thin liquidlike layer, with a thickness of few molecular layer. The temperature in the contact zone (T) must be lower than the melting temperature (T_m) and the lubricating thickness (h) has to be far smaller than the roughness (R). These conditions must be valid *everywhere in the contact zone*. Generally the friction coefficient in this regime is lower with respect to the dry friction conditions (see [64]).

8.4.2.3 Mixed Friction

Here we have that, *at some points in contact zone*, these conditions must hold: $T < T_m$ and $h < R$. Slider is now partly supported by the asperities and by the lubricating layer, resulting in:

$$F_T = A_c \cdot \left(\alpha\tau_c + (1-\alpha)\eta\frac{v}{h}\right) \tag{8.23}$$

where α is the not lubricating part of the surface, v is the slider relative velocity, and η the viscosity (see [83]). In this regime, there is also the formation of capillarity bridges between the asperities and the sliding surface, resulting in a drag force, but these effects are not easy to be modeled (see [64, 86]). This regime is the most important for winter sports, due to its "trade-off" between solid–solid contact and the capillary drag [87].

8.4.2.4 Hydrodynamic Friction

In this case along the *whole contact zone* the temperature is always above the T_m and the lubricating layer thickness is always higher than the asperities. No solid–solid adhesion phenomena occurs and the real contact zone is identical to the slider's surface area (A). One can define:

$$F_T = \tau_l A \quad \text{and} \quad \tau_l = \eta\frac{v}{h} \tag{8.24}$$

with τ_l shear stress developed from shearing of the liquidlike layer. Also in this regime the capillarity drag forces play a role on the final friction coefficient [64].

8.4.3 Ice Friction for a Speed Skate Blade: FAST 2.0i Model

Until now we have discussed the general theory of ice friction: in this section, a focus on a model of ice friction for a speed skate blade is provided. A good and complete numerical model (called FAST 2.0i) was developed by Lozowski, Szilder,

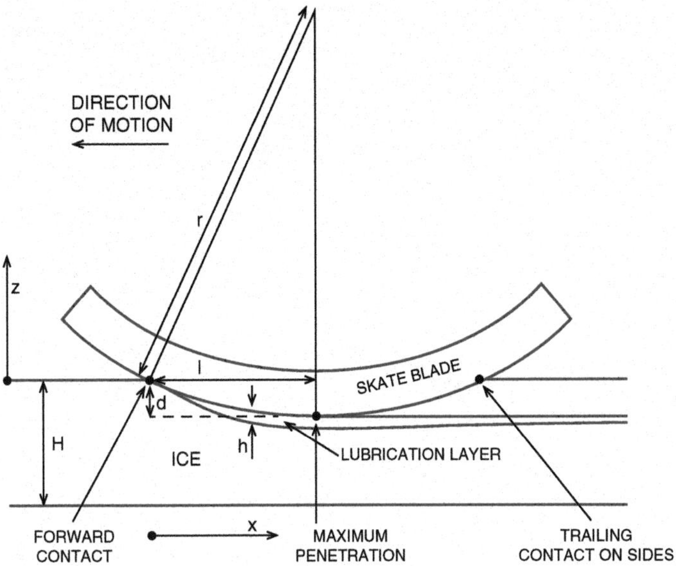

Fig. 8.15 Lateral view of a skate blade during ice contact. To show the parameters in a better way, dimensions and proportions are not strictly respected

and Maw (in the following paper the reader can see the evolution of the model: [63, 88, 89]). It works with a hydrodynamic friction regime, taking account of frictional melting, heat conduction, and lateral squeeze flow of the lubricating water.

We will now discuss the model presented by Lozowski et al. both for a vertically oriented (see [89]) and for an inclined (see [63]) blade sliding on a slab of ice with thickness H, like in Fig. 8.15. $h(x)$ represents the thickness of meltwater under the blade (exaggerated in Fig. 8.15).

8.4.3.1 Contact Geometry

If we consider the blade vertically oriented is quite easy to define the contact surface: the bottom contact is a rectangle of width w (around 1.1 mm in typical long track speed skate blade) and length l, with the two side contacts identical all along the contact surface.

Generally speaking, considering a constant longitudinal radius of curvature r we get that the side contacts are portions of a circular disk. Under this hypothesis, we can consider the distance from the forward contact point to the rear one equal to $2l$; at the point of maximum penetration the blade depth into the ice is equal to d.

In Fig. 8.16a, a transverse view of a speed skate blade varying the θ angle of inclination is provided: θ_c is the critical angle, at which the smaller interface between blade and ice vanishes. a_1 is the maximum transverse dimension of

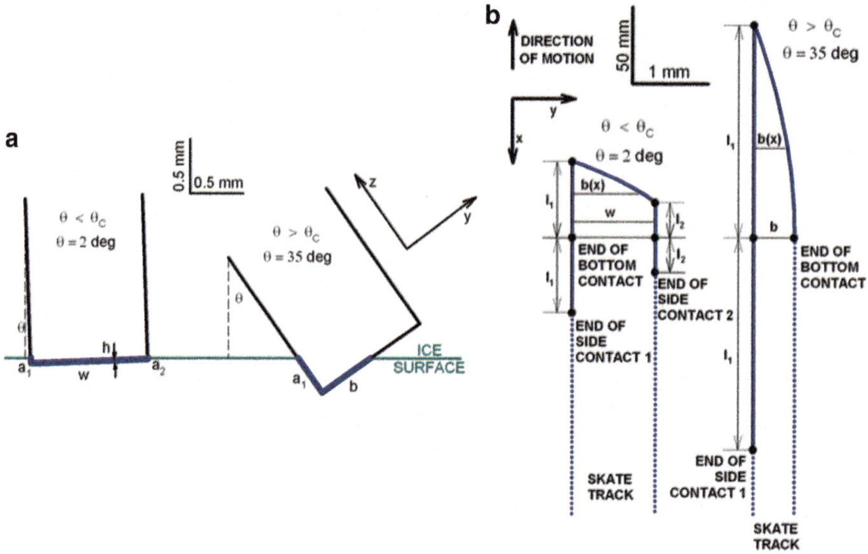

Fig. 8.16 (**a**) Transverse view of the blade, at the point of maximum penetration into the ice. The *blue lines* correspond to the meltwater lubrication layers. (**b**) The contact area between blade bottom and ice, seen from above. Adapted from [63]

the larger side of contact, a_2 the smaller side of contact, and b the bottom contact [63]. If θ is lower than the critical value, one can infer (see Fig. 8.16b):

$$\begin{aligned} l_1^2 &= 2a_1 r \\ l_2^2 &= 2a_2 r \\ a_1 &= a_2 + w \tan\theta \end{aligned} \tag{8.25}$$

Above the critical angle, a_2 is equal to 0 and w must be replaced by b. One can define the contact area using basic geometry (see [63]):

$$A = \frac{(l_1^3 - l_2^3)}{3r \tan\theta} \tag{8.26}$$

8.4.3.2 Normal Force

Considering the forces transmitted only by the bottom part of the blade, we can identify (see Fig. 8.17): the centrifugal force on the skater F_c, the weight of the skater mg, and the thrust applied by the skater T. We also assume that the vectorial sum of F_c and mg is oriented in the same direction of T, anti-parallel to the z-direction. Stating the relation $F_c = mg \tan\theta$, we have a scalar force balance equal to:

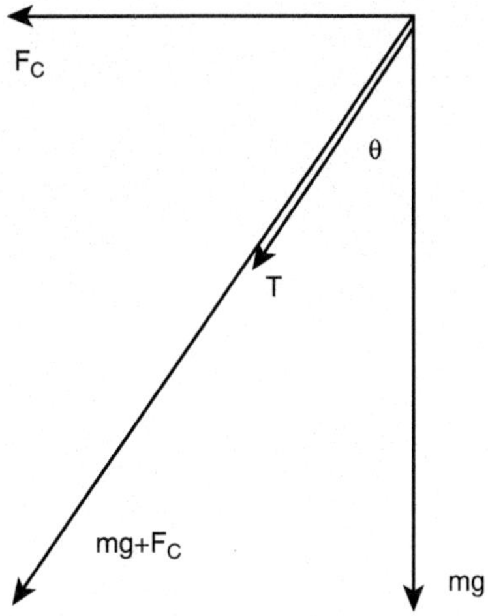

Fig. 8.17 The forces transmitted by the skater through the blade. Adapted from [63]

$$pA = \frac{mg}{\cos\theta} + T \tag{8.27}$$

where p is ice hardness (as measured by Poirier et al. [90]), which can be calculated in this way (see [89]): $p = 14.8 - 0.6T$, p is expressed in MPa and T, the temperature of ice surface, in Celsius degrees. It is strictly valid between -1 and $-15\,^{\circ}$C, but can be used in a wider range (see [63]).

At this point, considering the thrust contribution equal to zero[8] $T = 0$, one can calculate the geometrical parameters for different values of the inclination angle. For example, the critical angle is obtained evaluating and combining the above equations for $a_2 = 0$, leading to:

$$\theta_c = 0.5\sin^{-1}\left[\frac{9}{4r\,w^3}\left(\frac{mg}{p}\right)^2\right] \tag{8.28}$$

For motions with an inclination angle below the critical angle, it is possible to obtain the geometrical parameters through a suitable reshaping of the equations:

$$(2a_2r + 2wr\tan\theta)^{\frac{3}{2}} - (2a_2r)^{\frac{3}{2}} - \frac{3rmg\tan\theta}{p\cos\theta} = 0 \tag{8.29}$$

With this relation one can calculate numerically the a_2 coefficient and then solve Eq. (8.25), obtaining l_1, l_2, and a_1. For an angle that exceeds the critical value (so $a_2 = 0$, see [63]), one has to solve:

[8]This hypothesis is made to simplify the model, avoiding to introduce a biomechanical model for the skater.

$$
\begin{aligned}
b &= \left(\frac{9}{4r\sin(2\theta)}\right)^{\frac{1}{3}} \left(\frac{mg}{p}\right)^{\frac{2}{3}} \\
l_1 &= \frac{3mg}{2pb\cos\theta} \\
a_1 &= \frac{9}{8r}\left(\frac{mg}{pb\cos\theta}\right)^2
\end{aligned} \tag{8.30}
$$

The reader can refer to [63, 89, 91] for a more complete analysis of the model.

8.4.3.3 Ploughing Force

A skater leaves a well-known "skate track" after its transit, this is due to the formation of a kind of channel due to the ploughing of the blade on ice and it requires a force to "break" ice resistance. It is possible to use as a parameter the same hardness p used above, lacking in precision because of the anisotropy of ice resistance (see [89] for some observations) but simplifying the model. In the case of a vertical blade the force is modeled as $F_p = \text{pwd}$ (where d is the maximum penetration of the blade, derivable from $d = l^2/2r$) and for an inclined one it is easily $F_p = pb(a_1 + a_2)/2$. This is not enough to properly calculate the friction coefficient, which would be really lower than the one expected (see [34, 63]), that is to say that there are other resistance forces to be considered.

8.4.3.4 Shear Stress Force

An important part of this model is the inclusion of lubrication effects acting on the contact surfaces. For the sake of simplicity it is assumed that the contact zones are fully lubricated, so there is no direct contact between the surface asperities and the blade. Also the larger-scale asperities and waviness are not considered [63]. The assumption of a Couette flow is made, with zero velocity at the ice/liquid interface and skate velocity at the skate/liquid interface. The shear stress force in the lubricating fluid can be expressed as:

$$
F_s = \mu V w \int_0^l \frac{1}{h} dx \tag{8.31}
$$

where μ is the dynamic viscosity, V the skate velocity, and $h = h(x)$ the local thickness of lubricating layer (x is the longitudinal coordinate, increasing from the front to the rear part of the blade).

Several factors influences the shear stress force, we now give an overview on them.

8.4.3.5 Squeeze Flow

A lateral squeeze flow arises because of the force exerted by the bottom of the skate blade of the liquid. The modeling is simpler in the case of vertical blade, under the assumption of a symmetrical problem. Assuming low Reynolds number (no inertia of fluid), the equation of motion is (y is the lateral coordinate and z the vertical one):

$$\frac{\partial p}{\partial y} = \mu \frac{\partial^2 v}{\partial z^2} \tag{8.32}$$

With $v(0) = v(h) = 0$ and $(\partial v/\partial z)_{h/2}$, the solution is:

$$v(y,z) = \frac{1}{2\mu}\frac{\partial p}{\partial y}(z^2 - zh) \tag{8.33}$$

Introducing then the continuity equation for an incompressible fluid (W is the z-component of fluid velocity):

$$\frac{\partial v}{\partial y} + \frac{\partial W}{\partial z} = 0 \tag{8.34}$$

Using Eq. (8.33) in Eq. (8.34) and integrating between 0 and h, one can get

$$\frac{\partial^2 p}{\partial y^2} = \frac{12\mu \Delta W}{h^3} \tag{8.35}$$

where $\Delta W = dh/dt$ is the rate of thickening of the lubrication layer, due to squeeze flow (see [63]). With the two boundary conditions $(\partial p/\partial y)_0 = 0$ and $p(w/2) = 0$, one can get the pressure formula integrating two times:

$$p = \frac{6\mu \Delta W}{h^3}\left(\frac{w^2}{4} - y^2\right) \tag{8.36}$$

Through a force balance, one can write:

$$2\int_0^{w/2} pldy = mg \tag{8.37}$$

and finally imposing $\Delta W = V(dh/dx)$:

$$\frac{dh}{dx} = -\frac{ph^3}{\mu w^2 V} \tag{8.38}$$

For an inclined blade the formulation is more complicated, resulting in this formula:

$$\frac{dh(x)}{dx} = -\left(\frac{2}{3\beta(x)}\right)\left(\frac{ph^3(x)}{\mu w_t^2 x)V}\right) \tag{8.39}$$

defining $w_t(x) = a_1(x) + a_2(x) + b(x)$ total width of the lubricating layer and $\beta(x)$ function of the geometrical parameters (see [63]).

8.4.3.6 Heat Conduction and Frictional Melting

The heat flux depends on the difference of temperature between the lubricating layer and the ice surface, defining consequently also the direction of the heat exchange. Further the heat penetration involves only tens of micrometers, so one can argue that the inclination of the blade does not affect the heat exchange (as described in [63, 89]). The effects of the heat exchange on the variation of h can be described as:

$$\frac{dh}{dx} = -\frac{k_i \Delta T_i}{\rho_w l_f \sqrt{\pi \kappa_i x V}} \tag{8.40}$$

where ΔT_i is the difference between the pressure melting point temperature and the ice temperature, κ_i is the thermal diffusivity of ice, k_i is the thermal conductivity of ice, ρ_w is the density of liquid water at the melting point and l_f the specific latent heat of freezing.

8.4.3.7 Frictional Melting

An energy source near the ice-blade contact zone realizes the melting of some ice, creating part of the lubricating layer. As already presented by Lozowski et al. [89], this effect can be modelled (in longitudinal direction) as (see [63]):

$$\frac{dh}{dx} = \frac{\mu V}{h \rho_w l_f} \tag{8.41}$$

8.4.3.8 Total Equation of Lubrication

Now, without losing generality, one can use the above equations to compose the total equation for the lubrication phenomenon (neglecting y and z directions):

$$\frac{dh}{dx} = \frac{\mu V}{h \rho_w l_f} - \frac{k_i \Delta T_i}{\rho_w l_f \sqrt{\pi \kappa_i x V}} - \frac{2ph^3}{3\beta \mu V w_t^2} \tag{8.42}$$

which has to be integrated from $x = 0$ to $x = l_1$.

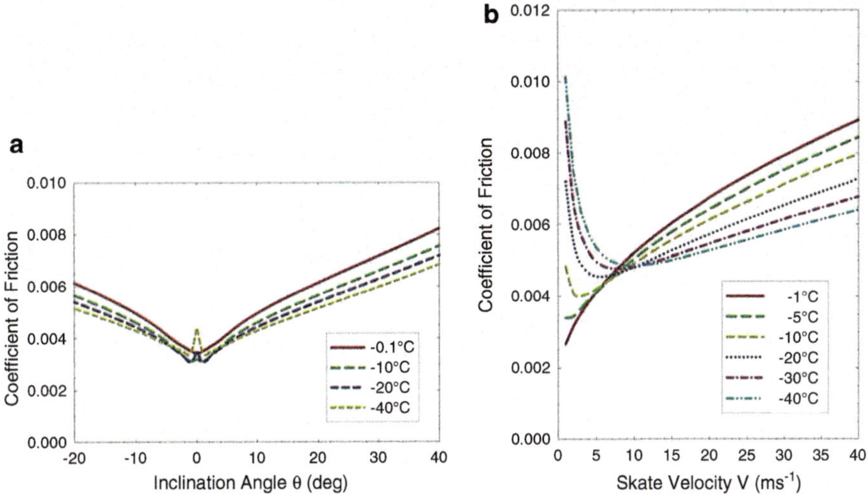

Fig. 8.18 (**a**) Relation between the coefficient of friction and the blade inclination angle (positive angles indicate inside edge), varying the temperature of ice; (**b**) average coefficient of friction during a stroke function of the skate velocity V (positive angles indicate inside edge), varying the temperature of ice. Both the plots refer to standard conditions: $T_i = -5\,^{\circ}\mathrm{C}$, skate speed $V = 12\,\mathrm{m\,s^{-1}}$, skater mass $m = 75\,\mathrm{kg}$, blade rocker radius $r = 25\,\mathrm{m}$, blade width $w = 1.1\,\mathrm{mm}$. Adapted from [63]

8.4.3.9 Results

For a complete discussion of results please refer to [63]. We here show the most relevant results obtained by the model:

- Except for a small peak near zero degrees of inclination angle, the coefficient of friction increases with inclination angle (Fig. 8.18a), even if this dependence is correlated with the ice temperature.
- Figure 8.18b illustrates the relation between the coefficient of friction and the skate velocity (results are averaged over a range of blade inclination angle varying from −20° and 40°, typical of a skating stroke). In general it is possible to notice that temperature dependence of the ice friction coefficient is different at low and high velocities. According to the model, at higher velocities a warmer ice presents an higher coefficient of friction, while at lower velocities a warmer ice is "faster" (low coefficient of friction). Hence, according to the model, very fast skaters (in high level competitions velocities around $16\,\mathrm{m\,s^{-1}}$ can be reached) will experience lower ice friction with colder ice, while slow skaters will experience lower ice friction with warmer ice [63]. Around 6 and $10\,\mathrm{m\,s^{-1}}$, it is possible to find optimum values for the average friction coefficient: these results are comparable with the ones of De Koning et al. [34].
- Results of the model were compared with the measurements performed by De Koning et al. [34], made with a skater of 72 kg (really close to the standard

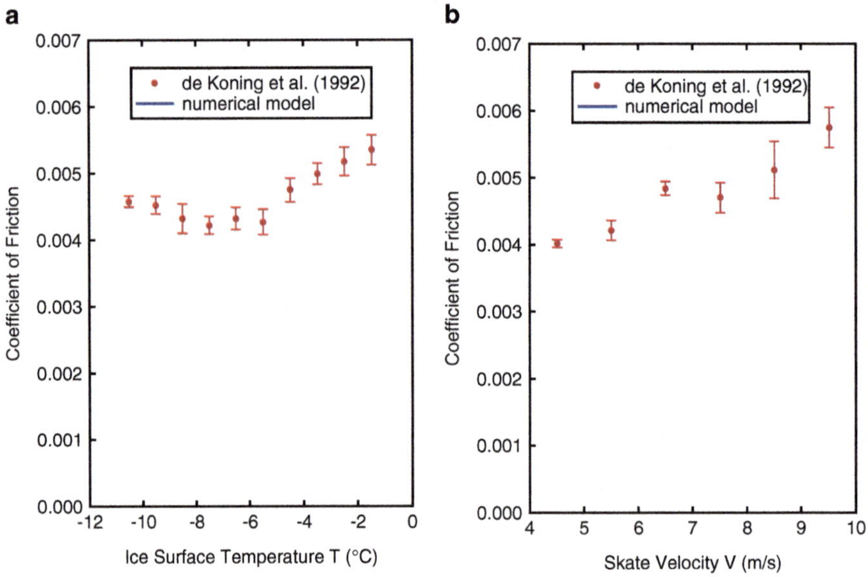

Fig. 8.19 (**a**) Average ice friction coefficient during a skating stroke, function of ice surface temperature with a skate speed of $8\,\mathrm{m\,s^{-1}}$; (**b**) average ice friction coefficient during a skating stroke, function of skate velocity with an ice surface temperature of $-4.6\,°\mathrm{C}$. Standard conditions are used. Adapted from [63]

mass of 75 kg here considered) and skate velocity of $8\,\mathrm{m\,s^{-1}}$. In Fig. 8.19a the model and experimental values of friction coefficient are remarkably similar when averaged over all ice surface temperatures. The model does not simulate the systematic behavior seen in the experiments, with a minimum at $-7\,°\mathrm{C}$. Also in Fig. 8.19b the results are qualitatively comparable, even though the model underestimates the increase of ice friction with increasing skating speed. Authors suppose the discrepancy depends on the hypothesis of neglecting the thrust force [63].

8.4.3.10 Conclusions

The FAST2.0i here presented is one of the most recent and complete models for ice friction of a speed skater. An interested reader can check the good results of this approach compared with numerical and experimental results in [63]. This model has some lacks and must respect numerous hypotheses. In particular, further developments may expand the validity of this formulation. The thrust is neglected, while it could be introduced using a biomechanical model of the skater. Again, a model of this kind can be useful to consider a potential moment exerted by the skater on the blade. Squeeze flow and heat flow models can be extended, e.g. including the y and z directions and the heat conduction between blade and ice (see also [89]).

References

1. Wikipedia -speed skating (May 2014). http://en.wikipedia.org/wiki/Speed_skating
2. D.L. Bird, Skating a brief history of ice and the national ice skating association of great Britain (May 2014). http://www.iceskating.org.uk/about/history
3. Wikipedia -Edinburgh skating club (May 2014). http://en.wikipedia.org/wiki/Edinburgh_Skating_Club
4. The edinburgh skating-club, with diagrams of figures and a list of the members. Edinburgh skating club (1865). http://books.google.it/books?id=8CQ7MwEACAAJ
5. Wikipedia -international skating union (May 2014). http://en.wikipedia.org/wiki/International_Skating_Union
6. The international skating union (ISU) (May 2014). http://www.isu.org/en/home
7. Special regulations and technical rules -speed skating and short track speed skating. As accepted by the 54th Ordinary Congress June 2012. International Skating Union (ISU) (2012)
8. Olympic -speed skating equipment and history (May 2014). http://www.olympic.org/speed-skating-equipment-andhistory?tab=equipment
9. Wikipedia -clap skate, May 2014. http://it.wikipedia.org/wiki/Clap_skate
10. G.J. Van Ingen Schenau et al., Skate, more particularly ice-skate for speed skating. EP Patent App. EP19,860,200,279, 1986. https://www.google.com/patents/EP0192312A2?cl=en
11. A. Shum, Clap skate with spring and cable biasing system. US Patent 6,007,075, 1999. http://www.google.com/patents/US6007075
12. T. Martin, Evolution of ice rinks. ASHRAE J. **46**(11), S24–S30 (2004)
13. G. McCourt, S.L. Goodrich, Calgary's olympic oval (Canada). Industria Italiana del Cemento **59**(9), 524–543 (1989)
14. S.S. Jensen, Ice skating rink at the brisbane bicentenary sports and entertainment complex: Boondall Queensland. Aust. Refrig. Air Cond. Heat. **40**(8), 23–24, 28 (1986)
15. K. Tusima et al., Development of a high speed skating rink by the control of the crystallographic plane of ice. Jpn. J. Tribol. **45**, 17–26 (2000)
16. C. Câmpian, M. Cristujiu, I. Benke, The supporting steel structure of the ice rink—city of Târgu Mureş, Romania Zs. Nagy, in *Structures & Architecture*, ed. by P.J.S. Cruz. ICSA 2010 – 1st International Conference on Structures & Architecture, 21–23 July 2010, Guimaraes (CRC Press, 2010), pp. 167–168
17. M. Schulitz, J. Güsgen, J. Kaußen, Ice skating and swimming stadium Lentpark in Cologne [Eis-und Schwimmstadion Lentpark in Köln] Stahlbau **83**(1), 57–60 (2014)
18. Anon, roofing of an artificial ice skating rink. [Couverture D'UNE Patinoire Artificielle Ueberdachung Der Kunsteisbahn Arosa (CH).] Acier English ed. **44**(4), 153–155 (1979)
19. www.scopus.com, Stadiums and arenas: a grandstand construction for German ice rink. Concr. Eng. Int. **11**(1), 42–43 (2007)
20. Z. Nagy, I. Mircea Cristutiu, Advanced nonlinear investigations of a 50 m span frame case study: the steel structure of the ice rink, city of Targu-Mures, Romania, vol. 2, cited by (since 1996)1. 2010, pp. 649–656
21. A. Zak, O. Sikula, M. Trcala, Analysis of local moisture increase of timber constructions on ice arena roof. Adv. Mater. Res. **649**, 291–294 (2013)
22. T. Davis, S. Kneisel, Building control systems at three new arenas. Can. Consult. Eng. (1995). www.scopus.com
23. H. Guo, S.C. Lee, L.Y. Chan, Indoor air quality in ice skating rinks in Hong Kong. Environ. Res. **94**(3), 327–335 (2004)
24. J. Matsuo, T. Nagai, A. Sagae, M. Nakamura, S. Shimizu, T. Fujimoto, F. Maeda, Thermal characteristics and energy conservation measures in an indoor speed-skating arena. Paper presented at the Proceedings of Building Simulation 2011: 12th Conference of International Building Performance Simulation Association (2011), pp. 2072–2079
25. G. Teyssedou, R. Zmeureanu, D. Giguère, Thermal response of the concrete slab of an indoor ice rink (rp-1289). HVAC R Res. **15**(3), 509–523 (2009)

26. G. Teyssedou, R. Zmeureanu, D. Giguère, Benchmarking model for the ongoing commissioning of the refrigeration system of an indoor ice rink. Autom. Constr. **35**, 229–237 (2013)
27. Real world physics problems: physics of ice skating (June 2014). http://www.real-world-physics-problems.com/physics-of-ice-skating.html
28. D.M. Fintelman, O. den Braver, A.L. Schwab, A simple 2-dimensional model of speed skating which mimics observed forces and motions, in *Multibody Dynamics 2011, ECCOMAS Thematic Conference*, 2011
29. H. Houdijk et al., Push-off mechanics in speed skating with conventional skates and klapskates. Med. Sci. Sports Exerc. **32**(3), 635–641 (2000)
30. G.J. Van Ingen Schenau, R.W. De Boer, G. De Groot, On the technique of speed skating. Int. J. Sport Biomech. **3**, 419–431 (1987)
31. G.J. Van Ingen Schenau, R.W. de Boer, G. de Groot, The control of speed in elite female speed skaters. J. Biomech. **18**, 91–96 (1985)
32. J.J. De Koning, Biomechanical aspects of speed skating. Ph.D. thesis, Vrije Universiteit, Faculty of Human Movement Sciences, Amsterdam, 1991
33. M.R. Yeadon, A method for obtaining three-dimensional data on ski jumping using pan and tilt cameras. Int. J. Sport Biomech. **5**, 238–247 (1989)
34. J.J. De Koning, G. De Groot, G.J. van Ingen Schenau, Ice friction during speed skating. J. Biomech. **25**, 565–571 (1992)
35. H. Jobse et al., Measurement of the push-off force and ice friction during speed skating. Int. J. Sport Biomech. **6**, 92–100 (1990)
36. J.J. De Koning et al., From biomechanical theory to application in top sports: the Klapskate story. J. Biomech. **33**, 1225–1229 (2000)
37. G.J. Van Ingen Schenau et al., A new skate allowing powerful plantar flexions improves performance. Med. Sci. Sports Exerc. **28**, 531–535 (1996)
38. J.J. De Koning, G. De Groot, G.J. van Ingen Schenau, Coordination of leg muscles during speed skating. J. Biomech. **24**(2), 137–146 (1991)
39. T.L. Allinger, A.J. Bogert, Skating technique for the straights, based on the optimization of a simulation model. Med. Sci. Sports Exerc. **29**(2), 279–286 (1997)
40. J.J. De Koning, G.J. van Ingen Schenau, Performance-determining factors in speed skating, in *Biomechanics in Sport: Performance Enhancement and Injury Prevention: Olympic Encyclopaedia of Sports Medicine*, ed. by V. Zatsiorsky, vol. IX (Blackwell Science Ltd, Oxford, 2000), pp. 232–246. doi:http://dx.doi.org/10.1002/9780470693797.ch11.
41. E. Otten, Inverse and forward dynamics: models of multi-body systems. Philos. Trans. R. Soc. Lond. **358**, 1493–1500 (2003)
42. P. Mantegazza, P. Masarati, *Analysis of Systems of Differential-Algebraic Equations (DAE)*. Lecture Notes of Graduate Course on "Multibody System Dynamics" (22 November 2012)
43. R.W. de Boer, K.L. Nislen, The gliding and push-off technique of male and female olympic speed skaters. J. Appl. Biomech. **5**(2), 119–134 (1989)
44. J.J. De Koning, G. de Groot, G.J. van Ingen Schenau, A power equation for the sprint in speed skating. J. Biomech **25**, 573–580 (1992)
45. J.J. De Koning, G.J. van Ingen Schenau, On the estimation of mechanical power in endurance sports. Sport Sci. Rev. **3**, 34–54 (1994)
46. J.J. De Koning, G.J. van Ingen Schenau, Performance-determining factors in speed skating, in *Biomechanics in Sport*, ed. by V.M. Zatsiorsky (Blackwell Science Ltd., Oxford, 2008)
47. G.J. Van Ingen Schenau, J.J. De Koning, G. De Groot, A simulation of speed skating performances based on a power equation. Med. Sci. Sports. Exerc. **22**, 718–728 (1990)
48. J.J. De Koning et al., Experimental evaluation of the power balance model of speed skating. J. Appl. Physiol. **98**(1), 227–233 (2004)
49. T.J. Barstow, P.A. Mole, Linear and nonlinear characteristics of oxygen uptake kinetics during heavy exercise. J. Appl. Physiol. **71**, 2099–2106 (1991)
50. C. Foster et al., Pattern of energy expenditure during simulated competition. Med. Sci. Sports Exerc. **35**, 826–831 (2003)

51. G.J. van Ingen Schenau, G. De Groot, A.P. Hollander, Some technical, physiological and anthropometrical aspects of speed skating. Eur. J. Appl. Physiol. Occup. Physiol. **50**(3), 343–354 (1983)
52. G.J. van Ingen Schenau, The influence of air friction in speed skating. J. Biomech. **15**(6), 449–58 (1982)
53. A. D'Auteuil, G. Larose, S. Zan, The effect of motion on wind tunnel drag measurement for athletes. Procedia Eng. **34**, 62–67 (2012)
54. A. D'Auteuil, G.L. Larose, S.J. Zan, Relevance of similitude parameters for drag reduction in sport aerodynamics. Procedia Eng. **2**, 2393–2398 (2010)
55. A. D'Auteuil et al., Detection of the boundary layer transition for non-circular cross-sections using surface pressure measurements, in *Proceedings of the 13th International Conference on Wind Engineering*, Amsterdam, 2011
56. M.M. Zdravkovich, *Flow Around Circular Cylinders Volume 1: Fundamentals* (Oxford Science Publications, New York, 1997)
57. S.J. Zan, K. Matsuda, Steady and unsteady loading on a roughened circular cylinder at Reynolds numbers up to 900,000. J. Wind Eng. Ind. Aerodyn. **90**, 567–58 (2002)
58. A. D'Auteuil, G. Larose, S. Zan, Wind turbulence in speed skating: measurement, simulation and its effect on aerodynamic drag. J. Wind Eng. Ind. Aerodyn. **104–106**, 585–593 (2012)
59. L.W. Brownlie et al., Reducing the aerodynamic drag of sports apparel: development of the Nike Swift sprint running and SwiftSkin speed skating suits, in *The Engineering of Sport 5* (ISEA, Sheffield, 2004), pp. 90–96
60. G.H. Kuper, E. Sterken, Do skin suits increase average skating speed? Technical Report. University of Groningen, 2008
61. L.W. Brownlie, C.R. Kyle. Evidence that skin suits affect long track speed skating performance. Procedia Eng. **34**, 26–31 (2012)
62. R. Rosenberg, Why is ice slippery? Phys. Today **58**(12), 50–55 (2005)
63. E.P. Lozowski, K. Szilder, S. Maw, A model of ice friction for a speed skate blade. Sports Eng. **16**(4), 239–253 (2013)
64. A.M. Kietzig, S.G. Hatzikiriakos, P. Englezos, Physics of ice friction. J. Appl. Phys. **107**, 08110 (2010)
65. P. Barnes, D. Tabor, Plastic flow and pressure melting in the deformation of ice I. Nature **210**, 878–882 (1966)
66. F.P. Bowden, T.P. Hughes, The mechanism of sliding on ice and snow. Proc. R. Soc. Lond. A **172**(949), 280–298 (1939)
67. F.P. Bowden, Friction on snow and ice. Proc. R. Soc. Lond. A **217**, 462–478 (1953)
68. M.J. Furey, Friction, wear and lubrication, in *Chemistry and Physics of Interfaces II* (American Chemical Society Publications, Washington, 1971)
69. V.F. Petrenko, R.W. Whitworth, *Physics of Ice* (Oxford University Press, Oxford, 1999), p. 373
70. P.V. Hobbs, *Ice Physics* (Clarendon Press, Oxford, 1974)
71. N.H. Fletcher, Surface structure of water and ice. Philos. Mag. **7**, 255–59 (1962)
72. N.H. Fletcher, Surface structure of water and ice: II. A revised model. Philos. Mag. **18**, 1287–1300 (1968)
73. R. Lacmann, I.N. Stranski, The growth of snow crystals. J. Cryst. Growth **13–14**(C), 236–240 (1972)
74. J.G. Dash, F. Haiying, J.S. Wettlaufer. The premelting of ice and its environmental consequences. Rep. Prog. Phys. **58**, 115 (1995)
75. L. Makkonen, Surface melting of ice. J. Phys. Chem. B **101**(32), 6196–6200 (1997)
76. N. Fukuta, J. Phys. (Paris) **48**, 503 (1987)
77. G.-J. Kroes, Surface melting of the (0001) face of {TIP4P} ice. Surf. Sci. **275**(3), 365–382 (1992). ISSN: 0039-6028. doi: http://dx.doi.org/10.1016/0039-6028(92)90809-K. http://www.sciencedirect.com/science/article/pii/003960289290809K
78. J.P. Devlin, V. Buch, Surface of ice as viewed from combined spectroscopic and computer modeling studies. J. Phys. Chem. **99**(45), 16534–16548 (1995)

79. Y. Furukawa, H. Nada, Anisotropic surface melting of an ice crystal and its relationship to growth forms. J. Phys. Chem. B **101**(32), 6167–6170 (1997)
80. N. Materer et al., Molecular surface structure of ice(0001): dynamical low-energy electron diffraction, total-energy calculations and molecular dynamics simulations. Surf. Sci. **381**(2–3), 190–210 (1997)
81. F.P. Bowden, Introduction to the discussion: the mechanism of friction. Proc. R. Soc. Lond. A Math. Phys. Sci. **212**, 440–449 (1952)
82. B.N.J. Persson, *Sliding Friction. Physical Principles and Applications*, 2nd edn. (Springer, Berlin, 2000)
83. B. Bhusam, *Introduction to Tribology* (Wiley, New York, 2002)
84. F.P. Bowden, D. Tabor, *The Friction and Lubrication of Solids* (Oxford University Press, Oxford, 1964)
85. V.F. Petrenko, R.W. Whitworth, *Physics of Ice* (Oxford University Press, Oxford, 1999)
86. S.C. Colbeck, Kinetic friction of snow. J. Glaciol. **34**(116), 78–86 (1988)
87. A.-M. Kietzig, S.G. Hatzikiriakos, P. Englezos, Ice friction: the effects of surface roughness, structure and hydrophobicity. J. Appl. Phys. **106** 024303 (2009)
88. A. Penny et al., Speedskate ice friction: review and numerical model -FAST 1.0, in *Physics and Chemistry of Ice*, ed. by W.F. Kuhs. R. Soc. Chem. (2007), pp. 495–504. doi:10.1039/9781847557773
89. E.P. Lozowski, K. Szilder, Derivation and new analysis of a hydrodynamic model of speed skate ice friction. Int. J. Offshore Polar Eng. **23**, 104–111 (2013)
90. L. Poirier et al., Getting a grip on ice friction, in *Proceedings of the Twenty-first (2011) International Offshore and Polar Engineering Conference*, Maui, HI, 19–24 June 2011
91. E.P. Lozowski, K. Szilder, S. Maw, A model of ice friction for an inclined incising slider, in *The Twenty-second International Offshore and Polar Engineering Conference. International Society of Offshore and Polar Engineers* (2012), pp. 1243–1251

Chapter 9
Ice Hockey Skate, Stick Design and Performance Measures

R.A. Turcotte, P. Renaud, and D.J. Pearsall

9.1 Skate Design and Performance Measures

Ice hockey has a long history dating back to the 1880s in Canada and Europe. The game has evolved, especially in the last two decades into a fast-paced power game. Ice hockey has become increasingly sophisticated in terms of technological innovations, equipment design and improvements in training, coaching and game strategies [1, 2]. This has led to a game where players accentuate speed and power to a greater extent than previous generations of hockey players. However, in nearly two decades of research in this area it is understood that although speed and power have been optimized, skate development has also had implications for the improvement of a range of skating skills, some of which require finesse and precision for optimal execution. The unique environmental conditions (e.g. low surface friction) of ice hockey demand a very distinctive skill set. The skating skills in particular can be subdivided into skills that require power (forward and backward skating) and skills that require precision and/or finesse during execution, such as changes in direction. The earliest records of the first known skate models are quite primitive in their design; but the goal behind them was the same: to allow a person to move and travel around faster on ice. Although today most ice hockey skates are used for leisure and sport, the earlier models of ice skates were fabricated in order to allow people to transport themselves across frozen waterways in cold climate cities [3]. The intent of this section of the chapter is to examine the importance of skate design in relation to comfort, fit and performance for skating in ice hockey.

R.A. Turcotte (✉) • P. Renaud • D.J. Pearsall
Department of Kinesiology and Physical Education, McGill University, Montreal, QC, Canada H2W 1S4
e-mail: rene.turcotte@mcgill.ca

F. Braghin et al. (eds.), *The Engineering Approach to Winter Sports*,
DOI 10.1007/978-1-4939-3020-3_9

The first archaeological evidence of skates dates back to the Bronze Age in Europe and northern Russia. Within these areas, there are many findings of animal bones with horizontal holes pierced perpendicular to the long axis to allow the installment of leather straps [3]. These types of bone skates did not have edges to them like later models, so the person wearing them had to propel themselves forward by using a stick between their legs; but surprisingly, some bone skates were still used up until the eighteenth century in Iceland, Gotland and parts of Europe. This innovation allowed the person using the bone skates to take advantage of a grip and glide technique, which could not be afforded by shoe soles. Bone blades, which were heavy and resulted in a slow skating speed, were eventually replaced by metal blades. This demonstrates the effect of skate design on skating performance [4]. These later models of skates which incorporated metal runners to the bottom of wooden blocks were also fastened to the feet by leather straps attached to the wooden blocks [3]. These metal runners allowed for better edge grip during the push off of the stride, as well as an enhanced glide length. These models also allowed the person using them to propel themselves forward by more effective use of their lower body muscles in their legs and hips. Later versions of these types of skates, from the fifteenth to the eighteenth centuries, were similar in their design but were reduced in weight by a significant amount of about 30 % [3]. This allowed the person to travel much more easily, and with less effort or more speed.

During the eighteenth century and thereafter, skates were fabricated with longer blades because of a better mechanical understanding that only the edges of the blades were touching the ice surface. That is, if the blade is longer, the contact area between it and the ice would be larger, and consequently, the blade would not be pushed into the ice as much, which minimizes resistance to allow for easier traveling [3]. A longer blade also provided a more stable standing surface on the ice; as well as creating deeper etchings into the ice which would allow channels for lateral grip and steering during the forward glide. From these blade innovations came forerunners of our modern day skate design, which were developed in the nineteenth century, with better integration of the boots and blade by way of metal blade "holders". This modification increased the skater's control which allowed for a safer and easier traveling experience; ultimately allowing the skater to take fewer strides over the same distance and conserving their energy at the same time [3].

The evolution of the design of the modern day ice hockey skate can be attributed to an amount of trial and error from the designers and engineers of the ice hockey industry. The materials used in today's skate productions have replaced metal holders and leather boots with synthetic materials [5]. The use of these synthetic materials, such as molded nylon, Kevlar® (DuPont) and graphite polymers, is to increase the skate's durability, reduce its weight and enhance foot protection from puck impacts. For these reasons, there is now widespread production of standard molded hard synthetic composite boots which incorporate all of the previous design aspects. Designers are also looking at factors affecting the skate wearer's comfort and fit with function [6]; to enhance the user's satisfaction. Optimizing these variables to make the whole skate is a challenge to sporting goods designers. The ice hockey skate design has three distinct parts to it: the boot, the blade holder and the blade itself (see Fig. 9.1).

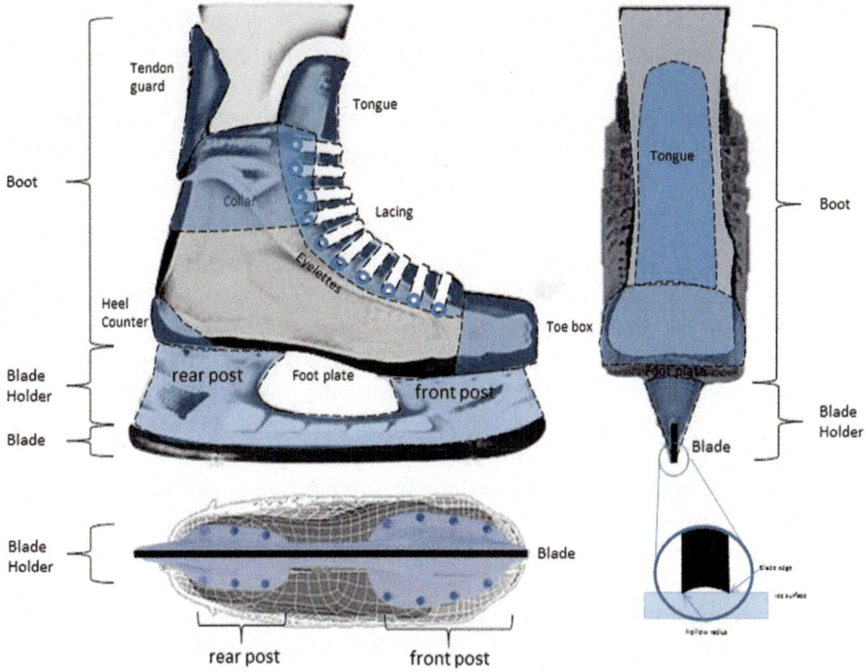

Fig. 9.1 The components of an ice hockey skate

The main complicating factor when evaluating the effectiveness of skate design is to ensure that the skate will be both effective and appropriate for the performance of a large variety of skills. Most of the work done in evaluating product design has consisted of a "reverse engineering" approach since many products have been conceived by intuition and evaluated post hoc. In order for a skate design to provide optimal function it would seem logical that a number of conditions must be met when evaluating the design. In keeping with the theme of comfort fit and performance a skate design can be evaluated in each of these categories.

For proper fit and comfort:

1. The skate should permit adequate kinaesthetic sense of joint position and limb orientation.
2. The skate should be as comfortable as possible by avoiding pinching of sensitive soft tissue areas overlying muscles and neurovascular structures.
3. The skate should accommodate for geometric anthropometrics and orientation of bony structures.

For performance:

1. The skate should allow for effective anterior-posterior and medial-lateral alignment, stability and range of motion.
2. The skate should provide effective forefoot and rear foot and medial-lateral leverage for more effective force production and better control of blade movement.
3. The skate should minimize the restriction of joint movements and facilitate coupled foot-ankle-knee-hip chain coordination.

Using a variety of technologies the hockey equipment industry has attempted to address many of these design issues. The following sections of the chapter will summarize past research and highlight some of the newest findings.

The industry's leading companies have been using a heuristic approach often based on subjective impressions of athletes using the products on ice and, in turn, subjective impressions/preferences in skate construction. For example, one popular notion that has led to lengthy discussions and subsequent design ideas was the idea that a stiff skate was crucial for proper support during skating tasks. Stiffer skates could also offer more injury protection from impacts [7]. Early research was conducted specifically to evaluate the effect of stiffness on fit and comfort. A wide range of stiffness within skate boots was made possible by the many types of materials available [8]. Logically, determining if a specific design resulted in stiffer properties was a first step. Turcotte et al. [8] constructed a jig that could measure the load imposed on the boot (in Newton) in comparison with the extension in cm (stiffness/rigidity modulus) when the skate boots were moved through full ankle and foot range of motions. Indeed it was determined that the Young's modulus varied substantially in upper boot stiffness across skate models and corresponded to differences in boot construction and materials. Subsequent study using human subjects [9] with boot motion and torques measured using a Biodex dynamometer confirmed that boot stiffness and construction properties have substantial effects on functional range of motion and work output.

Until that point in time however, no one had evaluated whether stiffer skates were comfortable and whether they fit properly. Indeed a complaint from many athletes field testing these types of skates was one of discomfort and in many cases a lack of mobility or movement constraints imposed by the excessively stiff skate construction. Thus, subsequent studies made use of pressure sensor technology to evaluate the fit of skates and subjects' rating of comfort of constructed skates [6].

One study in particular was designed to measure pressure (kPa) in different areas of the skate boot. This allowed for the determination of "hot spots" (high pressure areas) that existed with standard skate boots using the previously mentioned materials to design skate boots. Thus, it was now possible to quantify "fit" in relation to comfort and perceived function afforded [10]. In a second experiment, correlations of pressure points to subjects' perception of comfort while wearing skates confirmed that higher pressures tended to be negatively associated with comfort [6]. Conversely, athletes have also complained that skates can be too loose creating foot slipping within the boot, control instability and reduce push-off power. An optimal "snug" fit allows for a better foot/skate boot interface, which in turn increases the perception of foot orientation and kinaesthetic feedback. This is because a good fit likely helps to provide afferent information to the central nervous system (CNS), which optimizes kinematic adjustments during skating performance [11].

Certainly it would be possible to do more research on fit, in order to determine if there are precise pressure cut-off points at which comfort and proper fit can be quantified in various areas of the skate boot. One thing that is certain is that the new materials to design skates allow considerable flexibility in determining the precise shape of the boot and thus fit specifications and the new materials have also permitted the added value of superior protection from injury since the outer boot construction can be incredibly stiff without compromising movement or fit.

Perhaps the most interesting research conducted in the past two decades has consisted of a series of research studies evaluating the impact of skate design on skating performance. Pearsall et al. [1] attempted to identify the majority of tasks that make up the game of ice hockey and we also conducted a video analysis of over 10 NHL games to identify the frequency of execution of various tasks during a game [12]. What was evident from these studies was that ice hockey was a very open and complex sport with a variety of skating tasks and a large frequency of each of these different tasks occurring during the course of a game. These tasks require a player to stop, start and pivot in different directions, skate forwards and backwards, skate in non-linear trajectories in tight turns and frequent crossover manoeuvres in forward and backward directions, pivot in several different ways, use the skate to stabilize and balance and as a lever as a skater engages in puck battles in open ice and along the boards against opponents. Thus, a skate becomes a very important instrument for skilful task execution during the course of a game of ice hockey.

The scientific community created many hypotheses regarding specific design changes or design features that could be made to skates that could offer advantages to a skater attempting to execute a large variety of complex skating skills. One feature of a skate boot that was thought to be a possible advantage in many aspects of skating was "freedom of movement", especially in the lower limb (hip, knee, foot and ankle) [13]. Greater freedom of movement in the foot/ankle could in theory alter limb kinematics substantially and in turn result in greater degrees of freedom when attempting to execute skating tasks.

One of the first interventions used to create a skate with greater freedom of movement was to simply cut off the tendon guard of the traditional skate boot. Early experiments examining this modification resulted in a greater range of motion (7°) in dorsi/plantarflexion during the skating stride as measured with electrogoniometers compared to a standard skate boot [14]. For safety reasons such a modification is not feasible as a design feature in ice hockey skates but the use of the new materials in modern skate design makes it possible to achieve the same or similar advantages without removing the tendon guard. Indeed many experiments have been conducted in the ice hockey research laboratories in the last decade to evaluate a skate designed to improve mobility in the foot and ankle with the use of more flexible materials in the tendon guard that in effect allows a greater range of motion. In all studies conducted a consistently demonstrated result has been an increased range of motion of approximately 5–7° in dorsi/plantarflexion in these skates compared to standard unmodified modern skates [15].

Range of motion in the ankle is of particular importance. Restriction of this joint would prevent the ankle plantar flexors from generating optimal mechanical power output. Experiments were conducted comparing standard skates and modified skates to shoes on a Biodex™ machine and measured passive and active torques in dorsi-plantarflexion. These experiments demonstrated that the torques and the range of motion of the modified skates were greater in the modified skate than in the standard skate and, in addition, compared more favourably to the range of motion and the torques generated in shoes where there is far less movement restriction [9]. Other research in support of the importance of range of motion has shown that restriction of the ankle motion with standard figure skates in turn limits knee amplitude with a resultant decrease in lower limb work output of the knee and ankle [16].

A good example of the effect of skate design on skating mechanics is the well-known work of the Scandanavian researchers comparing standard speed skates to the now popular klap-skate in speed skating [10]. Although ice friction, body kinematics and muscle activation characteristics were similar when both skate types were compared, the klap-skate resulted in 5 % more velocity and a 35 W increase in power output [10], leading to profound performance differences in long track races. When the klapskate was first invented, people were reluctant to use it because there were kinematic skating technique adaptations needed to be made by the skaters in order to take advantage of its design. But after some groups trained only with these skates, they started to see results and many world records in speed skating were broken [10]; the entire speed skating community quickly followed suit and all of the elite long-track athletes are now using the klapskates. Whether the same benefits can be achieved in ice hockey has yet to be determined.

It has been shown that larger forces exerted from the leg muscles during the skating stride and ultimately onto the ice surface will lead to faster skating [17]. But how can ground reaction forces (GRF) applied to the body from the ice during skating be measured? Traditional force plates used in human gait analysis cannot be used in the ice environment. One solution has been to adhere strain gauge force transducers to the skate blade holders [18], see Fig. 9.2. Strain signals may be directly recorded to a data acquisition device worn in a backpack by the player or transmitted wireless to an RF or Bluetooth receiver and PC. The person can then perform many different ice hockey tasks, such as shooting, skating, and stopping while dynamic force measures are recorded. Resultant vertical, medio-lateral and total GRF during the skating may then be viewed graphically (Fig. 9.2,[18]).

During forward skating in a series of studies the overarching observation was that kinetics were not different when comparing a modified skate design to a standard skate. It is important to note that in the majority of the studies, the one consistent result is an increased range of motion with the modified skate design [15, 19]. Additionally, most experiments showed that power, force and work were not significantly different between skate designs during forward skating tasks. Kinetic measures were similar in these studies in the modified versus standard skates as was the time to completion of full out forward skating tasks.

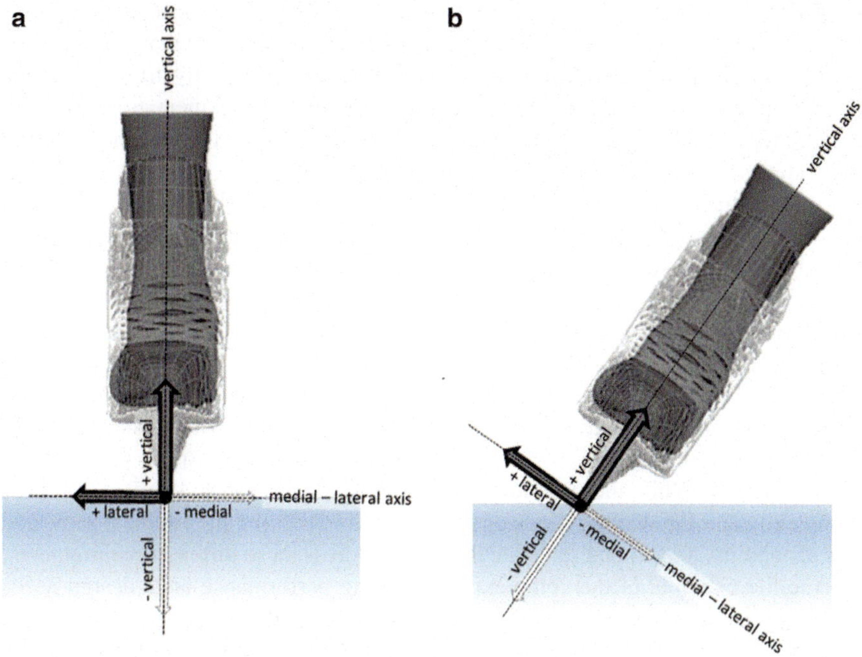

Fig. 9.2 Force transducer placement on the skate (**a**) with medio-lateral, vertical, and total forces for *left* and *right* skate strides (**b**) (adapted from [18])

Another recent study used pressure sensors to measure the Centre of Pressure (COP) in the heel, mid-foot and toe region of standard and modified skates. This study noted two important observations. The first observation was that most of the force generated in both skate types was generated primarily from the heel region. Greater force application was seen only in the toe region with the standard skate and it occurs only at the very end of the contact phase of the stride. A second important observation was that the excursion of the COP during all phases of the skating stride was diminished (by 65 vs 85 mm less forward) in the modified skate compared to the standard skate. The reason for this difference between skates is not completely clear at this time but may be related to a greater freedom of movement in the modified skate allowing for greater stability and balance because of a different kinematic organization of the forward skating stride. Although a full whole body kinematic analysis has not been completed, these results would imply that the kinematics of a skating stride in the modified skates is fundamentally different from standard skates. The modified skate likely allows for a different interaction of the hip knee and foot altering kinematics of the lower limb. Whether this is altered in a favourable fashion will need to be clarified with whole body kinematics analysis [20]. Thus, at least during forward skating other considerations may be important when determining the best method of skating to achieve maximum speed and efficiency.

The concept of improving "skating performance" during ice hockey skating tasks is a very complex one and is even more complicated due to the fact that a player must use his equipment and confront opponents in a variety of and vastly different skating conditions. All of these tasks are very different using different kinematic and kinetic paradigms. Not all of these tasks require full range of motion or maximum power generation.

Investigations have been conducted that illustrate this concept well. One such study used pressure sensors distributed in all areas of the foot including medial, lateral and plantar surfaces inside the skate boot to examine the pressures generated during the execution of tight turns. The most notable observation from this study was the very high pressures generated on the lateral (inner foot) and medial (outer foot) calcaneous of the skate boot in skilled skaters throughout tight turn execution [21]. These observations concluded that in the case of a tight turn, support was the primary requirement in the lateral and medial parts of the skate. Plantar pressures were diminished as pressures in the medial/lateral aspects of the skate boot increased. While these findings are perhaps not surprising, they highlight the complexity and the varying needs of skills performed during the course a game and this also points to the need for skate design to address these different requirements.

Another example of the complexity of skating performance and "design needs variation" is highlighted in another recent study where stops and starts were examined [22]. During the stop phase of a stop-and-go skill in ice hockey players, it was noted that peak and average vertical forces as well as the impulse generated in modified skates were significantly lower (by 10–20 % body weight) when compared to standard skate construction. Thus, skates that offered a greater range of motion in their design resulted in altered skater behaviour in modified skates. Specifically, lower vertical forces imply a different vector of approach during braking and thus, altered kinematics (Fig. 9.3). Presently, it is difficult to determine if this represents an advantage during the execution of this task and a whole-body kinematic analysis

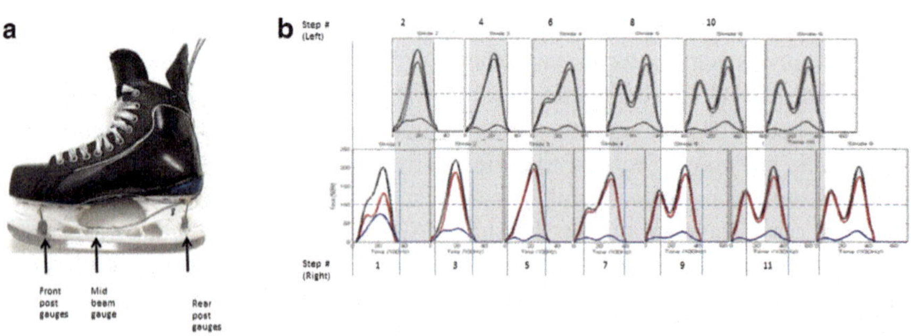

Fig. 9.3 Medial-lateral and vertical force vector components during upright (**a**) and inclined (**b**) skate orientations

will be necessary to determine how kinematics are altered and whether current design approaches represent an improvement/superiority of the modified design compared to the standard skate.

The design of the ice hockey skate has come a long way from its original model. Today, designers are trying to manufacture models which balance performance, comfort, durability, support, and weight. With the advancement in composite materials and testing protocols, one can only believe that the design of ice hockey skates will keep moving forward; with the hope of providing players with even better skating experiences and performances.

9.2 Stick Design and Performance Measures

As with many modern sports, ice hockey has become increasingly sophisticated in terms of technological innovations, equipment design and improvements in training, coaching and game strategies [1, 2, 23]. Ice hockey requires a different stick skill set due to the unique characteristics of the playing surface such as it is low surface friction and low temperature (−10 to −3 °C). In addition to ice skates, the other characteristic articles of the game of ice hockey are the puck and the stick, which are the primary object pursued and the primary tool to manoeuvre the puck, respectively.

Ice hockey pucks are made of natural rubber and fillers, usually carbon black [2]. Additives such as sulphur and anti-oxidants are used in the curing process and increase the puck's strength and hardness [24]. Carbon also helps improve the ability of the rubber to resist wear in hockey pucks. Pucks are often frozen before use in the game, thereby reducing puck bounce for better control and more uniform bounce and glide response. Controlled drop tests on to a rigid surface have shown the coefficient of restitution (e) of hockey pucks range between 0.45 and 0.55 at room temperature and 0.12 and 0.27 when frozen [2]. The dynamic stiffness of pucks is not constant, but becomes increasingly inelastic with higher collision speeds [25]. A cold puck yields a low coefficient of friction over an ice surface, providing for high and sustained puck glide velocities when passed over several metres from one player's stick to another team mate.

The ice hockey stick is so named because it was originally a "stick" carved from hard woods such as elm, birch, aspen or ash [26, 27]. The main stick skill categories are shooting, puck control, and passing/receiving [28] as well as checking and face offs [29]. The player controls the stick bimanually to extend his arm reach to the ice surface and the puck, as well as to amplify his trunk rotation and arm swing movements to propel the puck during blade-puck collisions towards the opposing team's net. Sticks can also serve to store and release shaft bend elastic energy to amplify the power of the shot. Successful skill execution depends on the mechanical interaction of the puck and the stick's blade.

The ice hockey stick is "L" shaped and its two main features are the shaft and blade (Fig. 9.4). The shaft must be a straight beam with a bevelled rectangular cross-sectional (3 cm wide by 2.5 cm thick) and range in length from 115, 137 up

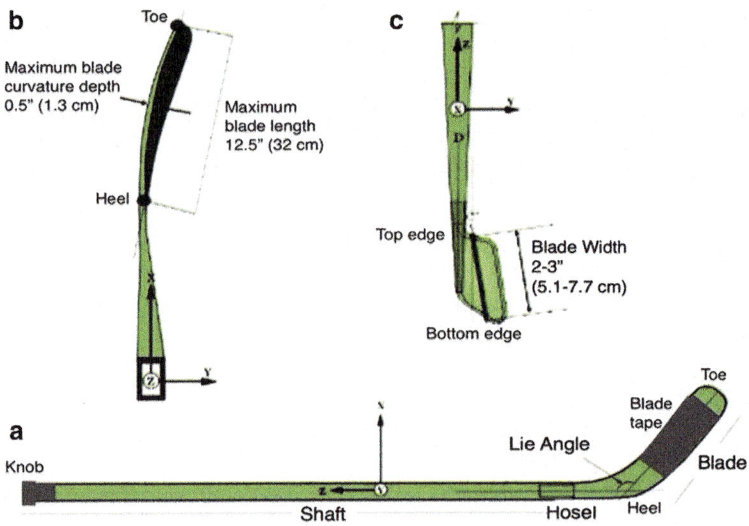

Fig. 9.4 The basic components of a hockey stick (**a**) and the blade as seen from top and (**b**) and side (**c**)

to 160 cm for Junior, Intermediate and Senior equipment size groups, respectively [30, 31]. The top end of the shaft is called the "butt" (handle) while the lower shaft end "hosel" tapers to unite with blade.

The blade is a lateral curved projection (with maximums of 32 cm length, and varying by 5–7.5 cm width and thickness from 5 to 10 mm) and is divided into heel, face and toe, the latter of which can be shaped round or square. The angle between the blade and the shaft is called the "lie angle" and varies from 137° to 131°. Low and high angles move the blade (when flush to ice surface) further or closer to the player, respectively. Players choose lie angles and stick lengths to match their standing height and preferred posture and skating technique. Furthermore, blade shape or pattern is characterized by its curve depth (0–19 mm), curve location (from heel to toe) and face angle (open to closed) orientation. A blade's curve is classified as a heel-, mid-, or toe-curve based on the location of the origin of the curve. To determine the curve type the blade is laid flat on the ice and viewed directly from above (Fig. 9.4). A variety of blade shapes with the face angle open or closed exist along the blade's long axis. Cloth tape is typically wrapped around the top of the shaft for better grip and around the blade for better puck control.

General stick dimensions are governed by regulations [30, 31]. Optimal length of a stick varies greatly when wearing skates and depends on the skater's posture while skating on the player's playing style. Some players prefer longer sticks to enable harder shots while others will use shorter sticks, an advantage for manipulation and puck control. As a general guide the upright hockey stick, from toe blade to butt, will extend upward to the chin or slightly but sometimes substantially below the chin (7.5 cm).

Players hold the stick either left (right hand gripping the top of the stick shaft) or right (left hand on the top of the stick shaft) and the concave curve will typically be oriented forward. Factors determining initial left-versus-right laterality selection vary with intrinsic and extrinsic context variables, but at higher skill levels one's stick side becomes strongly inter-related to playing position, such as left or right wing or playing side on the rink surface [32, 33].

As noted above, the blade's curvature cups the puck thereby providing greater control over puck movement and shot accuracy trajectory placement. There are numerous blade profile variations available. Players choose their blades based on their perception of optimal performance for puck control, passing/receiving the puck and shooting outcomes control.

9.3 Stick Skills

The hockey stick is used mainly to move the puck across the ice surface within the player's proximity and control, termed "possession". The manner and sequence of stick movements is often referred to as "stick handling". For this task, players will choose a stick based on weight distribution and effective inertia that affords them the best perception of "feel" [34]. With brief visual scans in combination with tactile and proprioceptive senses, players can successively manipulate the stick and blade orientation to push the puck in the desired vector. The stick is also used to pass and receive the puck to between your own team's players, as well as to steal, or "check" the puck away from the opposing player. Though these above represent a majority of the stick skills used in the game, the fundamental mechanics and human factors involved have received little scientific investigation.

Conversely, shooting tasks have been the most studied skills in ice hockey given that they are the primary tactical means of scoring [28]. Players commonly utilize a wide variety of shots during a typical game situation (e.g. slap, wrist, snap, backhand and sweep). The type of shot used is player and position specific [12, 35]. The ability to propel the puck at high speeds is the most common performance criterion measure. Puck speeds have been recorded using different techniques: high speed video, radar guns, light traps and accelerometers [36–43]. Puck velocities are typically greater for slap shots (80–160 km/h) than wrist shots (50–80 km/h) [41, 42].

All shot techniques involve a sequence of specific stick movement phases: preloading (stick blade-ice contact), loading (shaft deflection), blade-puck contact, and release (shaft recoil and release termination of puck contact) [41, 43–45]. The more vigorous slap shot includes prior stick back- and down-swing phases to amplify energy input to the puck projection. The mechanical factors of importance during the slap shot are stick speed prior to puck contact, pre-loading bending of the stick and puck-blade contact time [29, 43].

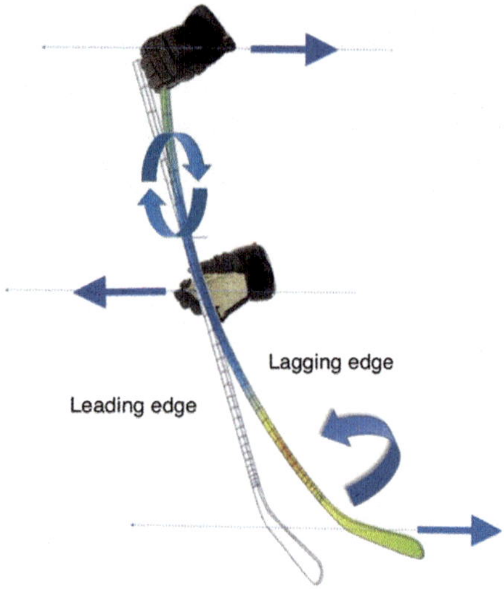

Fig. 9.5 Free body diagram of forces acting along the shaft during a slap shot. The bending moments deflect the shaft to store elastic energy that upon recoil contribute to greater puck projection velocity

The greater the stick blade's tangential velocity and shaft bend prior to puck contact, the potentially greater energy transfer to the puck. With slap shots, this was demonstrated by Lomond et al. [45] using 3D reconstruction from high speed video records (1000 Hz) of slap shots. Across a wide range of player skill levels from recreational to elite subjects, stick blade velocity and shaft bend were closely related to recorded puck speed. To achieve these top speeds requires optimal body segment "kinetic chain" movement coordination to both rotate and bend the stick during ice and puck contact events.

As noted previously, bending the stick's shaft allows stored elastic strain energy to be transmitted to the puck, which augments shot speed due to stick rotation is augmented by bending (or preloading) of the stick's shaft [2, 46]. In general, the player's choice of stick stiffness rating should match their strength and body mass, as well as personal preferences for proprioceptive response. To bend the stick, the player must use three points of contact: the (1) upper hand/top of stick and (2) blade-ice pressing backwards against the stick, and the (3) lower hand/mid stick shaft pressing forward (Fig. 9.5). This creates the transient bending moment to flex the stick shaft [47]. Initially, the stick blade must obliquely impact the ice surface at an angle of approximately 35° to the vertical to provide sufficient vertical reaction force (125 N or $\sim 1/5$ body weight) at point (2). Peak stick deflection angles may range up to 20° during slap shots by elite players. Too upright a blade's orientation yields insufficient ice contact reaction (20 N) due to low ice friction to bend the shaft [41, 45]. The latter is a common error shown by recreational players [42].

Given that elastic storage is transient and the total ice contact time is less than 60 ms, the stick's swing movement and load/unload response must be synchronized to yield the greatest energy propulsion of the puck. Skilled players have learned heuristically to better coordinate these events than novice players. In a study of using high-speed video recording (1000 Hz) and an accelerometer embedded within the puck, [43] identified that puck speed was highly correlated to both stick bend and total puck contact time (38 ms for elite vs. 27 ms for rec), but not the average acceleration force applied to the puck (64 g's). On further inspection, elite shooters better timed the bend/unbend(recoil) response of the sticks to be in phase with puck contact whereas recreational shooter lost contact with the puck well before stick recoil, thus losing this energy transfer to the puck. In summary, skilled players take better advantage of the stick's properties to project the puck.

Mechanical testing during stick construction and manufacturing is essential to determine a stick shaft's flexion rigidity and elasticity. Though industrial standards have not been established to quantify shaft stiffness or other properties [48], manufacturers categorized their sticks based on their own "flex" rating. Shaft stiffness may be estimated by controlled force and bend (or displacement) of the shaft [29, 49]. Typically, shaft (beam) stiffness is determined using a three-point bending test with central and/or cantilever loading. Using these techniques manufactures have produced sticks in a wide range of stiffness. In general, corresponding to player weight and strength, stick range from 40 to 50 flex for youth and junior players, 60–75 flex for intermediate players, and 85–110 flex for adults and more experienced players. It should be noted that cutting a stick's length to accommodate player's height and preferences effectively increases the flex rating by 5–10 points for each 2.5 cm shortened.

The resulting stiffness of stick shafts may be determined in quasi-static loading tests. For instance, using a cantilever length of 1.17 m between clamp and blade load points, the effective stiffness (Young's moduli) of composite sticks has been found to range from 31 to 42 GPa as compared to wood shafts at 10–12 GPa. The larger variation in elastic properties of composite materials depends in large part on the fiber material, the angle of the weave [50] as well as the geometry of the shaft cross-sectional area and wall thickness [51]. Failure modes of sticks vary. Prior wood model stick's durability were found to decrease with age and duration of use. The most common failure modes during shots occurred at the hosel junction of blade and shaft, mid-shaft break or delamination, and blade break or delamination [52]. Conversely, composite sticks retain greater mechanical flex stability with age; however, accumulation of microfibre fractures due to wear from cyclic shot shaft flexion and/or stress point risers from nicks due to stick-to-stick collisions with the opponent players ultimately leads to catastrophic structural failure during high shot loading.

The dynamic behaviour of stick (and resulting puck velocities post impact) are not easily extrapolated from static laboratory measures [29, 53]. For example, Bigford and Smith [48] demonstrated that the impact properties (i.e. the effective coefficient of restitution, COR) of the puck-to-blade are complex. Intrinsic stick shaft unloading (recoiling) response may contribute as much as 30 % to the outgoing puck velocity. However, impact characteristics were substantially affected by puck temperature, with COR varying from 0.27 at −4 °C to 0.39 at 22 °C, that in turn inversely affected the puck-to-stick reactive forces by 30 kN and 12 kN, respectively [48]. In addition, given that manufacturing process of carbon composite sticks permits full control of material and construction parameters, more sophisticated flexion-stiffness profiles along the shafts length are possible (as opposed to one uniform flex parameter). This has led to the introduction of sticks with varied flexion points such as mid- and low-kick point. By varying these properties, shot contact duration and release characteristics may be tuned to match player preferences and/or perceptions [54, 55].

In summary, this section has reviewed the conventional design and material practices as well as mechanical assessment of the ice hockey stick. Expanded fundamental research is required to better understand stick performance from shooting speed and accuracy [56, 57] as well as include other tangible skills such as stick handling [34].

References

1. D.J. Pearsall, R.A. Turcotte, S.D. Murphy, Biomechanics of ice hockey, in *Exercise and Sport Science*, ed. by W.E. Garrett, D.T. Kirkendall (Lippencott, Williams & Wilkins, Philadelphia, 2000), pp. 675–692
2. A. Haché, *The Physics of Hockey* (Johns Hopkins University Press, Baltimore, 2002)
3. F. Formenti, A.E. Minetti, Human locomotion on ice: the evolution of ice-skating energetics through history. J. Exp. Biol. **210**(Pt 10), 1825–1833 (2007)
4. B.G. Minkoff, J.V. Simonson, Ice hockey, in *Sport Injuries: Mechanisms, Prevention and Treatment*, 2nd edn., ed. by F.H. Fu, D.A. Stone (Lippincott Williams and Wilkins, Philadelphia, 1994), pp. 397–444
5. M. Jenkins, A.J. Subic (eds.), *Materials in Sports Equipment*, vol. 2 (Woodhead, Cambridge, 2007)
6. C.R. Gheorghiu, D.J. Pearsall, R.A. Turcotte, Quantifying fit in ice hockey skate boots. Paper presented at the ISBS-Conference Proceedings Archive, 2008
7. R.N. Humble, H. Smith, Skating, in *Athletic Footwear and Orthoses in Sports Medicine*, ed. by M.B. Werd, E.L. Knight (Springer, New York, 2010), pp. 247–266
8. R.A. Turcotte, D.J. Pearsall, D.L. Montgomery, An apparatus to measure stiffness properties of ice hockey skate boots. Sports Eng. **4**(1), 43–48 (2001)
9. D.J. Pearsall et al., Ice hockey skate boot mechanics: direct torque and contact pressure measures. Procedia Eng. **34**, 295–300 (2012). http://dx.doi.org/10.1016/j.proeng.2012.04.051
10. H. Houdijk et al., Push-off mechanics in speed skating with conventional skates and klapskates. Med. Sci. Sports Exerc. **32**(3), 635–641 (2000)
11. J. Kekoni et al., Mechanical sensibility of the sole of the foot determined with vibratory stimuli of varying frequency. Exp. Brain Res. **78**, 419–424 (1989)
12. D.L. Montgomery et al., Task analysis (hitting, shooting, passing and skating) of professional hockey players, in *Safety in Ice Hockey: 4th Volume*, ed. by D.J. Pearsall, A.B. Ashare (ASTM International, West Conshohocken, PA, 2004)
13. T. Upjohn et al., Three-dimensional kinematics of the lower limbs during forward ice hockey skating. Sports Biomech. **7**(2), 206–221 (2008)
14. C.J. Dewan, Biomechanics of the foot and the ankle during ice hockey skating. MA thesis, Unpublished Master's Thesis, M.Sc., McGill University, 2004
15. X. Robert-Lachaine et al., Impact of hockey skate design on ankle motion and force production. Sports Eng. **15**(4), 197–206 (2012)
16. M. Hagenauer, P. Legreneur, K.M. Monteuil, Influence of figure skating skates on vertical jump performance. J. Biomech. **39**(4), 699–707 (2006)
17. M.B. Bracko, Biomechanics powers ice hockey performance. Sports Med. **11**, 47–53 (2004)

18. T. Stidwill et al., Force transducer system for measurement of ice hockey skating force. Sports Eng. **12**(2), 63–68 (2009)
19. T.J. Stidwill, D. Pearsall, R. Turcotte, Comparison of skating kinetics and kinematics on ice and on a synthetic surface. Sports Biomech. **9**(1), 57–64 (2010)
20. C. Legnoc, Force and center of pressure measurement during forward skating with a standard hockey skate and a modified hockey skate. MA Thesis, Master's Thesis, McGill University, 2012
21. J.S.G. McGrail, Skate boot pressure analysis of elite and recreational ice hockey skaters during the execution of tight turns, Master Thesis, McGill University, 2006
22. S. Forget, Comparisons of player calibers and skate models during an ice hockey explosive transitional maneuver, Master Thesis, McGill University, 2013
23. D.J. Laliberte, Biomechanics of ice hockey slap shots: which stick is best? Sports J. **12**(1), 1 (2009)
24. T. Falconer, How hockey works, Equinox, January, 1994
25. T. Aspelmeier, F. Gerl, A. Zippelius, A microscopic model of energy dissipation in granular collisions, in *Physics of Dry Granular Media*, ed. by H.J. Herrmann, J.-P. Hovi, S. Luding, vol. 350 (Springer, Netherlands, 1998), pp. 407–412
26. G. Vaughn, *The Puck Starts Here: The Origin of Canada's Great Winter Game: Ice Hockey* (Goose Lane Editions, Fredericton, 1996)
27. B. Dowbiggin, *The Stick: A History, a Celebration, an Elegy* (MacFarlane Walter & Ross, Toronto, ON, 2001), pp. 27–37, 65
28. R. Renger, Identifying the task requirements essential to the success of a professional ice hockey player: a scout's perspective. J. Teach. Phys. Educ. **13**, 180–195 (1994)
29. G.W. Marino, Biomechanical investigations of performance characteristics of various types of ice hockey sticks, in *International Society of Biomechanics of Sports*, Konstanz, 1998, ed. by H.J. Riehle, M.M. Vieten (1998), pp. 184–187
30. IIHF, *Official Rule Book -Section 2*, p. 222 (2010–2014), http://www.iihf.com/iihf-home/sport/iihf-rule-book.html
31. NHL, *Official Rule Book -Section 3: Rule 10* (2010–2011), http://www.nhl.com/ice/page.htm?id=27011
32. D. Coates, S.A. Fard, Returns to handedness in professional hockey, *International Association of Sport Economists (IASE); Working Paper Series*, pp. 11–21 (2011)
33. J. Puterman, J. Schorer, J. Baker, Laterality differences in elite ice hockey: an investigation of shooting and catching orientations. J. Sports Sci. **28**, 1581–1593 (2010)
34. P. Hove, M.A. Riley, K. Shockley, Perceiving affordances of hockey sticks by dynamic touch. Ecol. Psychol. **18**(3), 163–189 (2006)
35. H. Pileggi et al., Snapshot: visualization to propel ice hockey analytics. IEEE Trans. Vis. Comput. Graph. **18**(12), 2819–2828 (2012)
36. J.F. Alexander, J.B. Haddow, G.A. Schultz, Comparison of the ice hockey slap and wrist shots for speed and accuracy. Res. Q. **34**, 259–266 (1963)
37. J.F. Alexander et al., Effect of strength development on speed of shooting of varsity ice hockey players. Res. Q. **35**, 101–106 (1964)
38. B. Roy, R. Doré, Kinematics of the slap shot in ice hockey as executed by players of different age classifications, in *5th International Congress on Biomechanics*, ed. by P. Komi (University Park Press, Baltimore, 1976), pp. 286–290
39. E.G. Chau et al., *Mechanics of Hockey Injuries* (American Society of Mechanical Engineers, New York, 1973), pp. 143–154
40. R. Doré, B. Roy, Influence de la Rigidité des Bâtons sur la Cinématique et la Cinétique des Tirs au Hockey sur Glace', Technical Report Ep 78-R-5, Montréal, Qc. Ecole Polytechnique De Montréal (1978)
41. D.J. Pearsall et al., The influence of stick stiffness on the performance of ice hockey slap shots. Sports Eng. **2**, 3–11 (1999)
42. T.-C. Wu et al., The performance of the ice hockey slap shot and wrist shots: the effects of stick construction and player skill. Sports Eng. **6**(1), 31–39 (2003)

43. A. Villaseñor-Herrera, R.A. Turcotte, D.J. Pearsall, Recoil effect of the ice hockey stick during a slap shot. J. Appl. Biomech. **22**(3), 200–209 (2006)
44. P.E. Fait et al., Increasing task complexity and ice hockey skills of youth athletes 1, 2. Percept. Motor Skills **112**(1), 29–43 (2011)
45. K.V. Lomond, R.A. Turcotte, D.J. Pearsall, Blade position and orientation during an ice hockey slap shot. Sports Eng. **10**(2), 87–100 (2007)
46. J.T. Worobets, J.C. Fairbairn, D.J. Stefanyshyn, The influence of shat stiffness on potential energy and puck speed during wrist and slap shots in ice hockey. Sports Eng. **9**(4), 191–200 (2006)
47. L. Zane et al., Grip force measures during ice hockey slap and wrist shots, in *Canadian Society of Biomechanics Conference*, 6–9 June 2012
48. R.L. Bigford, L.V. Smith, Experimental characterization of ice hockey pucks and sticks, in *Safety in Ice Hockey: 5th Volume*, ed. by R.M. Greenwald, A.B. Ashare (ASTM International, West Conshohocken, PA, 2009), pp. 186–197
49. E. Simard et al., Static and dynamic characteristics of composite one-piece hockey sticks, in *XXIInd International Symposium on Biomechanics in Sports*, Faculty of Health Sciences University of Ottawa, Ottawa, ed. by M. Lamontagne, D.G.E. Robertson, H. Sveistrup (2004), pp. 515–518
50. D. Russell, L. Hunt, Spring constants for hockey sticks. Dissertation, Kettering University, Flint, MI, 2009
51. G.W. Marino, J. Cort, Matching the implement to the player: hockey stick research and development, in *International Society of Biomechanics of Sports,* Ottawa, 2004
52. E.F. Hoerner, The dynamic role played by the ice hockey stick, in *Safety in Ice Hockey, ASTM STP1050*, ed. by C.R. Castaldi, E.F. Hoerner (ASTM International, Philadelphia, 1989), pp. 154–163
53. M. McQueen, J. McPhee, Design of an ice hockey stick test machine (P178), in *The Engineering of Sport 7* (Springer, Paris, 2008), pp. 199–206
54. A. Hannon et al., Dynamic strain profile of the ice hockey stick: comparisons of player calibre and stick shaft stiffness. Sports Eng. **14**(2–4), 57–65 (2011)
55. H. Soroush et al., Modelling of the deflection of an ice hockey stick during a slapshot, in *Canadian Society of Biomechanics Conference*, 6–9 June 2012
56. Y. Michaud-Paquette, D.J. Pearsall, R.A. Turcotte, Predictors of scoring accuracy: ice hockey wrist shot mechanics. Sports Eng. **11**, 75–84 (2009)
57. Y. Michaud-Paquette et al., Whole-body predictors of wrist shot accuracy in ice hockey: a kinematic analysis. Sports Biomech. **10**(1), 12–21 (2011)

Chapter 10
Curling

Norikazu Maeno

Curling is a winter sport played by two teams on an ice sheet. Each team is composed of four players and each player slides a granite stone towards a target, called "house" and marked on the ice sheet. At the conclusion of each end, points are scored for the stones resting closest to the center of the house. The result of a game is decided by the majority of points at the completion of scheduled ends of play. Usually a game consists of ten ends.

A stone delivered with twist slides with a curved path or "curls." The path may be altered after release by "sweeping" the ice surface in front of the stone to allow it slide farther and to decrease the degree of curl. In this sense curling is quite unique among many target-based sports and requires a variety of strategy and skill to choose the ideal path and placement of a stone. Techniques and skills of delivery, shots, sweeping, strategy of games, etc. are found in many books (e.g. [1–4]), manuals and textbooks prepared by numbers of curling clubs, schools, and associations all over the world.

10.1 Origin and History of Curling

The exact origin of curling is obscure but it is obvious that the game of curling appeared in Scotland in the early sixteenth century as evidenced by an unearthed curling stone engraved with the date 1511. Another evidence is the two oil paintings dated 1565, "Hunters in the Snow" and "Winter Landscape with a Bird Trap" by Flemish painter Pieter Bruegel the Elder. The two paintings depict peasants playing

N. Maeno (✉)
Professor Emeritus Hokkaido University, Hanakawa Minami 7-2-133, Ishikari, Hokkaido 061-3207, Japan
e-mail: maenony@ybb.ne.jp

F. Braghin et al. (eds.), *The Engineering Approach to Winter Sports*,
DOI 10.1007/978-1-4939-3020-3_10

a game similar to curling on frozen ponds. In those days Scotland and the Low Countries of Europe had strong trading and cultural links and it is possible to say that curling was rather popular and widespread in the cold Northern European countries. In Scotland the system and rules of curling were extensively developed and improved.

Scottish immigrants brought with them the game of curling to Canada in the eighteenth century and then to the United States in the nineteenth century. Curling was further developed in Canada and by the twentieth century the equipment and facilities were standardized. Now curling is played all over Europe and has spread to Australia, China, Japan, Korea, and New Zealand, and was officially included in the program of the Olympic Winter Games in 1998.

10.2 Ice Sheet

The ice sheet for a game of curling is defined precisely in the Rules of Curling and Rules of Competition [5] as a flat level rectangular ice surface, 146–150 feet (44.50–45.72 m) in length and 14 feet 6 inches to 16 feet 5 inches (4.42–5.0 m) in width (Figs. 10.1 and 10.2). At each end of the sheet a “hack,” foot-hold used to start the delivery of a stone, is fixed, and a target, “house,” is marked and distinguished by colour. The house consists of four concentric circles, 1 foot (0.31 m), 4 feet (1.22 m),

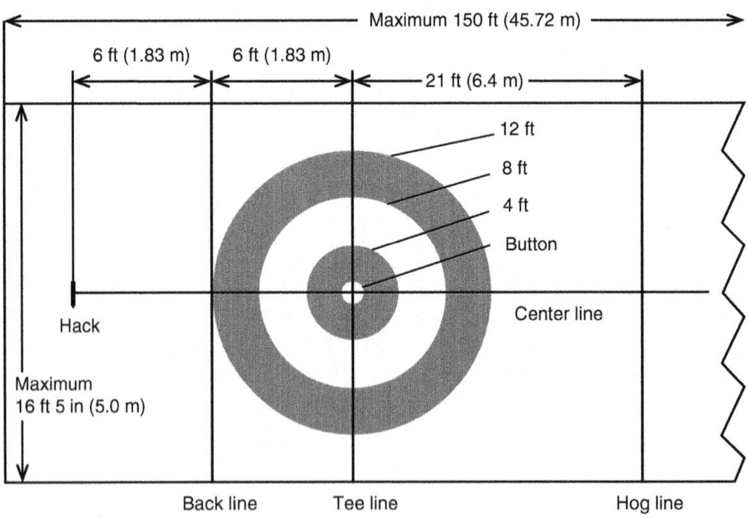

Fig. 10.1 Curling ice sheet. The distance between the two hog lines is 72 feet (21.94 m) and that between the two tee lines is 114 feet (34.74 m). More detailed dimensions in World Curling Federation [5]

Fig. 10.2 Photo of ice sheet

8 feet (2.44 m), and 12 feet (3.66 m) in diameter. The distance between the hack line and the "tee" line is 126 feet (38.41 m) and that between the "hog" line and the tee line is 93 feet (28.35 m).

The ice is usually made by freezing thin layers of water on the concrete floor in which a network of pipes is embedded. A chilled heat transfer fluid (brine) is pumped through the pipes to lower the temperature of the floor so that the water layer will freeze. The ice is built up to a thickness of a few centimeters by repeated supplies of water onto the surface. This procedure of freezing is vital to get clear ice since air, dust, and other foreign impurities in water, if any, can be removed from ice during freezing.

Before the game of curling the playing surface is prepared by spraying water droplets on the flat ice surface. The droplets freeze and form tiny ice bumps called "pebbles." Pebbles are extremely important for a curling ice sheet because pebbles make it possible for a curling stone to slide smoothly and curl. The size, form, and number density of pebbles are different in each curling rink but the height and diameter are roughly 1–2 and 3–10 mm respectively, and the number density is 1–10 pebbles per square cm (1×10^4 to $1 \times 10^5\ \mathrm{m}^{-2}$).

Pure water should be used for the preparation and maintenance of the ice sheet and pebbles. Any impurities in water, such as salts, acids, ions, and dust, may decrease the quality of ice. There are varieties of techniques to purify water, e.g., membrane filtration, ion exchange, reverse osmosis, distillation, deionization, and so on. Therefore it is not so difficult to obtain pure water suitable for curling ice. The purity of water may be easily checked by measuring the electric conductivity using simple conductivity meters. The electric conductivity of normal tap water is 10–100 mS/m (100–1000 μS/cm) at 25 °C and that of purified water is 0.01–1 mS/m (0.1–10 μS/cm) at 25 °C.

Details of technical information about curling ice can be found in the manual, Curling Ice Explained [6].

10.3 Stone and Broom

The curling stone is a circular shape and made of granite (Fig. 10.3). Its size is specified in the Rules of Curling and Rules of Competition [5]; weight including a handle and bolt 38–44 pounds (17.24–19.96 kg), maximum circumference 36 inches (91.44 cm), and minimum height 4.5 inches (11.43 cm). Its bottom is concave at the center so that the stone touches with ice at an annular band called "running band or edge," usually 0.13 m in diameter and 3–8 mm in width.

The interaction between the running band and pebbles on the ice sheet is a key process of curling. Combined with the characteristic property of ice their strange forms bring about the two important effects [7]. One is the reduction of friction coefficient due to the large force acting on pebbles. As the area of the running band is about 20 cm^2 the nominal pressure on the ice surface is roughly 0.1 MPa, but the actual pressure acting on each pebble tip is much larger, amounting to 0.4–8.1 MPa, and leads to the decrease in the friction coefficient of ice. It is the reason why curling stones slide so smoothly on the ice sheet.

The other effect is the mechanical deformation and fracture of ice pebbles, which occurs because the pressure created by the stone becomes comparable or larger than the yield or fracture stress of ice, usually a few MPa depending on temperature and strain rate [8]. Therefore it can be said that the slide of a curling stone on the ice sheet is not a smooth friction process but a complex process including smooth friction and mechanical abrasion.

Fig. 10.3 Curling stone. Curling stone (**a**) and its bottom and running band (**b**)

Abrasion of ice pebbles takes place because of large pressures as several MPa, or several 10 kg of loads acting on square centimeter area. It is often noticed that pebbles are deformed and broken and fine ice fragments and debris are formed on the ice sheet.

Brooms or brushes are used to sweep in front of a moving stone to clean and polish the ice surface. It is also used as a balancing aid during delivery of the stone. There are varieties of brooms; corn brooms, horsehair brooms, and more popular brooms having synthetic fabric over the pad surfaces.

Sweeping ahead of a stone raises the surface temperature of ice momentarily and allows the stone to travel farther and curl less or go straighter. Accurate measurements of the surface temperature of swept pebble tips are very difficult and not found in literatures, but we can apply varieties of common techniques to measure the average temperature rise of 1–2 °C by vigorous sweeping on the actual ice surface. The laboratory-based rubbing experiment [9] showed the maximum temperature rise about 0.8 °C at 2 mm below the ice surface.

10.4 Dynamics of a Curling Stone

The motion of a curling stone on an ice sheet is a two-dimensional motion of a rigid body so that the motion of the center of mass is exactly described with the following three equations,

$$m\frac{dV_x}{dt} = F_x \tag{10.1}$$

$$m\frac{dV_y}{dt} = F_y \tag{10.2}$$

and

$$I\frac{d\omega}{dt} = r \times F \tag{10.3}$$

where m and I are the mass and the moment of inertia of the stone, respectively, V (V_x, V_y) and ω are the translational and angular velocities, F (F_x, F_y) is the friction force caused by pebbles, and r is the radius of the running band. The air drag is negligibly small compared with the ice friction in the velocity range of curling game, usually less than 5 m/s. The ice friction F is the sum of friction forces caused by all the pebbles in touch with the running band,

$$F = \sum_{q=1}^{J} F_q = \sum_{q=1}^{J} \mu_q f \tag{10.4}$$

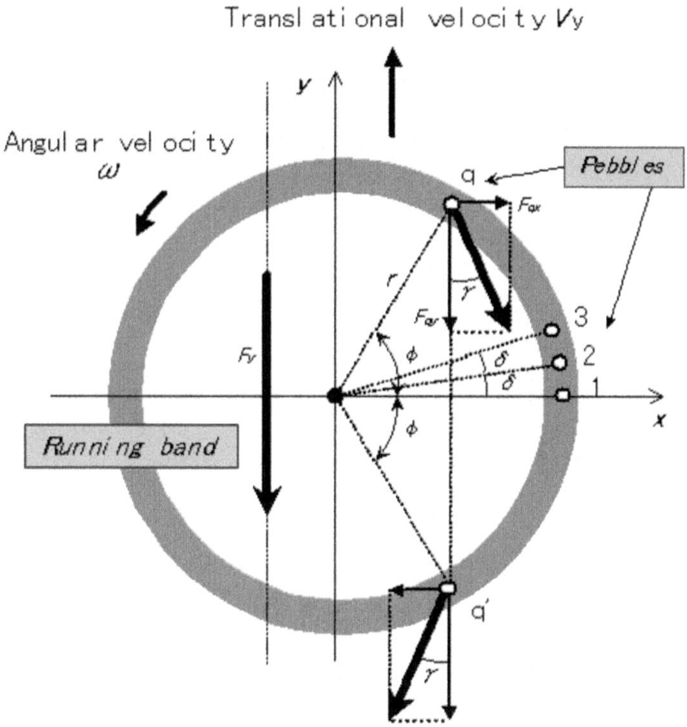

Fig. 10.4 Schematic diagram of a running band and friction forces. Pebbles are assumed to line up with an angle interval of δ and touch with the running band. Friction forces at the q-th pebble (angle ϕ) and those at the symmetric point (q') are shown, which are resolved into F_{qx} and F_{qy}. F_y is the total friction force in the y-direction expressed by Eq. (10.7), taking into account that friction force of each pebble on the left-hand side is larger than that on the right-hand side because the friction coefficient of ice increases with decreasing relative velocities

where J is the total number of pebbles, F_q is the friction force of each pebble, μ_q is the friction coefficient of ice, and f is the normal force exerting on each pebble (mg/J, g: the acceleration of gravity).

Assuming that pebbles line up with angle space $\delta = 2\pi/J$ on the running band (Fig. 10.4) we can write the relative velocity between the running band and the q-th pebble, U_q, as

$$U_q = [V_y^2 + 2r\omega V_y \cos\phi + (r\omega)^2]^{1/2} \tag{10.5}$$

where $\phi = (q-1)\delta$ is the angle. Writing the angle between U_q and y-axis as γ, the total friction force is decomposed into x- and y-components as

$$F_x = \sum_{q=1}^{J} F_{qx} = \sum_{q=1}^{J} \mu_q f \sin\gamma \tag{10.6}$$

and

$$F_y = -\sum_{q=1}^{J} F_{qy} = -\sum_{q=1}^{J} \mu_q f \cos\gamma \tag{10.7}$$

and the torque of rotation is

$$T = \sum (r \times F) = -\sum_{q=1}^{J} \mu_q fr \cos(\gamma - \phi) \tag{10.8}$$

The translational and angular velocities and the position at any instant can be calculated from Eqs. (10.1)–(10.3) by specifying the friction forces in Eqs. (10.4)–(10.8).

A simple computation of Eq. (10.6) gives $F_x = 0$. This means that there appear no effective lateral forces to make a stone to curl or slide in the x-direction. The strange fact may be more clearly understood in Fig. 10.4. Compare, for example, the x-component of friction force at the pebble, q, on the front running band with that at the symmetric pebble, q', on the rear band. They are equal in magnitude and opposite in direction so that they cancel each other. We find similar cancelation of lateral forces in all the combination of pebbles on the front running band ($q = 1, \ldots, J/2$) and the rear band ($q = J/2+1, \ldots, J$). Actual arrangements of pebbles may not be completely symmetric, but possible displacement of pebbles gives less than 10 % change so that there appear no effective lateral forces to make a stone to curl.

Computation of F_y in Eq. (10.7) gives a finite friction force leading to the deceleration of the stone motion with its line of action parallel to the y-axis, but actually the line of action of F_y deviates from the center of mass as shown in Fig. 10.4 because the friction coefficient of ice is larger at lower velocities so that friction force of each pebble on the left-hand side is larger than that on the right-hand side. The friction force, however, does not make a stone to curl, but only influences the spin or rotation motion through the torque of Eqs. (10.3) and (10.8).

The above consideration shows that the deflection of a curling stone cannot be explained by a simple left-right velocity difference or left-right friction force difference. It requires that F_x in Eq. (10.6) must have a finite value so that the friction force on the rear running band is larger than that on the front to explain the actual deflection, that is, a stone rotating clockwise curls towards right-hand side and anti-clockwise towards left-hand side. We will discuss physical models to produce the net lateral force in a later section.

For the practical purpose to calculate the motion of a curling stone without attachment to a specific physical mechanism, Maeno [10, 11] introduced a parameter, friction magnification factor A (≥ 1), which is defined as

$$\mu_q^R = A\mu_q^F \tag{10.9}$$

where μ_q^R and μ_q^F are the friction coefficients of the rear and front running bands, respectively. It was found that the magnitudes $A = 5\text{–}20$ can give realistic curl distances encountered in usual games. Figures 10.5, 10.6, and 10.7 give an example of calculation with $A = 10$ for a stone delivered from the hog line with the initial translational velocity 2.0 m/s and angular velocity 1.5 rad/s. The stone stops at 28.3 m with curl distance of 1.07 m. It is clearly shown that the curl distance increases more rapidly near the end of path. This corresponds to the rapid decrease of translational and angular velocities.

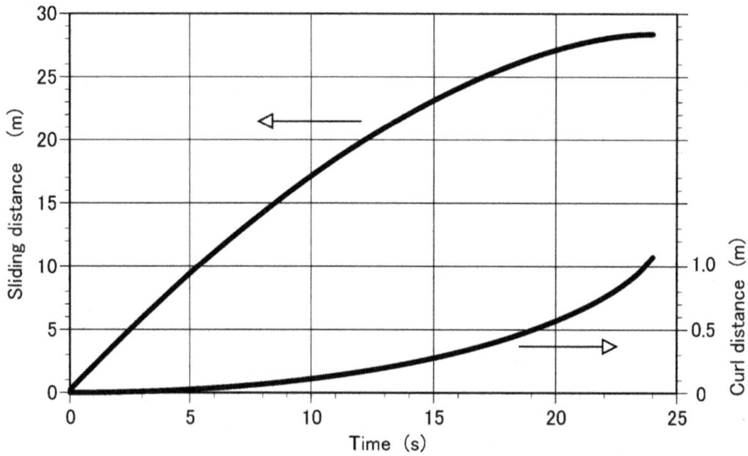

Fig. 10.5 Sliding and curl distances versus time. Initial translational velocity 2.0 m/s, angular velocity 1.9 rad/s, number density of pebbles $5 \times 10^4\,\text{m}^{-2}$, ice friction coefficient $\mu = 0.00148U^{-1/2}$ where U is in m/s, and friction magnification factor $A = 10$

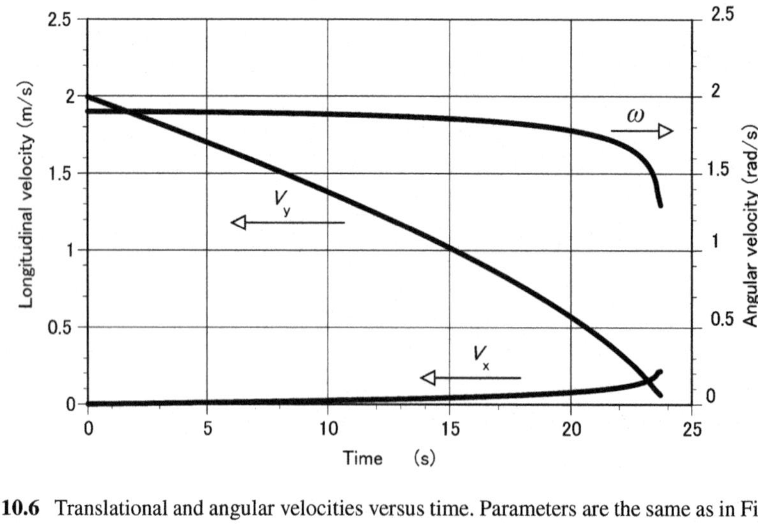

Fig. 10.6 Translational and angular velocities versus time. Parameters are the same as in Fig. 10.5

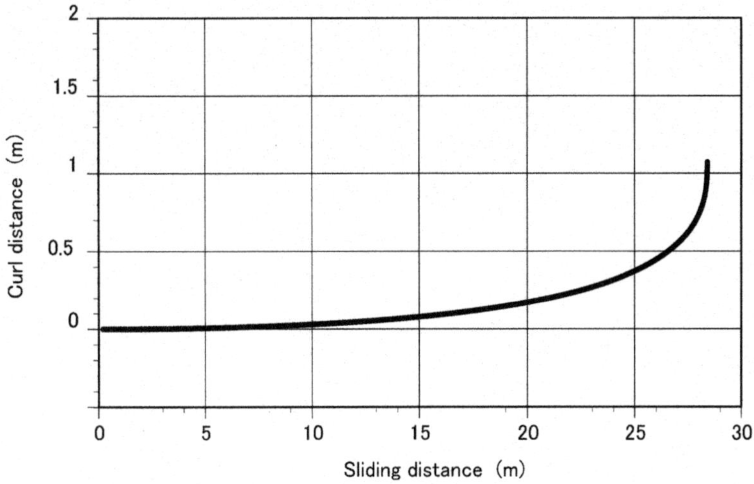

Fig. 10.7 Trajectory of the stone. Parameters are the same as in Fig. 10.5

It should be mentioned that in the above calculation the friction coefficient of ice was put to vary with the relative velocity as $U^{-1/2}$. The dependence is responsible for the rapid decrease of translational and angular velocities at the end of path. As summarized in Chap. 1 the physical mechanism of ice friction is water lubrication at sliding velocities above 1 cm/s. The friction force is caused by viscous flow of thin water layer, and the friction coefficient is described to vary with velocity as $\mu \sim U^{-1/2}$ at $U > 1\,\mathrm{cm/s}$ and $\mu \sim U^{1/2}$ at $U >$ a few m/s. The former relation is applied to our case of curling game.

10.5 Curl Ratio

The amount of curl or lateral deflection is an extremely important element in the strategy of curling, but most expressions to describe the property are unclear and obscure; the curl distance at stop is convenient but not satisfactory since it varies with different stopping distances, initial translational and angular velocities, and moreover it is a total deflection of the whole path and does not give the amount of curl of a stone in motion.

Maeno [10, 11] proposed a new quantity, curl ratio (C), to describe exactly the instantaneous amount of curl of a stone, which is defined as

$$C = \Delta X / \Delta Y \tag{10.10}$$

where ΔX and ΔY are, respectively, the curl and sliding distances in a short time Δt. C is a quantity to be defined at any instant, and if we know friction forces working in a short time, Eqs. (10.1) and (10.2) give the curl ratio:

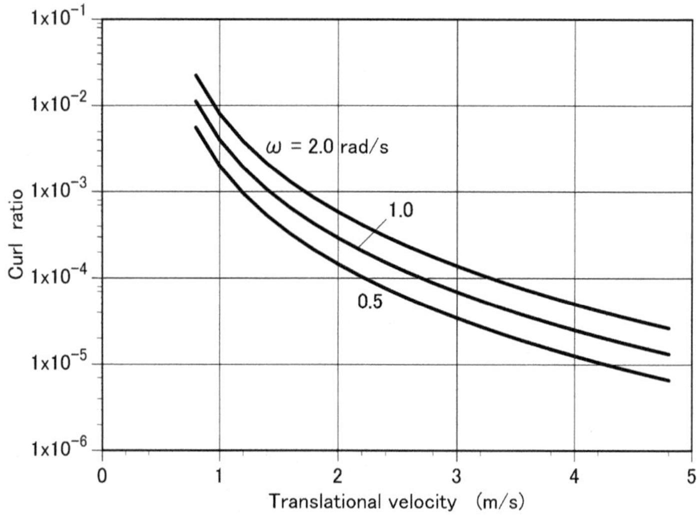

Fig. 10.8 Calculated curl ratio versus translational velocity. Ice friction coefficient $\mu = 0.007U^{-1/2}$, number density of pebbles $5 \times 10^4\,\mathrm{m}^{-2}$, $A = 10$ and $L = 0.5\,\mathrm{m}$

$$C = \left(\frac{F_x \Delta t}{2m}\right) \Bigg/ \left(V + \frac{F_y \Delta t}{2m}\right) = \frac{mF_x}{2F_y^2}\left(V + \sqrt{V^2 + \frac{2F_y L}{m}}\right)^2 \Bigg/ L \quad (10.11)$$

where L is the distance the stone slides in Δt, that is $L = (V + F_y\Delta t/(2m))\Delta t$. We can calculate C at any translational and angular velocities by specifying appropriate values of L or Δt. L is more convenient to specify since C thus estimated can be compared to actual observations and experiences.

Figure 10.8 gives the curl ratio calculated by Eqs. (10.6) and (10.11). Parameters were put as follows: number density of ice pebbles $n = 5 \times 10^4\,\mathrm{m}^{-2}$, ice friction coefficient $\mu = 0.007U^{-1/2}$, $A = 10$ and $L = 0.5\,\mathrm{m}$. The curl ratio decreases with the increasing translational velocity. In other words a stone curls more at lower translational velocities.

The calculation also shows that the curl ratio increases with angular velocity, but the result should not be considered very reliable since A was simply put constant in the calculation. There is a possibility that A might decrease with angular velocity and varies with the density and forms of pebbles.

Recently the curl ratio was measured by delivering a stone on an actual ice sheet maintained at $-3.1 \pm 0.3\,°\mathrm{C}$ [12]. Images of the sliding stone were taken with two CCD cameras from above at an interval of 0.1 s and analyzed to give the translational position of the center of mass and angle of rotation with accuracies of $\pm 1.7\,\mathrm{mm}$ and $\pm 0.0082\,\mathrm{rad}$ (0.47°), respectively. Figure 10.9 shows the curl ratio obtained by the analysis at $L = 0.5\,\mathrm{m}$ [13]. Larger values of curl ratio are found at smaller translational velocities. The result is in harmony with the calculated results.

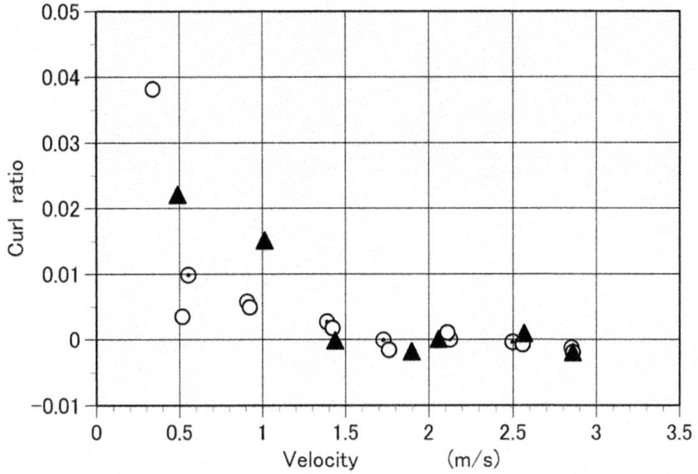

Fig. 10.9 Measured curl ratio versus translational velocity. Angular velocity: 0–1 rad/s (*dotted circle*), 1–2 rad/s (*open circle*), and 2–3 rad/s (*filled triangle*). Numerical data from Komagome et al. [13]

10.6 Sweeping

In order to discuss the motion of a curling stone in a variety of conditions we must have a general expression of the friction coefficient of ice because it varies with velocity, temperature, pressure, and so on. We may use the following expression for the present [10, 11]:

$$\mu = \mu_0 \left(\frac{P}{P_0}\right)^{-\frac{1}{4}} \left(\frac{T_m - T}{T_0}\right) \left(\frac{U}{U_0}\right)^{-\frac{1}{2}} \tag{10.12}$$

where μ_0, P_0, T_0, and U_0 are constants, T_m is the melting point of ice (273 K), and P and T are the pressure and temperature, respectively. This expression was derived by combining the results of theoretical and experimental researches [14–18], and discussed in some detail in Chap. 1. For convenience we put $P_0 = 0.1$ MPa, $T_0 = 5$ K and $U_0 = 1$ m/s, then μ_0 is the friction coefficient of ice at $P = 0.1$ MPa, $T = 268$ K (−5 °C) and $U = 1$ m/s. It should be noted in Eq. (10.12) that friction coefficient of ice decreases with increasing pressure, rising temperature, and increasing velocity, and gives great effects on the dynamics of curling stones.

Sweeping ahead of a stone raises the surface temperature of ice momentarily and allows the stone to travel farther and curl less or go straighter [9, 19]. These well-known effects are reasonably understood by the reduction of friction coefficient and curl ratio caused by the temperature rise. Figure 10.10 shows the effect of sweeping for trajectories of a stone delivered with the initial translational velocity 2.0 m/s and angular velocity 1.5 rad/s. Sweeping–1, sweeping–2, and sweeping–3 correspond,

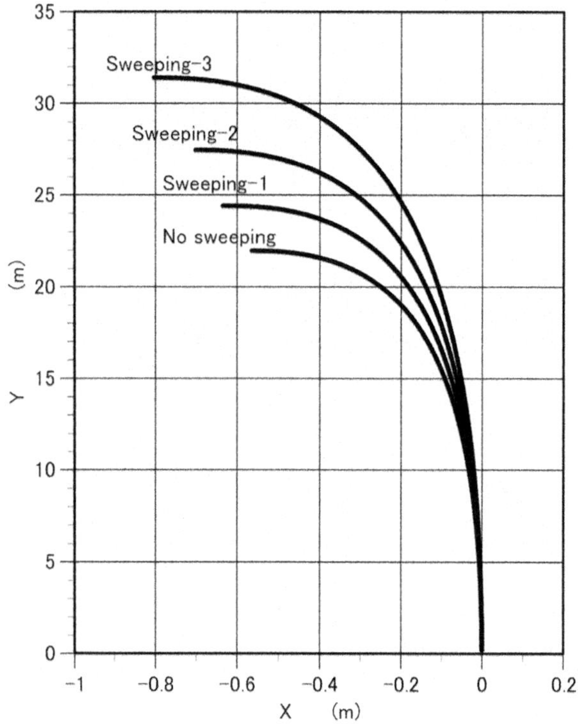

Fig. 10.10 Trajectories modified by sweeping. Initial translational velocity 2.0 m/s, angular velocity 1.5 rad/s, number density of pebbles $5 \times 10^4\,\text{m}^{-2}$ and $A = 10$. Ice friction coefficient $\mu_0 = 0.004$ and $P/P_0 = 1$ in Eq. (10.12). No sweeping, sweeping–1, sweeping–2, and sweeping–3 correspond, respectively, to $T_m - T = 5.0, 4.5, 4.0,$ and 3.5 K in Eq. (10.12). From Maeno [10, 11] with changes

respectively, to temperatures 0.5, 1.0, and 1.5 °C higher than not sweeping; the varieties are results of different speed, frequency, and pressure of the brush head movement curlers employ.

The increase of travel distance by sweeping is clear, but the effect of sweeping on the curl distance needs some careful evaluation since the curl distance at the stop is longer for more vigorous sweeping. Correct evaluation of sweeping should be made by comparison of the curl distance at the same travel distance. For example, at 20 m the curl distance without sweeping is 0.251 m, but it reduces to 0.178, 0.140, and 0.109 m for sweeping–1, sweeping–2, and sweeping–3, respectively.

10.7 Pebble Density

The number of pebbles in contact with a running band is important because it determines the pressure acting on each pebble. The minimum number that can support a standard curling stone without plastic deformation or fracture is estimated

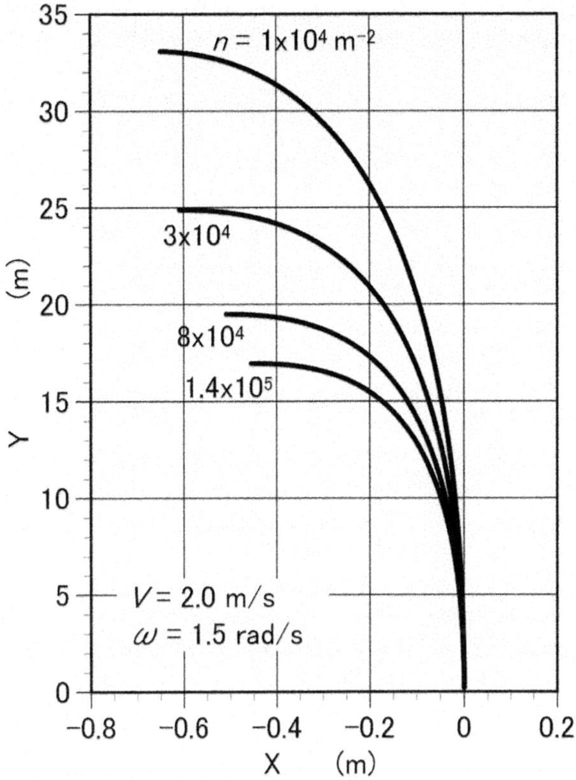

Fig. 10.11 Trajectories on an ice sheet of different pebble densities. Initial translational velocity 2.0 m/s, angular velocity 1.5 rad/s and $A = 10$. Ice friction coefficient $\mu_0 = 0.004$ and $(T_m - T)/T_0 = 1$ in Eq. (10.12). From Maeno [10, 11] with changes

as roughly 20 pebbles, which corresponds to the number density about 1×10^4 m^{-2} or one pebble per square centimeter. The maximum number depends on sizes of pebbles and running band; if we assume pebbles of 3 mm in diameter are closely arranged in a zigzag fashion on a standard running band it is roughly estimated as about 280 pebbles, which corresponds to 1.4×10^5 m^{-2} or 14 pebbles per square centimeter. The estimate implies that the variation in the number density of pebbles has physical meaning only in the range from 1×10^4 to 1.4×10^5 m^{-2}.

Figure 10.11 gives travel paths of a stone on an ice sheet of different number densities of pebbles. Initial translational and angular velocities are 2.0 m/s and 1.5 rad/s, respectively. At larger number densities the pressure on each pebble is smaller leading to the larger friction coefficient according to Eq. (10.12). As a result the sliding distance is shorter; the stopping distance becomes shorter as 33.0 m for 1×10^4 m^{-2}, 24.8 m for 3×10^4 m^{-2}, 19.5 m for 8×10^4 m^{-2} and 16.9 m for 1.4×10^5 m^{-2}. On the other hand, the curl distance is larger. At the point of 15 m, for example, it becomes larger as 0.046, 0.074, 0.120, and 0.177 m. The number of pebbles in touch with a stone can be changed by modifying the number density of pebbles in ice making or by using stones with running bands of different diameters and widths.

10.8 Level of Ice Sheet

We will mention briefly the level of an ice sheet since it is so important for the game of curling, unlike other sports played on ice as speed skating, figure skating, ice hockey, etc. We have discussed so far the motion of a stone only influenced by the friction force, but on a non-level ice sheet we must include a gravity force. Denoting the inclination angle of the ice sheet as α the component of the gravity force acting along the ice surface is $F_{\alpha} = mg\sin\alpha$. For example, its magnitude is 0.1 N for $\alpha = 0.0005$ rad (0.0286°), the sensitivity limit of normal spirit levels. If we assume the friction coefficient of ice as 0.001–0.01 the magnitude of friction force is $F_{\mu} = \mu mg = 0.2$–1.9 N. It is clear that the gravity force due to inclination cannot be ignored in the motion of a curling stone.

Figure 10.12 shows calculated trajectories of a stone delivered without rotation on an ice sheet inclined towards the positive direction of x-axis. The initial translational velocity is 1.5 m/s, and the inclination angle was varied from (a) 0.00001 rad (0.00057°) to (g) 0.0005 rad (0.0286°). In all cases the friction coefficient of ice is $0.004U^{-1/2}$ and a stone stops at 28 m. The trajectory (g) is for the largest inclination angle, 0.0005 rad, or 0.5 mm/m, which is the sensitivity limit of normal spirit levels. On this ice sheet the curl distance amounts to 2.4 m and is definitely unsuitable for normal curling games.

Floors of most buildings might be inclined much more so that it requires a great deal of technique and skill to make a sufficiently level and flat ice sheet. Many important experiences and techniques of flooding, cutting, etc. are explained deliberately and carefully in ice-make manuals [6], but there seem no clear definitions or rules about the maximum inclination angle permitted for curling.

For the present how level the ice sheet should be has not been established and the inclination angle may be too small to be measured with commercial levels. Therefore ice technicians are forced to make as level ice as possible by trial and error with the help of curlers delivering stones repeatedly to check the sliding and curl distances. The trajectories, b and c, in Fig. 10.4 might give some useful hints; the curl distances and inclinations of b and c are 0.24 m and 0.00005 rad (0.00286°) and 0.48 m and 0.0001 rad (0.0057°), respectively. The inclination angle around b and c may be considered a criterion. But the inclination corresponds to 0.25 mm over 5 m or 0.05 mm over 1 m, which cannot be measured easily by transit and other laser devices.

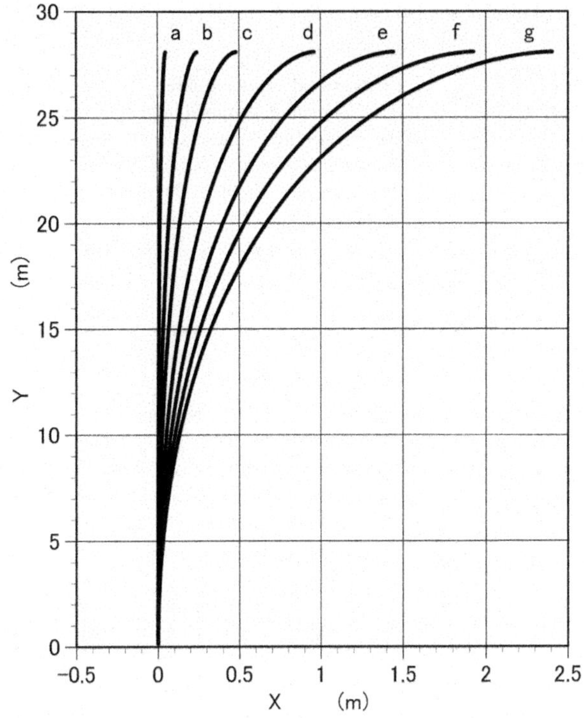

Fig. 10.12 Trajectories on an inclined ice sheet. The normal of the ice sheet is inclined towards the *x*-axis. Initial translational velocity 1.5 m/s with zero angular velocity, ice friction coefficient $\mu = 0.004U^{-1/2}$ and $A = 1$. Inclination angle: (*a*) 0.00001 rad (0.00057°), (*b*) 0.00005 rad (0.00286°), (*c*) 0.0001 rad (0.0057°), (*d*) 0.0002 rad (0.0116°), (*e*) 0.0003 rad (0.0171°), (*f*) 0.0004 rad (0.0229°), and (*g*) 0.0005 rad (0.0286°)

10.9 Mechanisms of Curl

It was shown in Sect. 10.4 that the lateral deflection or curl of a stone cannot be explained by a simple left-right asymmetry of velocity or friction force. Accepting that the stone curls because the friction force at the rear running band is larger than that at the front, the friction magnification factor, *A*, was introduced in Eq. (10.9). The physical mechanism to cause the front-rear asymmetry of friction force, however, has not been understood well [20] though several models have been proposed as discussed below. Table 10.1 summarizes the proposed mechanisms presented so far. Reports claiming the left-right asymmetry are not listed including the pioneering work of mechanics of curling by Harrington [21].

Macaulay and Smith [22] and Walker [23] remarked that the lateral deflection requires that the friction at the rear band should be greater than that at the front and that in consequence of retardation the pressure should be greater at the front than at the rear. Johnston [24] proposed that the deflection mechanism of a glass

Table 10.1 Physical mechanisms of curl

Authors	Models and remarks
Macaulay and Smith [22], Walker [23], Johnston [24]	**Pressure difference model**: The pressure at the front running band is assumed larger than the rear so that the friction coefficient is smaller. However, the friction force is not smaller but larger because the friction coefficient of ice decreases only slightly with pressure. The model cannot explain the curl correctly.
Shegelski et al. [25], Shegelski [26], Jensen and Shegelski [27]	**Water-layer model**: Water is assumed to be produced by frictional heating and dragged around by the rotation of a stone, lowering the friction coefficient of the front running band. However, the existence of such bulk water has not been confirmed and the transport mechanism is not obvious.
Denny [28]	**Snowplow model**: Picked up ice debris are assumed to be carried around by the rotation and accumulated at the front running band, leading to the reduction of the friction coefficient. However, the experimental results do not show the reduction of the friction coefficient of ice debris, and the transport mechanism is not clear.
Maeno [7]	**Evaporation-abrasion model**: The friction coefficient of the rear running band is assumed to be larger by cooling due to evaporation of pebble surface. At the rear band mechanical drag is added by ice fragments and debris formed by abrasion at the front band.
Nyberg et al. [29, 30]	**Scratch-guiding model**: Scratches formed on the surfaces of pebbles by asperities of the front running band are assumed to guide asperities of the rear band.

sliding on a table may be applied to the curling stone. If an empty top-down glass is rotated counterclockwise and allowed to slide on a smooth table, it deflects right because it tends to tip forward, resulting in the increase of pressure and friction force at the front. In the case of curling stone, he assumed that the similar increase in pressure at the front decreases the friction force F_F and leads to the deflection opposite to the glass.

However, his pressure-difference model cannot explain the curl of a stone on ice because it has been confirmed experimentally and theoretically that the friction coefficient of ice (μ) decreases with pressure (P) as $\mu \propto P^{-1/3}$ [15] or $\mu \propto P^{-1/4}$ [16], and therefore the friction force (F) increases as $F \propto \mu P \propto P^{2/3}$ or $F \propto \mu P \propto P^{3/4}$. That is, if the pressure at the front is larger than the rear the friction force is also larger, $F_F > F_R$, and the stone will curl in the same direction as a glass. In the case of ice the increase in pressure also leads to the increase in the friction force though the friction coefficient decreases.

The water-layer model was proposed by Shegelski et al. [25], Shegelski [26], and Jensen and Shegelski [27]. The model assumes that water layer produced by frictional heat is dragged around by the rotation of the stone and lower the friction coefficient of the front running band, leading to the friction force relation, $F_F < F_R$, so that a stone curls in the correct direction. The water they assumed seems to be bulk water, the existence of which, however, has not been confirmed experimentally. Furthermore the physical mechanism of transport of the water to the front half of the running band has not been clearly understood.

In the snowplow model Denny [28] assumed that intermediate-sized ice fragments and debris are carried around and accumulated at the front half of the running band, and that the friction coefficient is reduced due to friction of ice-ice rather than ice-granite. However, such reduction in friction coefficients has not been observed in the measurement of the ice-ice friction coefficient [31, 32]. Furthermore the accumulation of ice debris at the front half has not been observed, and the transport mechanism has not been confirmed.

The evaporation-abrasion model was proposed by Maeno [7] taking into account the two characteristic properties of curling, i.e., the forms of running band and pebble. Because of the peculiar form of the running band, each pebble rubbed by the front band will touch with the rear band after ten to hundred milliseconds. During the short time evaporation takes place on the tip surface to cool and lead to the increase in the friction coefficient of the rear band. Furthermore the peculiar form of pebbles is responsible for the abrasion or the mechanical deformation and fracture of pebbles due to high pressures exceeding the mechanical strength of ice. Ice fragments and debris produced by abrasion at the front running band will interact with the rear band producing mechanical drag to the rear band motion. The temperature decrease was calculated as about 0.7–1 °C at the sliding velocity 1–2 m/s leading to the increase of the friction coefficient by 19–26 %.

Much larger drag is expected to act to the rear band by the ice fragments and debris formed by the abrasion process. Ice debris and other obstacles on an ice sheet can be removed by sweeping, but the rear running band cannot avoid the ice debris produced by the front band. The ice fragments and debris produced by abrasion at the front running band may be a most essential drag force to cause curl since they are produced even if the stone does not rotate. It may give an explanation to the fact that a stone curls even if it is thrown without turns as evidenced in the next section.

Nyberg et al. [29, 30] proposed recently the scratch-guiding model. The model assumes that asperities of the front running band scratch the surfaces of pebbles, and then asperities of the rear running band encounter the scratches and guided by them, leading to curl. The process of scratch formation and guiding has not yet been formulated. The effect of scratches formed repeatedly on pebbles in various directions must also be taken account.

Considering the above discussions and the characteristic properties of ice at higher homologous temperatures (Chap. 1), it is reasonable to regard that the curl of a stone is caused by more than one physical mechanisms and that all the effects produced by these mechanisms can be reflected and expressed by a total or apparent friction coefficient. Their validity can only be verified by experiments though they are usually rather difficult to perform.

10.10 Measurements of Curl Distance

Finally we discuss the measurements of the curl distance. It was mentioned in Sect. 10.5 that the amount of curl can be accurately expressed with a physical quantity, curl ratio, which is defined by the ratio of the curl and sliding distances in a short time, Eq. (10.10). It gives the amount of curl of a stone at an instant of specific translational and angular velocities. In practice, however, the curl distance at stop is more useful, that is, the total lateral deflection from the start to stop.

Many curlers believe that the curl distance is rather insensitive to the angular velocity, but some curlers believe that it increases or even decreases with the angular velocity [1–4]. The uncertainty is attributed to the fact that most curlers judge the curl distance only by naked eyes and vision and there is no sufficient evidence to support one of the above beliefs because the accurate measurement of curl distance is very hard to make. We find only two research papers reporting the measurements of curl distance by Penner [20] and Jensen and Shegelski [27]. In both the measurements stones were thrown with various initial angular velocities along a center line in a local curling rink. The curl distance of stones which stopped within a house was measured together with the total number of rotations. The numerical data

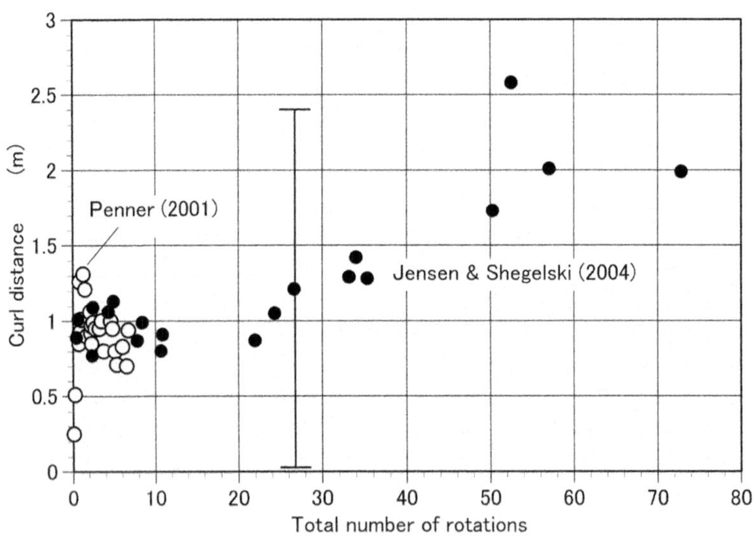

Fig. 10.13 Curl distance measured by Penner [20] and Jensen and Shegelski [27]. Numerical data: *open circle* from Fig. 8 in Penner [20], and *filled circle* from Fig. 7 in Jensen and Shegelski [27]. Curl distances were measured for stones which stopped within or close to the house. The initial translational velocity, ice temperature, pebble density and the accuracy of the measurements are not reported in the papers. The *vertical bar* is the possible maximum error ± 1.2 m for the curl distance measurements, which were estimated from the maximum skew angle 2.5° of shots from the y-axis [27]

of measured curl distance were read from the two figures in their original papers and plotted in Fig. 10.13. Scatter is fairly large but the general relation is clear, that is, the curl distance increases almost linearly with the total number of rotations, roughly from 1 to 2 m when the number of rotations increased from 0 to 70, in agreement with the second belief above. On the other hand, at smaller numbers of rotations, 0–10 rotations, where most curling games are played, the curl distance seems almost constant around 1 m in harmony with the first belief. But we notice also that it is not really constant but scattered in a range from 0.2 to 1.3 m.

The accuracy of the measurements is not given in both the papers, but it may give a serious effect to the measured data. The skew shot, that is, a stone thrown with a small angle against the center line, is one of the most difficult and significant factors to cause errors in the curl distance measurement. Jensen and Shegelski report that the skew angle was less than 2.5°, but the angle suggests a large error in the measured curl distance; 28 m times $\tan(2.5°) = \pm 1.2$ m, which is indicated by an error bar in Fig. 10.13. Taking account of large errors involved it can be concluded in Fig. 10.13 that the curl distance shows a general tendency of increase with the total number of rotations, amounting to 2 m at 70 turns, but any conclusions at smaller rotations are unreliable, that is, we have no grounds for the first belief.

Recently we have carried out a high-precision measurement of curl distance. The motion of a stone was recorded with two high-quality video-cameras and each image was digitally analyzed to give accurate translational and angular velocities together with the sliding and curl distances. Errors due to various sources were made as small as possible and corrections were made for skew shots, sliding and curl distances. Figure 10.14 shows an example of the results and more details were published by Hattori and others [33]. In the figure each data point gives the curl distance for a stone that was delivered with finite translational and angular velocities and stopped at a point 28.3 m apart from the start. The number of rotations during the slide was similarly estimated and corrected. We find clearly the negative proportionality, that is, the curl distance decreases almost linearly with the increasing number of rotations. This result gives an experimental evidence to the third belief of curlers. It should be emphasized, however, that the small change of curl distance with rotations could only be detected with this kind of high-precision measurement.

Finally we can summarize the experimental results obtained so far as follows: at extremely large numbers of rotations, 20 or larger, the curl distance increases with rotations. But it decreases slowly with rotations at smaller numbers of rotations where most curling games are played. It is now clear that a stone curls with a minimum rotation and that the degree of curl decreases with more rotations. We must note that the result has presented and pointed up a very perplexing and important subject that should be solved in the future.

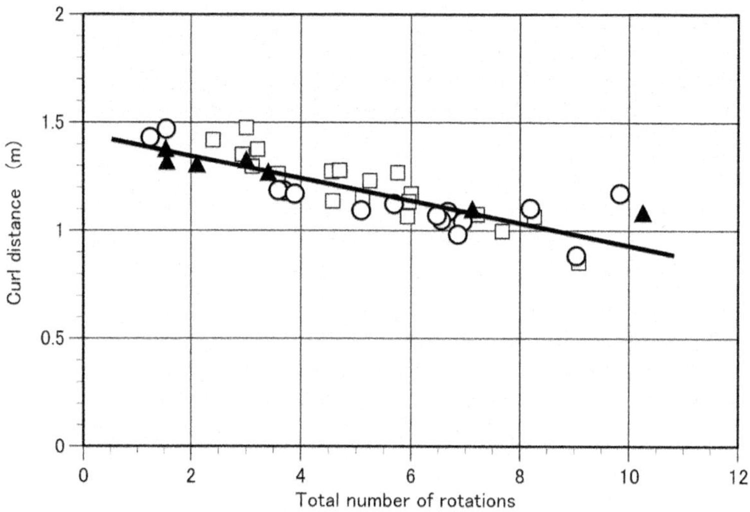

Fig. 10.14 Curl distance obtained by a high-precision measurement. The curl distance is a lateral deflection where the stone stops after sliding 28.3 m from the start. The total number of rotations refers to that during the 28.3 m slide. The difference of signs specifies different curlers. The ice surface temperature $-2.9\,^{\circ}$C and the number density of pebbles (4–6) $\times\ 10^4\ \mathrm{m}^{-2}$. The linear regression line is $D = 1.46 - 0.052N$, where D is the curl distance in m and N is the total number of rotations. Data from the report "Curling Experiment 201403 by Tokumoto, M., Kashiwazaki, K., Hattori, K. and Maeno, N.", presented at the second Curling Kenkyukai, held at Sapporo, June 5, 2014

References

1. W. Hansen, *Curling: The History, the Players, the Game* (Key Porter Books, Toronto, 1999)
2. B. Weeks, *Curling for Dummies* (CDG Books Canada, Toronto, 2001), 342 pp.
3. V. Pezer, *Smart Curling* (Fifth House, Calgary, 2007), 177 pp.
4. R. Howard, *Curl to Win* (HarperCollins New York, 2010) 192 pp.
5. World Curling Federation, *The Rules of Curling and Rules of Competition* (World Curling Federation, Perth, 2011)
6. L. Öhman, J. Minnaar, *Curling Ice Explained* (World Curling Federation, Perth, 2004)
7. N. Maeno, Curl mechanism of a curling stone on ice pebbles. Bull. Glaciol. Res. **28**, 1–6 (2010)
8. V.F. Petrenko, R.W. Whitworth, *Physics of Ice* (Oxford University Press, Oxford, 1999), 373 pp.
9. B.A. Marmo et al., Frictional heat generated by sweeping in curling and its effect on ice friction. Proc. Inst. Mech. Eng. L, J. Mater. Des. Appl. **220**(L4), 189–197 (2006)
10. N. Maeno, Dynamics and curl ratio of a curling stone. Sports Eng. **75**(1), 33–41 (2014)
11. N. Maeno, Erratum to: dynamics and curl ratio of a curling stone. Sports Eng. **75**(1), 43–44 (2014)
12. T. Nittono et al., Experimental study of curling: measurements of curl ratio of a curling stone. Seppyo **75**(3), 137–146 (2013)
13. T. Komagome et al., Motion analysis for curling stone, in *Proceedings of Symposium on Sports and Human Dynamics* **301**, 13–34 (2013)
14. P. Barnes, D. Tabor, J.C.F. Walker, The friction and creep of polycrystalline ice. Proc. R. Soc. Lond. **A324**, 280–298 (1971)

15. D.C.B. Evans, J.F. Nye, K.J. Cheeseman, The kinetic friction of ice. Proc. R. Soc. Lond. **A347**, 493–512 (1976)
16. P. Oksanen, J. Keinonen, The mechanism of friction of ice. Wear **78**, 315–324 (1982)
17. M. Akkok, C.M.McC. Ettles, S.J. Calabrese, Parameters affecting the kinetic friction of ice. ASME J. Tribol. **109**, 127–155 (1987)
18. N. Maeno, M. Arakawa, Adhesion shear theory of ice friction at low sliding velocities combined with ice sintering. J. Appl. Phys. **95**(1), 134–139 (2004)
19. J.L. Bradley, The sport science of curling: a practical review. J. Sports Sci. Med. **8**, 495–500 (2009)
20. A.R. Penner, The physics of sliding cylinders and curling rocks. Am. J. Phys. **69**(3), 332–339 (2001)
21. E.L. Harrington, An experimental study of the motion of curling stones. Trans. R. Soc. Can. **III**, 247–259 (1924)
22. W.H. Macaulay, G.E. Smith, Curling. Nature **125**(3150), 408–409 (1930)
23. G. Walker, Mechanics of sport. Nature **140**(3544), 567–568 (1937)
24. J.W. Johnston, The dynamics of a curling stone. Can. Aeronaut. Space J. **27**(2), 144–161 (1981)
25. M.R.A. Shegelski, R. Niebergal, M.A. Walton, The motion of a curling rock. Can. J. Phys. **74**, 663–670 (1996)
26. M.R.A. Shegelski, The motion of a curling rock: analytical approach. Can. J. Phys. **78**(9), 857–864 (2000)
27. E.T. Jensen, M.R.A. Shegelski, The motion of curling rocks: experimental investigation and semi-phenomenological description. Can. J. Phys. **82**(10), 791–809 (2004)
28. M. Denny, Curling rock dynamics: towards a realistic model. Can. J. Phys. **80**(9), 1005–1014 (2002)
29. H. Nyberg et al., The asymmetrical friction mechanism that puts the curl in the curling stone. Wear **301**, 583–589 (2013)
30. H. Nyberg, S. Hogmark, S. Jacobson, Calculated trajectories of curling stones sliding under asymmetrical friction: validation of published models. Tribol. Lett. **50**, 379–385 (2013)
31. A. Yasutome, M. Arakawa, N. Maeno, Measurements of ice-ice friction coefficients. Seppyo **61**(6), 437–443 (1999)
32. N. Maeno et al., Ice-ice friction measurements, and water lubrication and adhesion-shear mechanisms. Can. J. Phys. **81**, 241–249 (2003)
33. K. Hattori et al., High-precision measurement of curl distance by digital image analysis, in *Proceedings of Symposium on Sport and Human Dynamics*, **B-33**, 14–40, (2014)

Chapter 11
Why Did We Lose? Towards an Integrated Approach to Winter Sports Science

Dario Dalla Vedova

Life is short, art long, opportunity fleeting, experience treacherous, judgement difficult.
Hippocrates of Kos, V Cent. b.C.

Women's Downhill podium of Sochi in 2014 is likely to remain in the Olympic Winter Games history [1]. After 100 s at Rosa Khutor's slope, Dominique Gisin, Tina Maze, and Lara Gut arrived at the finish line wrapped in just 10/100 of a second. More than 2700 m of race, with several turns and jumps; and the difference between first and third place was less than 3 m, i.e. in the order of 1 %! Even more extraordinary was the result of the first two Athletes: exactly the same time, giving to the annals the first ex aequo gold medal of Winter Olympics. The same result, accurate to a hundredth of a second, was obtained by athletes differing in ages, nationalities, anthropometric characteristics, technical equipments, and race number. In Winter Sports and Olympics it was not the first time, in different disciplines. Even without these exceptional events, in modern sports differences between top athletes are almost minimal. It often happens that long races like the 50 km Cross Country Skiing of Vancouver 2010 end at photo finish: achieving the podium is often a matter of hundredths of second or centimetres. Athletes, coaches, trainers, researchers, engineers, and practitioners ask: where are the differences? It is still possible to improve [2, 3]?

The role of technology in Sport is almost irreplaceable, numerous studies analyze the impact that technological development has even in Olympic competitions [4–7]. Some wonder whether it is ethical to use technology to enhance athletic performance and, in some cases, whether a new type of technological doping is possible [8, 9]. But it is impossible to stop scientific progress; as long as there are challenges, man will always try to improve himself. So it still makes sense to try to understand why one finished first when athletes with different techniques and equipments, who compete in changing environments, arrive so close together? What conceptual

D. Dalla Vedova (✉)
CONI Servizi SPA. Medicine and Science Sport Institute. Largo Piero Gabrielli, 1 - 00197 Roma, ITALY
e-mail: dario.dallavedova@coni.it

F. Braghin et al. (eds.), *The Engineering Approach to Winter Sports*,
DOI 10.1007/978-1-4939-3020-3_11

frameworks shall be used to understand what happened in such a complex world [10]? Is it possible to establish causal links between final performances and variables studied by sports science [11]? What type of analysis is possible on race outcomes considering variability of individual performance and of external environment? How much the performance of high-level athletes should improve to have a reasonable probability to get a better race result [12]?

Athletes and Skimen are more accurate than any available sensor but it is not easy to objectify, generalize, and transfer their impressions. The attempt itself to correlate mechanical characteristics of sports equipment measured in laboratory and subjective judgements expressed by different users involves a very manifold process [13–20]. For example, the same snowboard deck responds in different way depending on users, type of slope, or snow temperature [21]. How to choose length, side cut, slab, or wax for a certain ski? How to take into account what happens when foils and blades "lose" a bit of edge and grip during a race? What happens to an athlete jumping or along a sharp turn on ice? Balance and posture of the athlete practising Winter Sport and how they are related to typical problems of Engineering cannot be neglected [22]. In addition skis, boots, helmets, suites also affect athlete's feedback, consequently, determining in some way final performance and ability for technical evaluations [23–30]. How many data are needed to be confident to really understand a phenomenon? The human tendency is to see what we want to see, so unconscious bias is inevitable. Without any ill intent, a researcher may look at the data so it supports the hypothesis, even if just barely. When only a few tests are possible, sometimes it happens to find regularities even where there are not [31–34]. What analytical approach should we use? [35].

Giving simple answers to these apparently simple questions is not always possible, sometimes it is not really possible to find unique solutions [36, 37]. The stopwatch alone is often too "poor" to read what happened on track and figure out factors which made the difference [38]. Sophisticated tools may be requested to analyze processes leading to victory.

Replication is the ability of another researcher to reproduce a finding, and it is the gold standard of Science. But in Winter Sports the nature of biological and environmental variables gets it very difficult. Often effects are so small that can be found only under ideal circumstances, and using highly specialized techniques. A research finding is less likely to be true when the studies conducted in a field are smaller; when effect sizes are smaller; when there is a greater number and lesser preselection of tested relationships; where there is greater flexibility in designs, definitions, outcomes, and analytical modes; when there is financial interest or prejudice, as it could happen in Winter Sports. Great efforts are constantly made by scientists trying to analyze and solve these problems [39, 40].

Researchers have to deal with another issue of primary importance when a test for applied Sport research is designed and executed: to match together internal and external validity [41]. Internal validity refers to control and measurement of involved variables, external validity is related to test conditions that should be as similar as possible to competition ones, including availability of high-level athletes. Especially in Winter Sports, it is difficult to match these two requirements. In lab tests or when the athlete wears a heavy and invasive data logger, very accurate data are acquired but far away from real racing condition. Analyzing competitions makes possible to see real phenomenons, but impossible to measure all variables determining the final outcome. Even when permitted by rules, it is difficult for an athlete to wear some measuring instruments while competing. It is hard to find consistent groups of high level athletes to perform reliable tests, have good data, perform powerful statistical analysis.

Finally, environments in which Winter Sports take place are sources of a lot of variability in final performance [42]. In some cases modern technologies allow to simulate technical variables in controlled and repeatable settings as it happens in wind tunnel or bobsleigh simulators. These methods have several advantages related to safety of athletes, affordability, ability to selectively control different parameters minimizing noise effects. On the other hand, it is impossible to simulate the real performance beyond a certain level of accuracy and reproduce any phenomena having effects on human performance in sport [43].

Only to have an idea of operative difficulties of applied Biomechanics in sports, the average number of tests acquired in all original researches published during an entire year of one of leading scientific Journal in field was 4; 50 % of all studies were about the analysis of a single test; the athletes studied were almost all men [37, 44, 45]. It is often difficult to log data for the entire length of a race, so the risk of losing important details exists when limiting analysis to single sections. In a Bobsleigh, Ski or Snowboard descent as well as in Ski Jump, the performance of each part is also function of what happened in the previous one. So the analysis of one or more isolated sector might not be enough to explain the final result [46, 47].

The engineering approach to Winter Sports is usually analytic, that is to say reductionist and mechanistic: the system Athlete equipment is divided into smaller sub-blocks considered independent of one another and easier to characterize and optimize. The same method is often used in quantifying human movements [48–50], or in analyzing different human tasks like management, problem solving, sports performance [51–55]. In this process it is assumed that a feature (improvement) found at a given level is still effective at higher levels, and that it is possible to sum together different effects obtained in distinct parts of the system. Sensitivity studies are performed through numerical models to assess weight of single variables [56, 57]. Geometries, behaviors, materials, structures, construction processes, tribology aspects, aerodynamics, chemical composition of waxes, skis, boards, boots, ice blades, suits, helmets, frames of Bobsleigh or Skeleton sleds are studied involving all branches of Engineering: mechanics, aerodynamics and spatial, chemistry, materials, medical, electronics, numerical, and computational. Borrowing a terminology of Bio-Medical Sciences, it is possible to experiment in vitro, in silico, and in vivo meaning in laboratory, with computers and on field.

As it will be seen in this book, this approach gave and continues to give excellent results in increasing performances and developing tools [4–6]. Sometimes rules are amended to limit technological exasperation, trying to increase Athletes safety and preventing injuries [5].

Moreover, people involved in Winter Sports research knows that it is relatively easy to measure characteristics and behavior of technical equipment, even modifying to make it lighter and stiffer to a degree 1 %. What is more, it is different to understand whether this intervention gives a gain (or a loss. . .) in a race. It is already very difficult to establish how mechanical and geometric properties of an equipment are related to perceived qualities by different athletes, on different conditions. There is not any guarantee to get unambiguous results [17].

Top level sport performance is something extremely complex, since many different variables working at different levels contribute to the final result often out of our control and forecast ability. With maximum simplification, the Winter Sports researchers have to face three different but strictly linked worlds: Physics, Environment, and Biology. Analysis of sports equipment is not easy, but it is even more difficult taking into account two elements between which they have to work: the legs of Athlete above and the snow or ice below [13, 58].

The physical problems are of extraordinary complexity and difficulty. It is easy to write complicated systems of differential equations, much more difficult to simplify and find coefficients making them useful to approximate what happens in reality. In top sport, there are examples of solutions which Physics has proven to be more efficient than those normally used in competitions. But Athletes are not able to adapt to them [59].

This shows that it is necessary to take into account Biology, or that everything has to be functional and useful to humans. Like all living systems, they require adequate tools to be modelled and analyzed [41, 53, 60–63]. Moreover, not all human characteristics are directly measurable as variables and, sometimes, can be thought only as hypothetical multifactorial constructs [64]. An example of these constructs is the "technique": What is it? How is it defined? How is it analyzed and judged? What are the key variables [35, 65]? Is the technique unique, or other techniques equally effective are possible?

Often performance indicators are used [34], but in turn they must refer to theoretical models and still remain doubt about how much a model itself is the proper one, or if something important was lost while selecting variables and simplifying the reality.

Finally, there is the environment. Even an indoor ice skating or curling track has its own variability linked to quality of ice and air. A fortiori, environment is of fundamental importance in outdoors winter disciplines.

So, under these circumstances, is it correct to model and simulate complex sports performance?

The Winter Sports engineer must take into account all these factors, remembering that the study of one cannot be separated from the others. Therefore he should work in a heterogeneous Team able to provide integrated and meanwhile differentiated answers; a team where Engineer, Biomechanics, Medical Doctor, Psychologist, Posturologist, Nutritionist, Trainer, Skimen and all the other professionals required from modern sport operate all together with the Athlete and the Coach. The Engineer should always maintain a global view of Athlete equipment system.

We would like to control what happens, calculate, simulate everything, understand what is needed to win, predict effects of a change. But this is not always possible. So what can we answer to Athlete and Coach asking us: "Why did we lose?".

11.1 Physics and Models

Theory is when one knows everything but nothing works. Practice is when everything works but nobody knows why. Finally, theory and practice go hand in hand: nothing works and nobody knows why. A. Einstein

Physics of Winter Sports is complex. Many races take place in three dimensions, at high speeds, under high loads of G-forces. The whole Athlete equipment system is often in nonlinear transient condition, great forces are generated in very short time and exerted on very strange, inconsistent, variables, yielding surfaces: snow and ice. Despite considerable efforts and progress made by Science, it is yet difficult to model what happens in contact regions during sliding as function of different speeds and vertical loads [66–70]. In addition, forces are transmitted on small surfaces, so generating enormous pressures, through blades or ski with their own mechanical characteristics. This information, in turn, is read by the Athletes' receptors and processed by their central nervous system (CNS) which, through complex phenomena of feedback and control, provides dynamic and appropriate responses to the external situation in a very short time [22, 71].

Competitions are held at different heights above sea level. Combined with techniques and metabolic demands of various disciplines, especially in endurance races, this has considerable practical implications on physiology of athletes, and so on their performance [42, 52, 72].

Aerodynamics plays a key role in a lot of winter disciplines being a major cause of drag of entire system. Studies in wind tunnels (WT) and with computational fluid dynamics (CFD) of positions, suits and undergarments, boots and skis, helmets and goggles, various appendices—including safety ones—are normal. Tests are performed looking for repeatability and validity, setting athletes at fixed positions to simulate race conditions, compensating for different Reynolds numbers and air properties, sometimes using instrumented dummies to avoid even small changes in position with possible influences on test results. But rarely systems run under stationary conditions, almost never straight. Athletes use a variety of customized accessories in combination with each other. It is not always possible to analyze and optimize individually aerodynamic properties of various elements or simply adding them together. C_{drag} coefficients do not remain constant while Reynolds number and airspeed change, sometimes it is not easy to quantify their absolute values [73, 74].

The human body is not a rigid frame, it is deformed in different ways to make movements, depending on technique, muscle stiffness, loads, speed [75–78]. Males and females,different athletes, have different body shape, size, and anthropometric characteristics. Aerodynamic data are transferable between athletes with care, so usually they need customized studies and optimizations, and this is possible only for top level.

Bobsled, Skeleton, and Luge run down in a wide range of speeds and in tracks with maximum width of 140 cm: effects of side walls and ground on Drag and Lift cannot be ignored in CFD simulation and WT test. During a descent rarely the system is in the center of the track and straightaway. Crew configuration, geometry of driver helmet, brakeman back position must be considered remembering that both athletes have to efficiently push and enter in bobsleigh at the start. So necessarily the shell is a compromise, real 3D aerodynamic field around it often is not symmetrical and stable, and therefore is more complex than that reproducible in laboratory or on computers [79–82].

Laboratories of manufacturers of sports equipment and universities study shapes, dynamic behavior and constructive processes of skis, metallurgical characteristics and tribological properties of steels, blades and foils, boots, bindings [83, 84].

Necessary and sufficient conditions are analyzed to get best performances. For Bobsleigh necessary conditions are ability to push at start and driving skills. They are necessary but not sufficient because experience and literature demonstrate that it is impossible to win without a good start, but between the first 10, 15 crews rarely push time is a discriminating factor as it happens for ice temperature or starting order [38, 46, 85, 86]. So a hard push is a necessary condition to win, but not enough. For equipments, (macro) necessary conditions are mechanical and aerodynamic properties: sliding, stiffness, handling, and low drag. Not even the best Athlete can win without adequate equipment and right setup. Studies on aerodynamics, ergonomics, materials, and chemistry to minimize all resistance and dissipation are fundamental and will be shown in this book.

Sophisticated models are used, differential equations systems are written to represent involved phenomena, sensitivity studies are carried out. This type of approach is very powerful because simplifying and modelling reality and environment under appropriate hypothesis allows parametric analysis to weight each variable with respect to the final performance, saving time, and resources [57]. Sometimes in the real world it is impossible to separate variables, values of certain parameters cannot be extreme because systems may become unstable and dangerous. Powerful computers perform millions of calculations, optimum research methods suggest most convenient areas of research to be explored. These mathematical models are then implemented with field data and final outcomes of races, so that they converge towards more lifelike as possible results [87, 88].

Nowadays there are very sophisticated simulation software to study a vehicle dynamic behavior. Characteristics of tires, suspension, grip, chassis, mechanical and aerodynamic, engine power, drives are used as inputs to simulate, study, and optimize cars in different load conditions and use. Corners, speed, loads, transient, driving styles, tires and terrain are measured and simulated to develop increasingly sophisticated tools and design better performing, ecological, safe cars. Optimization of tires—ground contact is done using kinematic, kinetics, elastic laws governing different systems, although in complex ways. In order to improve and correct mathematical models, it is possible to perform numerous tests under standardized conditions so providing repeatable data.

This is only partially possible in Winter Sports. With exception of Bobsleigh where frame and shell are rigid, having its own suspension system and a mass comparable to that of Athletes on board [87, 88], usually the system Athlete equipment weights about 100 kg, where human body is 80–90 %. So studies of mechanical behavior of skis, snowboards, and ice blades must consider all parts of equipment and inertial, postural, muscular, kinematic, dynamic, motor control characteristics of athletes [89, 90].

Management and interpretation of stiffness, degrees of freedom, and dynamic responses of system are not easy. Sometimes, especially with more experienced and adapted athletes, it is difficult to exploit technological innovations to 100 % in profitable way.

Often reality is more complex. Sometimes it is not even methodologically correct to follow this way, speculating just on a few variables, considering them independent of each other or directly correlated, with simply summable effects on final performance [91]. Athlete, snow, ice, environment, and equipment are very difficult to simulate beyond a certain level of accuracy mainly because the system is biological and with a lot of degrees of freedom (DoF); dynamic responses are not fully defined or simulated; the system is highly sensitive, continuously fatiguing and self-adapting to external environment which, at same time, constantly changes [48, 92].

Studying the Ski Jump, correlations between release speed and jump length are not univocal. Inferences that can be made about these two single variables are inconsistent, and pose some questions about simulation models only based on ballistic laws. Information related to takeoff and flight phases, level of athletes, jumping technique and strength expressed, hill type, and study conditions (training vs. competition) must be taken into account [78, 93]. Even different flight phases do not impact all in the same way on final result.

Study of forces between ski and snow during a turn should start from choice of model to use [58, 94, 95]. Different snow conditions require different models [96–99], coefficient values also depend on contact surface, humidity, and temperature [100]. Air trapped in snow works lifting snowboards and skis [101]. Similarly, it is very difficult to assess what happens when yaw, transverse components, high vertical loads, different sliding speed, vibrations, changes in temperature are present [38, 102].

Energy analysis represents powerful approach for downhill disciplines, where gravity acceleration works as primary propulsion. It is possible to write more o less simple equations describing laws of motion and speculate on dissipative forces, efficiency, speed, trajectories, split times [74]. Since it is not easy to obtain accurate data concerning all run, usually slope discretizations are made or single pieces selected. However it is not possible to study what happens in a sequence of turns without knowing what happened in a previous section, and without regarding the next one. So a performance cannot be predicted only analyzing a series of isolated intervals [74, 103]. At the same time, system complexity implies that none of performance prediction concepts commonly used is alone able to explain time differences between two tests [37]. In high-level competitions, final times often differ by no more than fractions of a second even though the performance of fastest athletes on individual sections of race can vary by as much as 10 %, reflecting personal skills and technical characteristics that a model hardly can simulate [104]. At the same time, this demonstrates that room for improvements exists.

More questions arise considering interaction between all these variables. Mass and its distribution, shape, position, sliding, handling, aerodynamics of system are not independent from each other.

Finally, some epistemological considerations born in Ecosystem Science, especially in great and long debate toward Life, Earth, and Atmosphere Sciences [105]. They are interesting for Winter Sport studies because many problems, tools, and models are quite similar as they deal with open and complex systems.

What we call data are inference-laden signifiers of natural phenomena to which we have incomplete and discrete access. Variables selection, observation, and measurement come from data sets or general laws, that is, inductions, deductions, and assumptions. Variables do not work all at the same level, relationship between different levels is not always clear; *scale effects* and non-additive properties may exist. The distribution of dependent variable is the most well-known aspect of systems; the distribution of independent variable is the least well known. Results of experiments and models are not always unique; different models could provide similar results. Therefore we can never be sure to know and properly consider all real phenomena. Models can be confirmed by the demonstration of agreement between observation and prediction, but confirmation is inherently partial. Complete confirmation is logically precluded by the fallacy of affirming the consequent and by incomplete access to phenomena. Models can show a good agreement between observations and predictions, but confirmation is always inherently biased. A model could be built to confirm his prejudices or support incorrect intuitions. Models can only be evaluated in relative terms, and their predictive value is always open to question. The primary value of models is heuristic [106–109].

How much data is needed to be sure if an observed phenomenon really exists? Are we dealing with random series or bias of observer looking for regularity? Nobel Prize was involved in these topics, proving that are not trivial issues and not unique to sport field [31, 32, 110, 111]!

Another major problem is practical application of general laws to single case. Where there is very fine scientific knowledge of the laws in question it might be on occasion that the conditional probabilities are known from extensive scientific experimentation. But then another aspect of the problem related to the application to the individual event will not be known from such scientific experimentation except in very unusual cases, and judgement will enter necessarily [112, 113]. Examples of validation and practical use of models applied to Winter Sports science will be shown in different chapters of the book, referring to various sports disciplines.

11.2 Environment and Variability

As far as the laws of mathematics refer to reality, they are not certain; and as far as they are certain, they do not refer to reality. A. Einstein

Like all great inductive philosophers from Hume to Popper, also RA Fisher—one of the fathers of modern statistics—was a skeptic, but he was a constructive one. Uncertainty and error were, for Fisher, inevitable. But "rigorously specified uncertainty" provided a firm ground for making sense of the world [114]. This sense is provisional, always open to corrections or changes but it is the only tool we have to acquire a reliable empirical knowledge. For example, before the new theories and methodologies of Fisher for assessing performance of different types of wheat, the standard experimental approach was theoretically and conceptually simple: try to make culture ground as homogeneous as possible by controlling all factors thought to influence results. Since it is assumed that all other variables were constant, any final diversity in harvest could be linked only to type of used seed. The same is done when evaluating effects of any change in sports field: best efforts are done to neutralize or block all other variables, assuming that they are all well known, that do not vary during test or not affect those on which it is acting. From long time researches were made into two most recognizable environmental factors that impact on sport performance: wind speed and altitude [115–117].

Unfortunately, this is very difficult to achieve in Winter Sports. Snow and ice change their status during races held on natural, artificial, hot, cold, wet or dry surfaces, in sun and shadow, chemically treated and machined by snow or ice groomers. Environmental factors are important even in indoor disciplines like Speed Skating: quality of ice depends on more or less demineralized water, but also by air conditioning system that—if it is crowded, if the air is moist and not properly circulating—can frost increasing friction [118, 119]. Ice is periodically handled with liquid water and appropriate tools during Bob, Skeleton, Luge or Skating races. Athletes practicing these disciplines know that final time they can achieve is also function of temperature of ice and blades [38, 120]. It is not possible to standardize the ice of a Bob track ruining with passage of crews, or snow status of a 50 km Cross Country race varying with season, sun exposure, daytime. During a test it is normal to have changes in temperature and wind [38, 99].

Still there is not a unique system to objectively evaluate and weigh the state of a track and any environmental variables affecting performance regardless of athlete, driver, and equipments. So it is impossible to control all sources of environment variability. How to be sure that assumed causal links are correct and unambiguous? The winner usually "reads" the environment better, and faster adapts himself [78]. So it is not strange that sometimes a ruined track is an advantage for certain athletes or equipment, some skis, snowboards, ski waxes or blades work better in certain range of temperature and humidity.

For studies with biological variables, randomization, control groups and so-called double blind procedures were born. Unfortunately, Engineering and Biomechanics of Winter Sports rarely are able to use all these tools. High-level athletes are often too few to randomly create control groups. How to make a blind or placebo test with a ski? Athlete, coaches, and researcher always know what is being tested, they have expectations and prejudices perhaps unconscious, they cannot control all boundary variables. Operational difficulties in field are substantial: normally it is possible to do and analyze only few tests, sometimes only one [44, 45, 121]. Instruments do not always work adequately, not all acquired data can be used due to their quality, technical problems, or required post-processing time.

One possible solution is inductive, i.e. to invert the situation. As it happens for Evidence Based Medicine, it means to make analysis and decision integrating best available external evidence from systematic research with individual expertise. This requires huge amount of data and powerful statistical tools. In Sports Medicine this approach is used looking for answers related to stretching, reliability of physical examination, injuries prevention and treatment [122].

When studying the Winter Sports, it is possible to start from competitions data such splits and final times of cross country ski, bobsleigh, luge, skeleton, ice skating races now widely available on the Web and, through powerful statistical analysis, try to infer some general laws related to optimal sports performance. Regression analysis is most widely used technique, regression type depending on investigated variables. If dependent variables are continuous, unbounded, and measured on the interval or ratio scale such as a run time, appropriate simple or multiple linear regression methods have been validated [38, 41, 46, 123]. In these models, several independent variables are weighted with appropriate coefficients, summed together and inserted into a system of equations for predicting a dependent variable. Additive (linear) models are used to separate different variables and analyze single effects assuming everything else constant. The models are called Mixed because they incorporate fixed and random effects and can be hierarchic or multi-level; evaluations and adjustments on residual error—differences between observed and predicted values—are made [124]. Linear Mixed Models are so very useful tools for typical problems of Science of Winter Sports. Like all instruments, it is important always to be aware of their potentialities and limits [125].

In Winter Sports these analysis techniques are used respectively for studies of endurance disciplines, as Cross Country Skiing, of downhill ones like Skeleton, or for ambiguous phenomena like "home advantage" [126]. This type of information is potentially very interesting for researchers wanting to limit all possible sources of variability in a test.

Cross Country times of Olympics, World Championships, and World Cups races held in 10 years on different tracks, with classic and free style technique, Sprint and Distance competitions, attended by over 500 athletes, men and women, were analyzed focusing on top ten athletes of each discipline [42]. Type of snow, race altitude, and different distances were fixed effects. Times between different skiers and races, same skier between one race and another, in different seasons, were variable effects. The main results showed a great variability due to ground (4–10 %), probably incorporating effects of snow type and altitude (2 %). After adjustments to normalize environmental parameters and remembering that in cross country speed and average power are closely correlated, results demonstrated a remarkable stability in performances of top ten athletes, with women slightly more variable compared to men. Similar findings are found in other endurance summer sports [127]. Performance predictability was high considering all athletes, very low for top ten athletes [42].

Starting, split and final times of top 20 male and female athletes of 5 years of Olympic, World Cup, and World Cup Skeleton races held on over ten different tracks were analyzed. Tracks were classified by athletes and coaches according to type and driving difficulty. Track, season, home advantage, first or second round were fixed effects. Athlete's skill and the variability of its results within same season and in different seasons, variability of final mean times in same track between first and second heat and in different seasons were random effects. Separate analyses were made for men and women, for 1–10 and 11–20 classified. Main findings were: less variability in performance of top athletes and—for same reason—less predictable top ranking; also for best athletes more variability in difficult tracks due to greater number of errors; home advantage for women but not for men; some possible effects of track degradation and ice softening between first and the second heat that is slower in average; times of first heats have greater variability than second one because in race day athletes do not perform a training runs; correlation between push and final time is low and related to difficulty of the track only for men; split times of some sectors of different tracks show higher correlations with respective final times, other sectors are more related only with track driving difficulty; race outcomes are largely unpredictable [86, 128]. There are many other tools for methodological and statistical analysis in Sport, Biomechanics, and Physical Activity [12, 41, 44, 45, 60, 61, 63, 129, 130].

Analyzing all races of 20 different Winter Olympics held along a period of 90 years between 1908 and 1998, and do not considering sports with subjective and aesthetic judgements as figure skating or freestyle skiing, home advantages for ski jumping, alpine skiing, short track and speed skating were found. Contrary to popular belief considering track knowledge very useful, small advantages (if any) have been reported for Nordic combined and skiing, bobsleigh, skeleton, luge, biathlon, speed skating [123, 131]. Other authors found small effects for speed skating [132]; in skeleton effects are not present for men but only for women, perhaps because of their lower technical level in average [86].

Finally, it must be remembered that in mass races as cross-country skiing, speed skating, short track or biathlon, final result also depends on what the opponents do. So also presence of other athletes is an important environmental factor.

11.3 Biology

> *Science is all about a normal distribution curve, but with elite athletes you don't always have the bell curve shape because you're dealing with the abnormal people, the outliers [on the normal distribution curve]. So gold medallists are often abnormal.* T.B. Smith, World Rowing Magazine, Dec. 2010

During Winter Olympics of Nagano 1988, the great Norwegian Cross Country Champion Bjørn Erlend Dæhlie—beaten in number of medals at Winter Olympics only by compatriot Ole Einar Bjoerndalen at Sochi 2014—won three gold medals in 10 km, 50 km, 4 × 10 km relay and a silver one in pursuit 10/15 km. Surprisingly he was only 20th in the race of 30 km: what happened [133]? Chronicles told of a strong snowfall before the race and of a wrong choice of wax. Considering the extraordinary results obtained by same Athlete in other Olympic events of those days, it is reasonable to give a simple answer to question why Bjørn Dæhlie has lost in 30 km: the wax. Now let us look at the story from another point of view: it is possible to say that Myllylä Mika, winner of that race, was first only thanks to choice of right wax? When we study complex systems and phenomena, there is a strong asymmetry between what we can understand and inferences we can make from available data [108, 125, 134]. Evidently, there are necessary conditions to win, but completely different matter is identifications of sufficient conditions.

In modern sports, no Athlete can excel without the best available equipments, customized and tuned as best as possible. But it is not so easy to understand why winner arrived first. Biology differs from Physics because of less rigid Laws [135]. Evolution produces organisms and sub-living organisms, so biological Laws without exceptions are few. Biology and Physics are often closely tied together in a very complex, nonlinear way through difficult to understand interactions and retroactive processes. Sport is played by humans, open systems continually exchanging energy and matter with external environment, looking for dynamic homeostatic balance without ever being in chemical, physical, and thermodynamic equilibrium [136].

Athletes have anthropometric and biomechanical characteristics very different from each other, highly optimized and adapted technical movements, their trainings are customized, as they use equipments and specific sports technology sewn on them like clothes from a tailor.

Furthermore, human body has a lot of DoF, redundant considering motor tasks capable to perform. The relationship between movements and innervational impulses which evoke them is extremely complex and is, moreover, by no means univocal. The fundamental equation for the movement of a limb in gravitational field under the influence of a muscle becomes extremely complicated as considerations of the mechanical effect of one muscle upon others and the moment of inertia of the system becomes a variable term [137].

The human system is equipped with its own variability developed from Evolution to adapt to a variety of different situations. With the same feet we walk, run, jump, skate, ski, climb or swim—not to mention what we are able to do with our hands. There are several sources of variability when studying movements of athletes: biovariance that affects human performance repeatability; continuous changes in musculoskeletal system and in external forces and moments acting on system preventing exact reproduction of technical movements; biological measurement errors influencing experimental results [137, 138].

Skeletal muscles of human bodies can work either as agonists or antagonists in movements. This means that there is not only one way to do a specific motor task, thus not always possible to define unique motor patterns. Since beginning of last century, some authors spoke of *motor equivalence*, i.e. different but all equally functional coordinative movement patterns can be used to produce similar values of performance [139]. As a result, movements themselves depend more frequently on the field of external forces than on the central control [137]. Trying to understand the behavior of so complex chains with few discrete data, as occurs with using kinematic data or electromyography, we can never be completely confident to have "captured" the phenomenon. In fact, Motor Control scientists are necessarily limited to observations on motor outputs from which infer underlying processes involved, since it is not possible experimentally to separate and study all different sub-systems individually [140]. The final result is like an iceberg: only ice outside water is visible, but the biggest part is hidden underneath.

An extraordinarily interesting field of investigation for Winter Sports is related to human equilibrium and sports equipments, as demonstrated by many studies on dynamic balance in high level athletes [141]; on injury prevention [142–144]; on vibrations control, fatigue, and performance [99, 145–148]; or by controversial debate on balance bracelets [149–151].

From long time, Motor Control scientists are studying the static and dynamic balance of man, i.e. ability to maintain the projection of center of gravity within support base taking into account movements, mutual position of all body segments, system of reaction forces, and trying to minimize the sway. Balance is not regulated by a single sense organ, as it happens for sight, hearing, touch, taste, and smell. A set of organs and systems must work together in an incredible synergy of mechanical and biological aspects [62]. Information relating both outside (exteroception) and inside (proprioception) world are needed. The role of proprioception in perception and control of human gait is essential but, at same time, very difficult to analyze and even to conceptualize [152].

To maintain balance, somatosensory, vestibular, and visual apparatus must be coordinated. Somatosensory system is the part of nervous system that provides information coming from skin through tactile and kinaesthetic subsystems, receiving and processing information about relative position and movement of the body segments. Receptors that detect these stimuli are in muscles, tendons, and joints and analyze tactile sensations of vibration, temperature, pain, and pressure arising from mechanoreceptors in skin. This is a key source of proprioceptive information [153] and, consequently, very important for human balance [23, 154].

The complex process of object recognition through touch, reading body position and expressing strength, is called *haptic* perception. Gibson [155] defined it as the *sensitivity of the individual to the world adjacent to his body*, emphasizing the relationship between haptic perception and body movements. The haptic approach in Winter Sports is studied and already found interesting practical applications in snowboard [71, 154].

Therefore it is not surprising that tapes applied to the skin or tight clothing compressing the legs may change balance, proprioception, perceived feelings [24, 27, 156]. The effect of compressive garments is a fantastic example of complexity of living systems: they decrease cross-sectional area of venous system, improving hemodynamic, reducing cardio-respiratory and neuro-metabolic stress, activate tactile mechanoreceptors, attenuating muscles oscillations, increase sensory and proprioceptive feedback closely connected with the ability to maintain balance on skis [25, 30]. A pressure of 20–40 mmHg exerted on lower part of legs allows athletes to maintain a deeper tuck position in Skiing with positive effects on aerodynamics, increasing muscle activation, decreasing oscillatory movements, without negative effects on maximum voluntary isometric strength, lowers level of perceived exertion, potentially increasing performance [30, 157]. All elements of technical equipment have an important role in proprioception and postural aspects and require specific adaptations. High-level skiers not wearing their boots may be disadvantaged in execution of specific tests [26, 158].

Bernstein already realized that motion control is not continuous but discrete, since brain needs time to transform sensory coordinates into motor ones providing appropriate responses. In general, the muscular reference system has more than one dimension so the transformation problem has many solutions, and nature developed several strategies to simplify neuro computation. A solution is to read continuous discretely cutting a lot of information. Neuroscientist Llinás [159] says we think at 40 Hz and move at 10 Hz. That is, elementary time to process brain data is about 25 ms; a basic operation of motion control requires about 100 ms. Different receptors acquire and transmit information in discreet way, every sensory system working in an optimal range of frequencies integrated with others to ensure necessary quality and timing [160].

All this information has important implications for study of technical equipments of Winter Sports considering athletes' feedback, required handling and precision, what happens with harder or softer snow and ice, damaged by passage of the other athletes, fatigue and stress due to vibrations. Control of vertical loads is one of the most important features of technical movements in alpine skiing and snowboarding, where athlete has carefully to cross gates and turn while subjected to high accelerations, to gravitational and inertial forces oscillating transmitted from ground.

Damping and sliding ability are very important qualities for athletes. They are functions of several variables: technique, position, maximum strength, power, balance, materials, type of snow, speed and frequency of the vibrations. In legs of athlete, passive vibrations cause activation of a large number of motor neurons with high activity of some specific muscle groups [161], which consequently implies premature fatigue [27, 162].

Studies carried out with skiers placed on vibrating platforms showed that damping ability of the whole system presents both intra-individual differences in function of the position and inter-individual related to level of Athlete. Differences between skiers of different levels are amplified above certain thresholds of frequencies and amplitudes. It is suggested that this capacity is linked both to ability to achieve high frequencies of active damping with rhythmic muscles contractions and more efficiency regulating stiffness of the muscle-tendon system [147].

Tactile and proprioceptive receptors provide data only of internal body movements, between different body segments, of forces and pressures exerted on skin. This is not enough for realization of complex movements for which the brain must know absolute movements of the head and body in space. For this, there are vestibular receptors and sight. Labyrinth is an inertial sensor working as a triaxial accelerometer without need for external anchors; three planes defined by its semicircular canals form a local Euclidean reference system. But Labyrinth is a biological sensor and may not properly work. It happens, for example, with thermic stimulation or pressure compensation [163–165]. This could be of great interest for athletes doing test in winter sports, suggesting to pay attention that helmets should not obstruct ears and provide adequate protection for low temperatures.

As it happens for accelerometers, inertial sensors give ambiguous information: they are related to head local reference system and cannot discriminate between inertial and gravitational accelerations, are subject to drift, may give illusory sensations. These uncertainties are resolved with integration of visual and acoustic information.

Sense of sight is crucial [160, 166–169]. In Winter Sports visibility, light, and shadows can quickly change influencing Athletes' equilibrium.

Sounds produced during movements contribute to additional sensory feedback. Because close relations between actions and corresponding sounds exist, acoustic analysis is used both in Biomechanics and Engineering. Athletes, coaches, and engineers correlate noise produced by bobsleigh with mechanical characteristics; rustling produced by ski contains a lot of information about quality of materials, type of snow, force-time curves at the basis of movements. Acoustic information can help understanding very complex parameters of movements [47, 170].

Labyrinths, eyes, and ears are all located in the head having, therefore, a "privileged balance." Numerous studies in animals and humans confirmed importance of head movements stabilization. Migratory birds use body and neck to keep it as much as possible in uniform motion, athletes try to achieve same result while running and jumping [22, 171–173].

Equilibrium, warm-up, and technical execution are another critical issue for fields of our interest. Professional athletes involved in sports that require execution of fine motor skills usually practice for a considerable length of time before competing in an event. Why is such practice necessary? Is it merely to warm up muscles, tendons, and ligaments, or does the athlete's sensorimotor network need to be constantly recalibrated? It is assumed that human sensorimotor system is characterized by high noise level and high learning rate. The combination of these two properties makes system continually recalibrated to be able to perform fine tasks. The higher the level of an athlete's skills, the more important this would be [174]. A question arises: Can we trust impressions given by an athlete when it is already so difficult for him to manage his internal balance?

More often, in execution of movements, there is not enough time to acquire and process all necessary information; then complex anticipatory strategies are at work [22].

Other causes of variability related to biological mechanisms exist making them difficult to understand and study. For example, changes in body temperature, heart rate, flexibility and stiffness of joints, muscle strength, ability to provide short-term power, perceived exertion of fatigue linked to circadian rhythms and recovery of jet lag after long flights [175, 176]. Intrinsic variability of movements of each individual and, on different level, possibility that so different athletes get very similar final performances extraordinarily reflect the so-called equifinality of Biology and Cybernetics [177]. This is the principle that—unlike what happens in closed systems where a final state is determined uniquely by initial conditions and working laws—in open systems the same state can be reached starting from different initial conditions and in different ways.

Nonlinear control and complex phenomena are present, dynamical systems where the state of each moment affects successive instants. For these reasons, it is often necessary to give up generalizations and focus on individual, top Athletes trying to provide them best responses for training, techniques, injury prevention, and personal factors limiting their performance.

When trying to manage complexity related to biological variability, it is important to remember that usually top level athletes provide less variable performances, and men are more consistent than women [12, 42, 46, 86].

Remarkable synthesis of Engineering and Biology of Human Body is present in works of engineer and neurophysiologist Berthoz and his colleagues. As already noticed by Bernstein, we have no instrument directly measuring speed of our movement. So our perception of motion is the result of extremely complex phenomena and interactions between our external and internal world differing in respective geometries and degrees of freedom. They started from study of the five traditional senses to argue that must be added another: the "sense of movement" [22].

11.4 What Winter Sports Engineer Can Do?

It's always advisable to perceive clearly our ignorance. C. Darwin, The Expression of Emotion in Man and Animals

In a world where mechanical, fluid dynamics, chemical, environmental, and biological processes interact with each other in so complex and nonlinear ways, where variables are often partly outside control and not all simply summable together, where tests are not always repeatable, methodologies often complex and expensive, what a Winter Sports engineer can do? How can he be confident to capture real causes and effects [37]? A key point is search for causal links between technical characteristics of equipment, individual functional aspects, and final performance.

Due to performance enhancement, prevention, and market reasons, in Sport research great efforts to tie mechanical objective characteristics of equipments as footwear and subjective perceived qualities are made [21, 178, 179]. When it is impossible to identify clear and helpful law, very sophisticated techniques like neural network are used [180, 181]. Same and customized methods are used for Winter Sport equipments comparing measurable properties, objectifying sensations, emphasizing features, finding appropriate evaluation benchmarks such as quality function deployment (QFD) [13–15, 17, 20, 84]. The "smallest worthwhile enhancement" of equipments for top athletes is assessed [12, 42, 86, 130]. This process is necessary at all levels, both for Olympic Champion selecting new ice blades for next season and for magazine tester evaluating new skis by different manufacturers.

It is known that findings are not unique but related to many factors such as: test conditions, experience and technical characteristics of tester, different components of equipment, number of cycles, sequence of tests, definition of evaluation criteria, bias or external pressures, economic aspects [13, 15–19].

Experience and studies show that it is easier to find relationships between some technical characteristics and subjective judgements. For example, testers of different skill level link all positively low flexural rigidity and high torsional stiffness of skis with performance; instead, only high-level skiers appreciate their damping characteristics at different frequencies providing more accurate ratings [14, 144]. In snowboard, bending and torsional stiffness distribution influence feel and performance [15]. Athletes of Ice Hockey prefer flared blades [19].

As a result of closely related coexistence of physical, biological, and environmental phenomena, in any type of engineering study applied to Winter Sports subjective judgements of the Athlete and Coach must be considered. Therefore it is important to evaluate this factor from very early stages of design research to develop appropriate testing, analysis, and evaluation tools.

It may seem strange that, almost two centuries after first pioneering applications of photography to study of human movement made by JE Marey and E Muybridge [182], motion analysis is still one of the most powerful and used tools in sport analysis. Videos have unique features: easy and cheap to obtain, do not require special interventions on athletes so can be used also to analyze opponents, reproducing sense of sight, i.e. our main way to access the outside world. High definition, speed and 3D video, accurate GPS, and dedicated software allow simple sophisticated analysis of fusion motion capture (FMC), 3D movements reconstructions, qualitative and quantitative evaluations of forces and technique, inverse dynamic, analysis of injuries mechanism [47, 103, 183–189]

Miniaturization of data logger, inertial measurement units (IMU), wearable devices and sensors provide direct measurements on athlete during motion on track without affecting his technique and allowing individualized skill assessments [190]. Surface electromyography (sEMG) records muscle data analyzing level, sequence, and time of activation. These methods allow to define patterns of motor activation, assessing accessories as boots, and comparing technical skiing movements between inclined treadmill and slope [191–194].

Due to potentially infinite human movements, invariant phenomena were researched as starting point for any theoretical speculation since very early biomechanical studies [48, 137]. Regularities are investigated at different levels, from execution of a single technical movement to global race profile of winners.

Ability of "human eye" to discriminate and select information in very noisy contexts is unique. In our lives full of software and PC, most familiar examples are anti-spam techniques. Although it may seem incredible, powerful software still cannot solve these problems, seemingly so simple for us. Bernstein [137] used small lights to record trajectories of anatomical landmarks during human locomotion of adults, children, and elderly approximating them with sums of sinusoids. Lashley analyzed recurrent features of human writing [195, 196]. Modern neuroscience postulates ability of the human brain to model Newton's laws [197]. Conceptually these principles are still valid and used for identification and quantification of human gait made by man and computers [53].

Technical movements possess their own intrinsic variability and can change as a result of fatigue, technique, other adjustments, equipment, phase of race, environment conditions. To achieve high reliability in study of kinematic parameters related sports movement, different numbers of trials are needed depending on the consistency of selected parameter to analyze. A single kinematics measure may not be representative of average pattern of movement investigated [198]. It is therefore necessary to carefully select kinematic variables to be analyzed staying, as far as possible, close to reaction point and using direct kinematic parameters rather those derived.

In cross-country skiing asymmetric oscillators were found that adequately describe leg kinematics accounting high percentage of average variance of movement. Consequently, leg kinematics can be viewed as resulting from a single nonlinear dynamical system resembling a limit-cycle [92]. Models of this type are also useful to analyze damping properties and stiffness of system, with interesting implications for data analysis.

In Alpine Skiing and Snowboard, movements of athletes are obviously more related to external environment. Slope, gates layout, type, and snow conditions are key variables that cannot be neglected. Sophisticated mathematical methods such as principal component analysis (PCA) are used with same goal: analysis of variables looking for regularities. A vector of n-dimensional posture is used to define position and principal movements (PMs). In skiing first 4 PMs, i.e. first four configurations of vector, contain over 95 % of system variance [199]. Interesting aspect is that these abstract mathematical tools are remarkably similar to the way used by coaches to analyze and evaluate movements of athletes. Incorporating many dimensions in a single indicator, they may represent equivalent of holistic approach to technique analysis while being uniquely and quantitatively definable.

In Endurance sports, optimal pacing strategy of winners is researched, and athletes do hard train to be able to replicate it in a race, at the right time of the season [200–205].

Other tools and procedure for scientific analysis of Winter Sports will be presented in this book with practical applications for different sports.

11.5 Conclusions

I know of an uncouth region whose librarians repudiate the vain and superstitious custom of finding a meaning in books and equate it with that of finding a meaning in dreams or in the chaotic lines of one's palm. J. L. Borges, The Library of Babel

Main purpose of these pages is to stimulate reflections towards an integrated approach to Winter Sports. A lot of analytical and practical tools are presented in the other sections of the book.

Science and technology are in constant and fast evolution; every day, new discoveries always move a bit more forward boundaries of our knowledge. With this amount of information and sophisticated tools now available, various branches of science are even more specialized while scientific approach tends to be reductionist.

But it must not be forgotten that the center of sport performance is the man with his muscles, brain, and heart. The athlete is not only the main actor but, provocatively flipping point of view, also the major limiting factor. This makes work of whole team around the Athlete so stimulating: result found for an athlete with some equipment, in a certain moment of his career, might no more work well in next seasons or for his team mate. Physical theory and methods are well coded, tested, and validated, margins remain for models, materials, analysis, and results. Materials are constantly evolving, technology and electronics make all more miniaturized and cheap, accessible to increasing number of researchers [190], analysis software are ever more powerful, searches and scientific publications are increasing.

Although technology plays a crucial role in modern sports, it is not easy to quantify its impact on performance. Study on aerodynamics and materials gave remarkable results, some technological innovations such as carving skis or speed skating claps have even changed athletes movements and training. But final performance and possible improvements may be in other variables [201]. The average level of athletes practicing professional and Olympic sport is increased. Their selection process, specialization, physical qualities, training and nutrition techniques made enormous progress [2, 3]. Especially in winter sports, biological and environmental variability make difficult to compare performances between different years and discriminate technology improvements. So a lot of results cannot be definitive and general. Occasionally rules limit performance for security reasons. Changing few words in technical regulation, as tissue allowed for race suits, it is enough to start again all previous studies to find new solutions.

Books do not contain all the answers. This happens because who discovers something really working maybe not releases it to public, or because "eye" and practical experience are irreplaceable. A wide part of knowledge of Winter Sports is generated by athletes and coaches and often remains in their hands. They are usually more interested in practical "know-how" rather than in scientific verifications having a holistic view of system Athlete–equipment–environment [35]. According to previous pages, this approach is correct from their point of view. On the other hand, there is a risk to be exposed to pseudoscientific beliefs, not being able to uniquely define and transfer parameters of interest, formulate models and laws of general validity. This happens because different athletes fit to same equipment in different ways; very similar results are obtained with various techniques and equipment. So, with scientific research, it is very important to find contact points between practitioners and researchers [65, 206, 207].

Even from a conceptual point of view, debate among researchers is far from over: shall we orient our analysis more on performance or technology, focusing more on *what* or on *how* [91, 199, 208–210]? Coach and Engineer must learn to speak the same language or, at least, to find a common ground [113].

Usually, sport scientist has a reductionist approach, i.e. logs and analyzes pre-selected variables quantifying important aspects of movement. Coaches and Athletes tend to have a more holistic approach, they analyze posture and movement of entire body in qualitative and often subjective way. A combination of movements and relative positions of body segments is defined "technique" [35, 65, 206]. Both approaches have plus and minus. Measuring only few variables, very accurate data are obtained but never being sure to have recorded all really important information. Many degrees of freedom of human body and potentially infinite causes of variability related to the external environment make it sometimes very difficult to understand causal links of studied variables. Results may be valid only for personal characteristics of an athlete, and not generalizable or transferable beyond a certain level. Time required to process all acquired data can be a bottleneck making results less effective, because we are no longer able to give immediate feedback and contextualized to the testing situation.

Expert coaches quickly provide useful answers even only briefly looking at the athlete. This method is appreciated because it is practical, allowing immediate verification and modification of gesture when motor memory is still fresh. However, the judgement of the coach is subjective, it could be suffering from prejudices, not follow scientific criteria, not be most effective for a particular athlete, unable to formulate laws and models of general validity.

One of the main challenges of modern sports science is putting together these two aspects of human ways of reading and interpreting reality. A contact point to bridge the gap then can be pattern analysis, looking for invariants and limit cycles. Modern technology helps because it allows field recordings and analysis, is cheap and so more accessible, always looking for "new eyes for the coaches" [211, 212].

The ways to achieve these goals are not always simple. Studying an athlete who has just become World Champion in Giant Slalom, none of current predictors of performance has been singularly able to explain differences between his different runs [37] while not all actions made by skier are directly correlated with final time [36]. So there is still room for improvement, allowing the victory to more skilful athletes, faster to understand and adapt to situations, environments, tools, and regulations [78].

Several practical messages can be taken. First is, at same time, most obvious but also most difficult to achieve: higher level athletes are able to provide more stable performance. If possible, it is always better to do test with them! Performances of amateur, student, or young athlete are variable, easily showing fatigue or adaptations to a specific test. Numerous environmental and motivational factors should be taken into account as well as the presence or absence of other athletes to deal with.

A second consideration is related to the smallest worthwhile enhancement useful to increase the likelihood of improving the race result [12]. Often effects of tested equipment are smaller than normal biological variability of athletes, therefore it is not easy to discriminate results. Perhaps some improvements exist, but our methodology, technology, and statistics tools do not allow us to see them.

Between high level athletes is very difficult to find single functional or technical factor able to explain their top performance [37, 41]. On average, better athletes finish all very close even with different anthropometric characteristics, techniques, and equipments [42, 46, 86].

Given the problems of objectification in technical judgement, preliminary definition of evaluation criteria of sports equipment is crucial. Differently defined qualities could lead to different final results. Some mechanical and geometrical characteristics are more easily correlated with inter-subjective impressions of athletes and testers, others are more critical and therefore should be used with caution or avoided [13–16, 18, 19]. Evaluation methods should be defined and shared in very early stages of research to enable development and use of appropriate tools.

Although effects of "home advantage" are not always clear, testers should be familiar with tracks to reduce possible sources of performance variability. An easy slope decreases chances of error and potentially enhances technical equipment qualities. Conversely, driving skills needed in difficult tracks could partially hide some outcomes. Environment light should be as good as possible to avoid possible influences on athletes' sight and balance.

Performance analysis of single sector rather than entire race may enable new possibilities for study of several variables related to performance [19, 213]. In turn, also this approach has some problems. It is common experience that top athletes get very similar final times while split ones may differ from each other, reflecting personal skills or equipment technical properties [46, 104].

Knowledge of testing slope or performance profiles (pacing strategy) is also important to understand whether a single piece is critical for final outcome [46, 202–204]. In this case, it would be better to avoid certain types of analysis, start in a lower part of track or run fractionated test to not physiologically overload athletes. It is important to remember that—unlike an engine—depending on the required effort, an athlete doing a high intensity performance needs adequate recovery time before being able to make another similar test. This time is in the order of hours, if not days. Meanwhile many internal and external variables can change.

Physical and cognitive performance of athletes change as function of nutrition, quality of sleep, time of day, circadian rhythms, effects of "home advantage" [123, 131, 132, 175, 214]. All these details are not easy to evaluate but should be considered when planning a test, trying to match all other practical needs. For safety and performance reasons, it would be better to avoid test immediately after long travels involving jet lag.

General conditions, heating and clothing should be controlled before testing. Athletes with postural, muscular, or visual dysfunctions may not make the most of equipment or supply these problems with experience and technique, so confusing results. Too much warm-up before testing may result in tired performance; not enough warm-up increases injury risk and can lead to less accurate technical execution, adding variability in results.

More or less compressing garments, especially on legs, can change balance and perceived exertion with potential benefits for performance [30], plates and boots have important role on postural control also influencing body equilibrium [26, 158].

Testing with women entails to consider menstrual and, therefore, performance variations related to hormonal cycle [215, 216]. Emotion, stress, motivation, time of season should also be considered both in males and females.

Engineer and researchers of winter sport should know different practical issues and methodological tools available and use the most appropriate and effective. At the same time, it is hoped that Engineer, experienced and continually updated in his specific field of interest, always keeps his mind open, aware of many aspects of modern sport.

We are open systems, continuously exchanging matter and energy with outside. Our body, our metabolism, our balance are result of million years of evolution. We are successful in many different environments and situations. Laws and deterministic methods valid for closed systems are only partially useful, under specific conditions.

A Siberian husky will always be faster than man running on snow, a shark in swimming. Human body is less optimized for a certain motor task, and for this reason extremely versatile and adaptable. With same limbs it is capable of fine movements and great forces production. Ergo even tools, means and technology used for studies of humans must be versatile. The hope is that curiosity, flexibility, and skill will accompany Engineer of winter sports towards new and exciting challenges, bridging the gap between field experience and science, looking for appropriate scientific hypotheses, thinking new models, testing and rejecting not useful ones, processing results and—if necessary—correcting, increasing general knowledge in order to get new answers for the Athlete and Coach who ask: "Why did we lose?" Or for the specular, less common but perhaps more challenging question: "Why did we win?"

References

1. Olympic.org -Official website of the Olympic movement (2014), http://www.sochi2014.com/en/alpine-skiing-ladiesdownhill
2. G. Berthelot et al., The citius end: world records progression announces the completion of a brief ultra-physiological quest. PLoS One **3**(2), e1552 (2008)
3. F. Radicchi, Universality, limits and predictability of gold-medal performances at the Olympic Games. PLoS One **7**(7), e40503 (2012)

4. F.K. Fuss, A. Subic, S. Ujihashi, *The Impact of Technology on Sport II* (CRC Press, Boca Raton, 2007)
5. S. J. Haake, The impact of technology on sporting performance in Olympic sports. J. Sports Sci. **27**(13), 1421–1431 (2009)
6. N. Balmer, P. Pleasence, A. Nevill, Evolution and revolution: gauging the impact of technological and technical innovation on Olympic performance. J. Sports Sci. **30**(11), 1075–1083 (2012)
7. L. Foster, D. James, S. Haake, Influence of full body swimsuits on competitive performance. Procedia Eng. **34**, 712–717 (2012)
8. A. Miah. Rethinking enhancement in sport. Ann. N. Y. Acad. Sci. **1093**(1), 301–320 (2006)
9. D. James, The ethics of using engineering to enhance athletic performance. Procedia Eng. **2**(2), 3405–3410 (2010)
10. W. Sands, J. McNeal, Predicting athlete preparation and performance: a theoretical perspective. J. Sport Behav. **23**(3), 289–310 (2000)
11. M.A. Bunge, *La Causalitá: il Posto del Principio Causale Nella Scienza Moderna* (Bollati Boringhieri, Torino, 1970)
12. W.G. Hopkins, J.A. Hawley, L.M. Burke, Design and analysis of research on sport performance enhancement. Med. Sci. Sports Exerc. **31**(3), 472–485 (1999)
13. P.A. Federolf et al., Subjective evaluation of the performance of alpine skis and correlations with mechanical ski properties, in *The Engineering of Sport 6* (Springer, New York, 2006), pp. 287–292
14. C. Fischer et al., What static and dynamic properties should slalom skis possess? Judgements by advanced and expert skiers. J. Sports Sci. **25**(14), 1567–1576 (2007)
15. A. Subic, P. Clifton, J. Beneyto-Ferre, Identification of innovation opportunities for snowboard design through benchmarking. J. Sports Technol. **1**(1), 65–75 (2008)
16. A. Subic et al., Investigation of snowboard stiffness and camber characteristics for different riding styles. Sports Eng. **11**(2), 93–101 (2009)
17. P. Clifton, Investigation and customisation of snowboard performance characteristics for different riding styles. Ph.D. thesis, RMIT University, 2011
18. P. Clifton et al., Identification of performance requirements for user-centered design of running shoes, in *5th Asia-Pacific Congress on Sports Technology -Procedia Engineering*, vol. 13, 2011, pp. 100–106
19. P. Federolf, B. Nigg, Skating performance in ice hockey when using a flared skate blade design. Cold Reg. Sci. Technol. **70**, 12–18 (2012)
20. F. Giubilato, N. Petrone, V. Franch, A method for the quantitative correlation between quality requirements and product characteristics of sport equipment. *6th Asia-Pacific Congress on Sports Technology (APCST). Elsevier, Procedia Engineering*, vol. 60, 2013
21. P. Clifton et al., Effects of temperature change on snowboard stiffness and camber properties. Sports Technol. **2**(3), 87–96 (2009)
22. A. Berthoz, *Le Sens Du Movement* (Odile, Paris, 1997)
23. E. Moberg, The role of cutaneous afferents in position sense, kinaesthesia, and motor function of the hand. *Brain* **106**(1), 1–19 (1983)
24. G. G. Simoneau et al., Changes in ankle joint proprioception resulting from strips of athletic tape applied over the skin. J. Athl. Train. **32**(2), 141–147 (1997)
25. P. Schaff, W. Hauser, Ski boots versus the knee joint. What produces the forward leaning position of the ski boot? Sportverletz. Sportschaden **4**(1), 1–13 (1990)
26. F. Noé, D. Amarantini, T. Paillard, How experienced alpine-skiers cope with restrictions of ankle degrees-of-freedom when wearing ski-boots in postural exercises. J. Electromyogr. Kinesiol. **19**(2), 341–346 (2007)
27. S. Perrey, Compression garments: evidence for their physiological effects, in *The Engineering of Sport 7*, ed. by M. Estivalet, P. Brisson, vol. 2 (Springer, Paris 2008), pp. 319–328
28. C. Hrysomallis, Balance ability and athletic performance. Sports Med. **41**(3), 221–232 (2011)
29. D.E. Lieberman, What we can learn about running from barefoot running: an evolutionary medical perspective. Exerc. Sport Sci. Rev. **40**(2), 63–72 (2012)

30. B. Sperlich et al., Is leg compression beneficial for alpine skiers? BMC Sports Sci. Med. Rehabil. **5**, 18 (2013)
31. A. Tversky, D. Kahneman, Belief in the law of small numbers. Psychol. Bull. **76**(2), 105–110 (1971)
32. A. Tversky, D. Kahneman, Judgment under uncertainty: heuristics and biases. *Science* **185**(4157), 1124–1131 (1974)
33. R.K. Guy, The strong law of small numbers. Am. Math. Mon. **95**(8), 614–617 (1988)
34. M. Hughes, I. Franks, *The Essentials of Performance Analysis: An Introduction* (Routledge, New York, 2008)
35. S. Loland, The mechanics and meaning of alpine skiing: methodological and epistemological notes on the study of sport technique. J. Philos. Sport **19**(1), 55–77 (1992)
36. P. Federolf et al., Impact of skier actions on the gliding times in alpine skiing. Scand. J. Med. Sci. Sports **18**(6), 790–797 (2008)
37. J. Spörri et al., Turn characteristics of a top world class athlete in giant slalom: a case study assessing current performance prediction concepts. Int. J. Sports Sci. Coach. **7**(4), 647–660 (2012)
38. M.M. Morlock, V.M. Zatsiorsky, Factors influencing performance in bobsledding: influences of the bobsled crew and the environment. Int. J. Sport Biomech. **5**, 208–211 (1989)
39. J.P.A. Ioannidis, Why most published research findings are false. PLoS Med. **2**(8), e124 (2005)
40. Vv. Aa, Challenges in irreproducible research. Specials & supplements archive. Nature **456** (2013). http://www.nature.com/news/reproducibility-1.17552
41. G. Atkinson, A.M. Nevill, Selected issues in the design and analysis of sport performance research. J. Sports Sci. **19**(10), 811–827 (2001)
42. M. Spencer et al., Variability and predictability of performance times of elite cross-country skiers. Int. J. Sports Physiol. Perform. **9**, 5–11 (2014)
43. R. Rodano, Critical issues in applied sport biomechanics research, in *Proceedings of 20th International Symposium on Biomechanics in Sport*, 2002
44. D.R. Mullineaux, R.M. Bartlett, S. Bennett, Research design and statistics in biomechanics and motor control. J. Sports Sci. **19**(10), 739–760 (2001)
45. K. Hébert-Losier, M. Supej, H.C. Holmberg, Biomechanical factors influencing the performance of elite alpine ski racers. Sports Med. **44**(4), 519–533 (2013)
46. G.P. Bruggemann, M. Morlock, V.M. Zatsiorsky, Analysis of the Bobsled and Men's Luge Events at the XVII Olympic Winter Games in Lillehammer. J. Appl. Biomech. **13**(1), 98–108 (1997)
47. M.A.D. Brodie, A. Walmsley, W. Page, Fusion motion capture: a prototype system using inertial measurement units and GPS for the biomechanical analysis of ski racing. Sports Technol. **1**(1), 17–28 (2008)
48. G. Taga, A model of the neuromusculoskeletal system for human locomotion. I. Emergence of basic gait. Biol. Cybern. **73**, 97–111 (1995)
49. G. Taga, A model of the neuromusculoskeletal system for human locomotion. II. Real-time adaptability under various constraints. Biol. Cybern. **73**, 113–121 (1995)
50. D.V. Knudson, C.S. Morrison, *Qualitative Analysis of Human Movement* (Human Kinetics, Champaign, 2002), pp. 73–189
51. E.A. Fleishman, M.K. Quaintance, L.A. Broedling, *Taxonomies of Human Performance: The Description of Human Tasks* (Academic Press INC, Orlando, 1984)
52. A. Dal Monte, M. Faina et al., *Valutazione dell'atleta: analisi funzionale e biomeccanica della capacità di prestazione* (UTET, Torino, 1999)
53. N.F. Troje, Decomposing biological motion: a framework for analysis and synthesis of human gait patterns. J. Vision **2**, 371–387 (2002)
54. M.D. Hughes, R.M. Bartlett, The use of performance indicators in performance analysis. J. Sports Sci. **20**, 739–754 (2002)
55. M. Hughes, Notational analysis -a mathematical perspective. Int. J. Perf. Anal. Sport **4**(2), 97–139 (2004)

56. Y.L. Zhang, M. Hubbard, R.K. Huffman, Optimum control of bobsled steering. J. Optim. Theory Appl. **85**(1), 1–19 (1995)
57. K. Hosokawa, Y. Sato, T. Sakata, Study on modification of ski referring characteristic of ski-turn (change of width and shoe center), in *The Engineering of Sport 6* (Springer, New York, 2006), pp. 213–218
58. W. Niessen et al., Structural dynamic analysis of alpine skis during turns, in *Science and Skiing*, ed. by E. Müller, H. Schwameder, E. Kornexl, C. Raschner (E & FN Spon, London, 1997), pp. 333–348
59. J.D. Barrow, Rowing and the same-sum problem have their moments. Am. J. Phys. **78**, 728–732 (2010)
60. D. Gordon et al., *Research Methods in Biomechanics* (Human Kinetics, Champaign, 2004)
61. C. Gratton, I. Jones, *Research Methods for Sport Studies*, 2nd edn. (Routledge, New York, 2010)
62. F. Danion, M.L. Latash, *Motor Control. Theories, Experiments and Applications* (Oxford University Press, Oxford, 2011)
63. J.R. Thomas, J.K. Nelson, S.J. Silverman, *Research Methods in Physical Activity*, 6th edn. (Human Kinetics, New York, 2011)
64. G. Atkinson, Sport performance: variable or construct? J. Sports Sci. **20**(4), 291–292 (2002)
65. A. Lees, Technique analysis in sports: a critical review. J. Sports **20**, 813–828 (2002)
66. M. Gorlin et al., Sliding friction and boundary lubrication of snow. J. Tribol. **109**, 614–617 (1987)
67. J.L. Giesbrecht, Polymers on snow. Towards skiing faster. Ph.D. thesis, ETH, Zurich, Switzerland, 2010
68. L. Poirier et al., Experimental analysis of ice friction in the sport of bobsleigh. Sports Eng. **14**, 67–72 (2011)
69. L. Poirier, E.P. Lozowski, R.I. Thompson, Ice hardness in winter sports. Cold Reg. Sci. Technol. **67**(3), 129–134 (2011)
70. N.A. Nordin, P. Styring, Superhydrophobic ski bases for reduced water adhesion. Procedia Eng. **72**, 605–610 (2014)
71. M. Jacobs, Design and recognition of tactile feedback patterns for snowboarding. Ph.D. thesis, RWTH Aachen University, 2008
72. A. Dal Monte, *La valutazione funzionale dell'Atleta* (Sansoni, Firenze, 1983)
73. E. Spring et al., Drag area of a cross country skier. Int. J. Sports Biomech. **4**, 103–113 (1988)
74. M. Supej et al., Aerodynamic drag is not the major determinant of performance during giant slalom skiing at the elite level. Scand. J. Med. Sci. Sports **23**(1), e38–e47 (2013)
75. A.J. Ward-Smith, D. Clements, Experimental determination of aerodynamic characteristics of ski jumpers. Aeronaut. J. **86**, 384–391 (1982)
76. K. Watanabe, I. Watanabe, Aerodynamics of ski-jumping: effect of 'Vstyle' to distance, in *XIVth Congress of the International Society of Biomechanics*, 1993, pp. 1452–1453
77. W. Muller, D. Platzer, B. Schmolzer, Dynamics of human flight on skis: improvements in safety and fairness in ski-jumping. J. Biomech. **29**, 1061–1068 (1996)
78. E. Muller, H. Schwameder, Biomechanical aspects of new techniques in alpine skiing and ski-jumping. J. Sports Sci. **21**, 679–692 (2003)
79. P. Dabnichki, E. Avital, Influence of the position of crew members on aerodynamics performance of two-man bobsleigh. J. Biomech. **39**, 2733–2742 (2006)
80. O. Lewis, Aerodynamic analysis of a 2-man bobsleigh. MA thesis, TU Delft, 2006
81. G. Gibertini et al., Aerodynamic analysis of a two-man bobsleigh, in *6th World Congress of Biomechanics* (Springer, Berlin/Heidelberg, 2010), pp. 228–231
82. A. Winkler, A. Pernpeintner, Automated aerodynamic optimization of the position and posture of a bobsleigh crew. Procedia Eng. **2**(2), 2399–2405 (2010)
83. A. Lüthi et al., Effect of bindings and plates on ski mechanical properties and carving performance, in *The Engineering of Sport 6* (Springer, New York, 2006), pp. 299–304
84. N. Petrone, The use of an Edge Load Profile static bench for the qualification of alpine skis. Procedia Eng. **34**, 385–390 (2012)

85. C. Zanoletti et al., Relationship between push phase and final race time in skeleton performance. J. Strength Cond. Res. **20**(3), 570–583 (2006)
86. N. Bullock et al., Characteristics of performance in skeleton World Cup races. J. Sports Sci. **27**(4), 367–372 (2009)
87. F. Braghin et al., A driver model of a two-man bobsleigh. Sports Eng. **13**(4), 181–193 (2011)
88. F. Braghin, F. Cheli, S. Melzi, E. Sabbioni, Experimental assessment of bobsleigh dynamics and ice-skate contact forces. In *Topics in Modal Analysis II*, Volume 6 (Springer, New York, 2012), pp. 487–498
89. M.L. Latash, F. Lestienne, *Motor Control and Learning* (Springer Science Media, Inc., New York, 2006)
90. V.M. Zatsiorsky et al., *Biomechanics in Sport. Performance Enhancement and Injury Prevention.* An IOC Medical Commission Publication, vol. 9 (Blackwell Science Ltd., Oxford, 2000)
91. P.S. Glazier, M.T. Robins, Comment on 'Use of deterministic models in sports and exercise biomechanics research' by Chow and Knudson. Sports Biomech. **11**(1), 120–122 (2012)
92. F. Cignetti et al., A limit-cycle model of leg movements in cross-country skiing and its adjustments with fatigue. Hum. Mov. Sci. **19**(4), 590–604 (2010)
93. P. Komi, M. Virmavirta, Ski-jumping take-off performance: determining factors and methodological advances, in *Science and Skiing*, ed. by E. Müller, H. Schwameder, E. Kornexl, C. Raschner, (E & F Spon, London, 1997), pp. 3–26
94. P. Federolf et al., Deformation of snow during a carved ski turn. Cold Reg. Sci. Technol. **46**(1), 69–77 (2006)
95. T. Yoneyama et al., A ski robot system for qualitative modelling of the carved turn. Sports Eng. **11**(3), 131–141 (2009)
96. A.A. Renshaw, C.D. Mote, A model for turning snow ski. Int. J. Mech. Sci **31**, 721–736 (1989)
97. Y. Hirano, N. Tada, Mechanics of a turning snow ski. Int. J. Mech. Sci. **5**(36), 421–429 (1994)
98. Y. Hirano, Numerical simulation of a turning alpine ski during recreational skiing. Int. J. Med. Sci. Sports Exerc. **28**(9), 1209–1213 (1996)
99. B. Nemec, O. Kugovnik, M. Supej, Influence of the ski side cut on vibrations in alpine skiing, *Science and Skiing II* (Meyer & Meyer, Aachen, 2001)
100. D.A. Moldestad, Some aspects of ski base sliding friction and ski base structure. Engineering Thesis, Dep of Structural Engineering, Norwegian University of Science and Technology, Trondheim, 1999
101. Q. Wu et al., Lift mechanics of downhill skiing and snowboarding. Med. Sci. Sports Exerc. **38**(6), 1132–1146 (2006)
102. A. Koptyug et al., Cross-country ski vibrations and possible mechanisms of their influence on the free gliding. Procedia Eng. **34**, 473–478 (2012)
103. M. Supej, O. Kugovnik, B. Nemec, *Advanced Analysis of Skiing Based on 3D Kinematic Measurements* (Meyer & Meyer Sport, New York, 2005)
104. M. Supej, M. Cernigoj, Relations between different technical and tactical approaches and overall time at men's world cup giant slalom races. Kinesiologia Slovenica **12**(1), 63–69 (2006)
105. C.D.W. Canham, J. Cole, W.K. Lauenroth, *Models in Ecosystem Science* (Princeton University Press, Princeton, 2003)
106. N. Oreskes, K. Shrader-Frechette, K. Belitz, Verification, validation, and confirmation of numerical models in the earth. Sciences, New Series **263**(5147), 641–646 (1994)
107. N. Oreskes, Why believe a computer? Models, measures, and meaning in the natural world. In *The Earth Around Us: Maintaining a Livable Planet* (W.H. Freeman & Company, New York, 2000), pp. 70–82
108. N. Oreskes, K. Belitz, Philosophical issues in model assessment, in *Model Validation: Perspectives in Hydrological Science, Cap 3* (Wiley, New York, 2001), pp. 23–41
109. N. Oreskes, *The Role of Quantitative Models in Science*, ed. by C.D.W. Canham, J. Cole, W.K. Lauenroth (Princeton University Press, Princeton, 2003)

110. T. Gilovich, R. Vallone, A. Tversky, The hot hand in basketball: on the misperception of random sequences. Cogn. Psychol. **17**(3), 295–314 (1985)
111. M. Bar-Eli, S. Avugos, M. Raab, Twenty years of ‘hot hand’ research: review and critique. Psychol. Sport Exerc. **7**(6), 525–553 (2006)
112. P. Suppes, Conflicting intuitions about causality. Midwest Stud. Philos. **9**(1), 151–168 (1984)
113. R.A. Matthews, Facts versus factions: the use and abuse of subjectivity in scientific research. ESEF Working Paper 2/98, The European Science and Environment Forum Cambridge (1998)
114. H.M. Marks, Rigorous uncertainty: why RA Fisher is important. Int. J. Epidemiol. **32**, 932–937 (2003)
115. A.V. Hill, The air-resistance to a runner. Proc. R. Soc. Lond. B: Biol. Sci., **102**(718), 380–385 (1928)
116. B.B. Lloyd, Theoretical effects of altitude on the equation of motion of a runner, in *Exercise and Altitude*, ed. by R. Margaria (Excerpta Medica Foundation, Amsterdam, 1967), pp. 65–72
117. S.C. Hollings, W.G. Hopkins, P.A. Hume, Environmental and venue-related factors affecting the performance of elite male track athletes. Eur. J. Sport Sci. **12**(3), 201–206 (2012)
118. J.J. De Koning, G. De Groot, G.J. van Ingen Schenau, Ice friction during speed skating. J. Biomech. **25**(6), 565–571 (1992)
119. A.M. Kietzig, S.G. Hatzikiriakos, P. Englezos, Ice friction: the effects of surface roughness, structure and hydrophobicity. J. Appl. Phys. **106**(2) 024303 (2009)
120. S.C. Colbeck, L. Najarian, H.B. Smith, Sliding temperatures of ice skates. Am. J. Phys. **65**(6), 488–492 (1997)
121. M. Virmavirta, P. Komi, Takeoff analysis of a champion ski jumper. J. Coach. Sport Sci. **1**, 23–27 (1994)
122. D. MacAuley, M.T. Best, *Evidence-Based Sports Medicine* (BMG Books, Johannesburg, 2002)
123. N.J. Balmer, A.M. Nevill, M. Williams, Home advantage in the Winter Olympics (1908–1998). J. Sports Sci. **19**(2), 129–139 (2001)
124. W.G. Hopkins et al., Mixed linear modelling: a very short introduction. Sportscience **17** (July 2013)
125. N.E. Breslow, D.G. Clayton, Approximate inference in generalized linear mixed models. J. Am. Stat. Assoc. **88**(421), 9–25 (1993)
126. A.M. Nevill, R.L. Holder, Home advantage in sport: an overview of studies on the advantage of playing at home. Sports Med. **28**(4), 221–236 (1999)
127. T.B. Smith, W.G. Hopkins, Variability and predictability of finals times of elite rowers. Med. Sci. Sports Exerc. **43**(11), 2155–2160 (2011)
128. N. Bullock, W.G. Hopkins, Methods for tracking athletes’ competitive performance in skeleton. J. Sports Sci. **27**(9), 973–940 (2009)
129. G. Atkinson, A.M. Nevill, Statistical methods for assessing measurement error (reliability) variables relevant to sports medicine. Sports Med. **26**(4), 217–238 (1998)
130. W.G. Hopkins, Measures of reliability in sports. Medicine and science. Sports Med. **30**(1), 1–15 (2000)
131. S. Bray, A. Carron, The home advantage in alpine skiing. Aust. J. Sci. Med. Sport **25**(4), 76–81 (1993)
132. R.H. Koning, Home advantage in speed skating: evidence from individual data. J. Sports Sci. **23**(4), 417–427 (2005)
133. SR Olympic Sports (2014), http://www.sports-reference.com/olympics/athletes/da/bjorn-daehlie-1.html
134. J.D. Sterman et al., The meaning of models. Science, New Series **264**(5157), 329–331 (1994)
135. L. Geymonat et al., *Storia del Pensiero Filosofico e Scientifico. Vol IX, Il Novecento* (Garzanti, Milano, 1972)
136. L. Von Bertalanffy, *General System Theory: Foundations, Development, Applications* (George Brazilier, Inc., New York, 1968)
137. N.A. Bernstein, *The Coordination and Regulation of Movement* (Pergamon Press, New York, 1967)

138. H. Hatze, Motion variability. Its definition, quantification and origin. J. Mot. Behav. **18**, 5–16 (1986)
139. K.S. Lashley, Basic neural mechanisms in behavior. Psychol. Rev. **37**, 1–24 (1930)
140. J.S. Kelso, K.G. Holt, A.E. Flatt, The role of proprioception in the perception and control of human movement: toward a theoretical reassessment. Percept. Psychophys. **28**(1), 45–52 (1980)
141. C.D. Davlin, Dynamic balance in high level athletes. Percept. Mot. Skills **98**(3c), 1171–1176 (2004)
142. A. Natri et al., Alpine ski bindings and injuries. Sports Med. **28**(1), 35–48 (1999)
143. M. Burtscher et al., Effects of modern ski equipment on the overall injury rate and the pattern of injury location in Alpine skiing. Clin. J. Sport Med. **18**(4), 355–357 (2008)
144. G. Bottoni et al., Effect of knee braces on balance ability wearing ski boots (a pilot study). Procedia Eng. **72**, 327–331 (2014)
145. J. Mester, Movement regulation in alpine skiing, in *Science and Skiing*, ed. by E. Müller, H. Schwameder, E. Kornexl, C. Raschner (E & FN Spon, London, 1996), pp. 333–348
146. O. Kugovnik, B. Nemec, Analyses of vibration and shocks during the parallel turn in alpine skiing, in *Proceedings of ISBS'98* (UKV -Univesitatsverlag Konstanz, Konstanz, 1998), pp. 164–167
147. C. Haas, C. Simon, D. Schmidtbleicher, Simulation of vibrations in alpine skiing, in *ISBS-Conference Proceedings Archive*, vol. 1, Issue 1, 1998
148. G.C. Foss, B. Glenne, Reducing on-snow vibration of skis and snow-boards. Sound Vib. **12**, 5 (2007)
149. R.L. Bratton et al., Effect of ionized wrist bracelets on musculoskeletal pain: a randomized, double-blind placebo-controlled trial. Mayo Clin. Proc. **77**, 1164–1168 (2002)
150. E. Bringman, C. Kimura, P. Scho, Effects of an ionic bracelet on physical, cognitive, and integrative tasks. Pac. Northwest J. Undergrad. Res. Creat. Act. 2 (2011). Article 4. http://commons.pacificu.edu/pnwestjurca/vol2/iss1/4
151. J. Porcari, Can the Power Balance © bracelet improve balance, flexibility, strength and power? J. Sports Sci. Med. **10**, 230–231 (2011)
152. C.T. Haas, S. Turbanski, D. Schmidtbleicher, Improved sensorimotor control is not connected with improved proprioception. Br. J Sports Med. **39**(Suppl 1), 388 (2005)
153. D.F. Collins, A. Prochazka, Movement illusions evoked by ensemble cutaneous input from the dorsum of the human hand. Physiology **496**(3), 857–871 (1996)
154. D. Spelmezan et al., Tactile motion instructions for physical activities, in *Proceedings of the SIGCHI Conference on Human Factors in Computing Systems* (2009), pp. 2243–2252
155. J.J. Gibson, *The Senses Considered as Perceptual Systems* (Houghton Mifflin, Boston, 1966)
156. V. Ibegbuna et al., Effect of elastic compression stockings on venous hemodynamics during walking. J. Vasc. Surg. **37**(2), 420–425 (2003)
157. D.P. Born, B. Sperlich, H.C. Holmberg, Bringing light into the dark: effects of compression clothing on performance and recovery. Int. J. Sports Physiol. Perform. **8**(1), 4–18 (2013)
158. F. Noé, T. Paillard, Is postural control affected by expertise in alpine skiing? Br. J. Sports Med. **39**(11), 835–837 (2005)
159. R. Llinás, The noncontinuous nature of movement execution, in *Motor Control: Concepts and Issues*, ed. by D.R. Humphrey, H.J. Freund (Wiley, New York, 1991), pp. 223–242
160. M.S. Redfern, L. Yardley, A.M. Bronstein, Visual influences on balance. J. Anxiety Disord. **15**(1), 81–94 (2001)
161. M. Cardinale, J. Lim, Electromyography activity of vastus lateralis muscle during whole-body vibrations of different frequencies. J. Strength Cond. Res. **17**(3), 621–624 (2003)
162. D. Burke et al., The responses of human muscle spindle endings to vibration of non-contracting muscles. J. Physiol. **261**(3), 673–693 (1976)
163. S.S. Maxwell, Labyrinth and equilibrium II. The mechanism of the dynamic functions of the labyrinth. J. Gen. Physiol. **2**(4), 349–355 (1920)
164. L.M. Nashner et al., Organization of posture controls: an analysis of sensory and mechanical constraints. Prog. Brain Res. **80**, 411–418 (1989)

165. G. Bottini, Modulation of conscious experience by peripheral sensory stimuli. Nature **376**, 778–781 (2013)
166. F. Lestienne, J. Soechting, A. Berthoz, Postural readjustments induced by linear motion of visual scenes. Exp. Brain Res. **28**, 363–384 (1977)
167. B. Kerr, S.M. Condon, L.A. McDonald, Cognitive spatial processing and the regulation of posture. J. Exp. Psychol. Hum. Percept. Perform. **11**(5), 617–622 (1985)
168. G.J. Andersen, B.P. Dyre, Spatial orientation from optic flow in the central visual field. Percept. Psychophys. **45**(5), 453–458 (1989)
169. T. Poggio, E. Bizzi, Generalization in vision and motor control. Nature **431**(7010), 768–774 (2004)
170. A.O. Effenberg, H. Mechling, Bewegung hörbar machen -Warum? Psychologie und Sport **5**(1), 28–38 (1998)
171. T. Pozzo, A. Berthoz, L. Lefort, Head stabilization during various locomotor tasks in humans. Exp. Brain Res. **82**, 97–106 (1990)
172. T. Pozzo, Y. Levik, A. Berthoz, Head and trunk movements in the frontal plane during complex dynamic equilibrium tasks in humans. Exp. Brain Res. **106**, 327–338 (1995)
173. B. Armstrong, P. McNair, D. Taylor, Head and neck position sense. Sports Med. **38**(20), 101–117 (2008)
174. R. Ajemian et al., Why professional athletes need a prolonged period of warm-up and other peculiarities of human motor learning. J. Mot. Behav. **42**(6), 381–388 (2010)
175. G. Atkinson, T. Reilly, Circadian variation in sports performance. Sports Med. **21**(4), 92–293 (1996)
176. B. Drust et al., Review: circadian rhythms in sports performance -an update. Chronobiol. Int. **22**(1), 21–44 (2005)
177. http://en.wikipedia.org/wiki/Equifinality (2014)
178. B.T. Bates et al., An assessment of subject variability, subject-shoe interaction and the evaluation of running shoes using ground reaction force data. J. Biomech. **16**(3), 181–191 (1983)
179. B.M. Nigg, M.A. Nurse, D.J. Stefanyshyn, Shoe inserts and orthotics for sport and physical activities. Med. Sci. Sports Exerc. **31**, S421–S428 (1999)
180. J.G. Barton, A. Lees, Comparison of shoe insole materials by neural network analysis. Med. Biol. Eng. Comput. **34**(6), 453–459 (1996)
181. M.J. Rupérez et al., Artificial neural networks for predicting dorsal pressures on the foot surface while walking. Expert Syst. Appl. **39**(5), 5349–5357 (2012)
182. R. Baker, The history of gait analysis before the advent of modern computers. Gait Posture **26**, 331–342 (2007)
183. A. Lüthi et al., Determination of forces in carving using three independent methods, in *Proceedings of 3rd International Congress on Science and Skiing*, Aspen, CO, 2004
184. Y. Ohgi et al., Measurement of jumper's body motion in ski jumping, in *The Engineering of Sport 6* (Springer, New York, 2006), pp. 275–280
185. M.A.D. Brodie, Development of fusion motion capture for optimisation of performance in alpine ski racing. Ph.D. thesis, Massey University, Wellington, NZ, 2009
186. T. Bere et al., Mechanisms of anterior cruciate ligament injury in world cup alpine skiing. A systematic video analysis of 20 cases. Am. J. Sports Med. **39**(7), 1421–1429 (2011)
187. F. Meyer, Biomechanical analysis of alpine skiers performing giant slalom turns. Ph.D. thesis, UNIL-University of Lausanne, 2012
188. M. Gilgien et al., Determination of external forces in alpine skiing using a differential global navigation satellite system. Sensors **13**, 9821–9835 (2013)
189. M. Gilgien, Characterisation of skiers' mechanics, course setting and terrain geomorphology in World Cup Alpine Skiing using global navigation satellite systems: injury risk, performance and methodological aspects. Ph.D. dissertation from the Norwegian School of Sport Sciences, 2014
190. M. Kranz et al., The mobile fitness coach: towards individualized skill assessment using personalized mobile devices. Pervasive Mob. Comput. **9**(2), 203–215 (2013)

191. H.E. Berg, O. Eiken, Muscle control in elite alpine skiing. Med. Sci. Sports Exerc. **31**(7), 1065–1067 (1999)
192. P.A. Tesch, Aspects on muscle properties and use in competitive Alpine skiing. Med. Sci. Sports Exerc. **27**(3), 310–314 (1995)
193. N. Petrone, G. Marcolin, M. De Gobbi, M. Nicoli, C. Zampieri, Acquisition of EMG signals during slalom with different ski boots, in *Science and Skiing IV* (Meyer & Meyer Sport, Aachen, 2009), pp. 399–409
194. F. Panizzolo, N. Petrone, G. Marcolin, Comparative analysis of muscle activation patterns between skiing on slopes and on training devices, in *8th Conference of the International Sports Engineering Association (ISEA). Elsevier,* Procedia Engineering, vol. 2 (2010), pp. 2537–2542
195. K.S. Lashley, The problem of cerebral organization in vision, in *Biological Symposia, vol. VII Visual Mechanism* (Jacques Cattel Press, New York, 1942), pp. 301–322
196. R. Schmidt, T. Lee, *Motor Control and Learning. A Behavioral Emphasis*, 5 edn. (Human Kinetics, Champaign, 2011)
197. J. McIntyre et al., Does the brain model Newton's laws? Nat. Neurosci. **4**, 693–694 (2001)
198. A. Salo, P.N. Grimshaw, J.T. Viitasalo, Reliability of variables in kinematic analysis of spring hurdles. Med. Sci. Sports Exerc. **29**(3), 383–389 (1997)
199. P. Federolf et al., The application of principal component analysis to quantify technique in sports. Scand. J. Med Sci Sports. **24**(3), 491–499 (2012)
200. R. Tucker, M.I. Lambert, T.D. Noakes, An analysis of pacing strategies during men's world-record performances in track athletics. Int. J. Sports Physiol. Perform. **1**, 233–245 (2006)
201. T.D. Noakes, M. Lambert, R. Human, Which lap is the slowest? An analysis of 32 world record performances. Br. J. Sports Med. **43**, 760–764 (2008)
202. C.R. Abbiss, P.B. Laursen, Describing and understanding pacing strategies during athletic competition. Sports Med. **38**(3), 239–252 (2008)
203. T. Muehlbauer, S. Panzer, C. Schindler, Pacing pattern and speed skating performance in competitive long-distance events. J. Strength Cond. Res. **24**(1), 114–119 (2010)
204. J.J. De Koning et al., Regulation of pacing strategy during athletic competition. PloS One. **6**(1), e15863 (2011)
205. B. Roelands et al., Neurophysiological determinants of theoretical concepts and mechanisms involved in pacing. Sports Med. **43**(5), 301–311 (2013)
206. I. Reade, W. Rodgers, N. Hall, Knowledge transfer: How do high performance coaches access the knowledge of sport scientists? Int. J. Sports Sci. Coach. **3**, 319–334 (2008)
207. I. Reade, W. Rodgers, K. Spriggs, New ideas for high performance coaches: a case study of knowledge transfer in sport science. Int. J. Sports Sci. Coach. **3**, 335–354 (2008)
208. J.W. Chow, D.V. Knudson, Use of deterministic models in sports and exercise biomechanics research. Sports Biomech. **10**, 219–233 (2011)
209. D.V. Knudson, J.W. Chow, Response to the comment on 'Use of deterministic models in sports and exercise biomechanics research'. Sports Biomech. **11**(1), 123–124 (2012)
210. M. McNamee (ed.), *Philosophy and the Sciences of Exercise, Health and Sport: Critical Perspectives on Research Methods* (Routledge, London, 2004)
211. A. Dal Monte, Nuovi occhi per l'allenatore, in *Scuola Dello Sport.* Speciale Seminario, vol. 1 (1982), pp. 13–14
212. D. Dalla Vedova et al., Nuovi occhi per l'allenatore: 25 anni dopo. Scuola dello Sport **25**(71), 3–11 (2007)
213. P.A. Federolf, Quantifying instantaneous performance in alpine ski racing. J. Sports Sci. **30**(10), 1063–1068 (2012)
214. S.L. Halson, Sleep in elite athletes and nutritional interventions to enhance sleep. Sports Med. **44**(1), 13–23 (2014)
215. M.A. Boutilier, L. San Giovanni, *The Sporting Woman* (Human Kinetics Publishers, Inc., Champaign, 1983)
216. B.L. Drinkwater, *Women in Sport.* The Encyclopaedia of Sports Medicine: An IOC Medical Commission Publication, vol. 8 (Wiley, New York, 2008)

Index

F. Braghin et al. (eds.), *The Engineering Approach to Winter Sports*,
DOI 10.1007/978-1-4939-3020-3

Printed by Printforce, the Netherlands